FUTURE HIGH ENERGY COLLIDERS

The Santa Barbara Symposium on *Future High Energy Colliders* brought together many physicists who will have a major impact on the future direction of the field. The group picture shows some of the participants.

FUTURE HIGH ENERGY COLLIDERS

Santa Barbara, California   October 1996

EDITOR
Zohreh Parsa
Brookhaven National Laboratory

American Institute of Physics

AIP CONFERENCE
PROCEEDINGS 397

Woodbury, New York

Authorization to photocopy items for internal or personal use, beyond the free copying permitted under the 1978 U.S. Copyright Law (see statement below), is granted by the American Institute of Physics for users registered with the Copyright Clearance Center (CCC) Transactional Reporting Service, provided that the base fee of $10.00 per copy is paid directly to CCC, 222 Rosewood Drive, Danvers, MA 01923. For those organizations that have been granted a photocopy license by CCC, a separate system of payment has been arranged. The fee code for users of the Transactional Reporting Service is: 1-56396-729-4/ 97 /$10.00.

© 1997 American Institute of Physics

Individual readers of this volume and nonprofit libraries, acting for them, are permitted to make fair use of the material in it, such as copying an article for use in teaching or research. Permission is granted to quote from this volume in scientific work with the customary acknowledgment of the source. To reprint a figure, table, or other excerpt requires the consent of one of the original authors and notification to AIP. Republication or systematic or multiple reproduction of any material in this volume is permitted only under license from AIP. Address inquiries to Office of Rights and Permissions, 500 Sunnyside Boulevard, Woodbury, NY 11797-2999; phone: 516-576-2268; fax: 516-576-2499; e-mail: rights@aip.org.

L.C. Catalog Card No. 97-73333
ISBN 1-56396-729-4
ISSN 0094-243X
DOE CONF- 961038

Printed in the United States of America

CONTENTS

Preface vii

Perspectives on Future High Energy Physics 1
  N. P. Samios

View from Washington: Perspectives on Future High Energy Physics 9
  M. Krebs

The Standard Model and Beyond (In Search of "New Physics") 11
  W. J. Marciano

Higgs Physics: An Historical Perspective 27
  S. Willenbrock

A Simplified Summary of Supersymmetry 41
  J. F. Gunion

Strongly Interacting New Physics 65
  T. Appelquist

LEP Status Report 81
  L. Rolandi

The TeV 2000 Report: Electroweak Physics with High Luminosity at the Fermilab Tevatron 95
  D. Amidei

Status of LEP2 and LHC 113
  E. Keil

Precision Physics at LHC 129
  I. Hinchliffe

Very Large Hadron Colliders 139
  M. Harrison

e^+e^- Physics 143
  H. Murayama

Physics at LHC and NLC 157
  F. E. Paige

The TESLA Superconducting Linear Collider 173
  R. Brinkmann

Scaling Linear Colliders to 5 TeV and Above 191
  P. B. Wilson

The Problems and Physics Prospects for a $\mu^+\mu^-$ Collider 203
  D. B. Cline

The Physics Capabilities of $\mu^+\mu^-$ Colliders 219
  V. Barger, M. S. Berger, J. F. Gunion, and T. Han

Linear Electron–Electron Colliders 235
  C. A. Heusch

Ultimate Luminosities and Energies of Photon Colliders 259
  V. Telnov

Lepton–Hadron Collider Physics 275
  S. Ritz

Perspectives on Future High Energy Physics 287
  B. Richter

Schedule 357

List of Participants 361

Author Index 368

PREFACE

A "Future High Energy Colliders" Symposium was held October 21–25, 1996 at the Institute for Theoretical Physics (ITP) in Santa Barbara. This was one of the three symposia[1] hosted by the ITP and supported by its sponsor, the National Science Foundation as part of a five month program on "New Ideas for Particle Accelerators." The long term program and symposia were organized and coordinated by Dr. Zohreh Parsa of Brookhaven National Laboratory/ITP.

The purpose of this symposium was to discuss the future direction of high energy physics by bringing together leaders from the theoretical, experimental, and accelerator physics communities. Their talks provided personal perspectives on the physics objectives and the technology demands of future high energy colliders. Collectively, they formed a vision for where the field should be heading and how it might best reach its objectives.

The design and construction of particle accelerators used in high energy physics are motivated and constrained by: 1) the forefront physics questions being asked; 2) the availability of technology needed to build and operate these machines, as well as capabilities to detect and analyze the collisions; and 3) the cost of the machine and the availability of construction funds from home and foreign governments. High Energy Physics in the United States is now at a crossroads, where its future will depend on: participation in foreign projects, upgrading, and utilizing existing facilities; and new construction initiatives.

Some of the underlying physics motivations and technical issues had been addressed at earlier workshops, such as Snowmass 1996. The Santa Barbara "Future High Energy Colliders" Symposium's novel aim was to begin the process of reaching a consensus on how to attain the vision.

The goal of Elementary Particle Physics is to understand the nature of matter and the forces acting on it. Experiments over the last two decades have convincingly shown that the strong, electromagnetic, and weak forces are all closely related and are simply described by the "Standard Model." In particular the anticipated sixth quark, top, has been recently found at Fermilab, and the predicted properties of the Z boson, one of the carriers of the weak force, have been tested to better than 1%. There is now little doubt that the Standard Model is a very good description of the basic forces responsible for all atomic and nuclear physics. While the Standard Model is a great success, there remain many open questions in Elementary Particle Physics. Perhaps the most urgent is to understand how masses of the particles originate. To that end, new physics beyond what is currently known is required. The simplest possibility, the "Higgs Mechanism," predicts the existence of a fundamental Higgs Boson. Finding that elusive particle, or whatever new physics is actually responsible for mass generation, motivated the Superconducting Supercollider (SSC) and remains the primary goal of the next generation of colliders. A number of other interesting and more elaborate models have been proposed, but there is as yet no direct experimental evidence supporting any of them. Nevertheless, consistency of the Standard Model requires that the new physics responsible for mass generation occur at an energy scale of less than about 1 TeV (i.e., within the range of the next generation of accelerators). In addition to the origin of mass, there are other compelling questions. For example, the observed matter-antimatter asymmetry in the universe is not understood. Also, astrophysical observations suggest that between 90% –99% of the matter that makes up the universe is invisible. There are a number of possible candidates for this "dark matter," but none have been proven experimentally to exist.

Currently, the operation of the Tevatron proton–antiproton collider at the Fermi National Accelerator Laboratory and the SLC electron–positron collider at the Stanford Linear Accelerator

---

[1] In addition to this symposium, a week long symposium was held on "New Modes of Particle Acceleration—Techniques and Sources" August 19–23, 1996. Some of the highlights of that meeting included Novel Modes of laser, plasma, wakefield accelerations, techniques, and power sources. A third symposium on "Beam Stability and Nonlinear Dynamics" was held on December 3–5, 1996 and dealt with some of the fundamental theoretical problems associated with accelerator physics.

Center are at the energy frontier of the field. The use of both proton–antiproton and electron–positron collisions is important in order to provide complementary information. At the energy frontier new particles never before observed are discovered and studied, providing unique insight into the laws of nature. Recently, the LEP electron–positron collider at the CERN laboratory in Geneva began upgrading its energy, expanding the energy frontier in electron–positron collisions by about a factor of two. Sometime near the year 2005 the LHC proton–proton collider in Geneva will operate with world record beam energy about seven times that of the Fermilab Tevatron.

The Large Hadron Collider (LHC) at CERN is intended to address the question of the origin of mass. Because we know the energy scale associated with mass generation, we can be reasonably confident that the LHC will discover this new physics, whether it is a Higgs boson or something quite different.

Although the LHC should elucidate the origin of mass, it probably cannot answer all the outstanding questions. For that reason, a collaboration involving institutions from Germany, Japan, United States, and many other countries has been developing a design for a high energy e^+e^- collider, the Next Linear Collider (NLC). This would have a somewhat lower energy reach than the LHC, at least initially, but it would be able to make many interesting unique measurements that would complement those at the LHC. Together the LHC and the NLC should clarify the origin of mass and address many other open questions. They might also find the explanation for the dark matter in the universe and open new frontiers, such as the much anticipated spectrum of heavy particles predicted by supersymmetry, an elegant extension of the underlying structure of space–time. If supersymmetry is found, it will become the focus of high energy physics. Studying its spectrum of particles and their properties will be a major experimental enterprise.

Also under discussion at Brookhaven National Laboratory, Fermilab, Lawrence Berkeley National Laboratory, and many Universities is a muon collider. The feasibility of building an accelerator in which muons collide with antimuons is currently under serious investigation. Since muons (at rest) decay in about two millionth of a second; building such a collider would be a major technological achievement. A survey of the physics that could be studied at such a machine overlaps with e^+e^- collider capabilities but also includes novelties, such as the possibility for fusion of the colliding beams to produce Higgs particles. This process would permit for example, the precise direct measurement of the mass and lifetime of the Higgs particle as well as its decay properties. However, the main enthusiasm for the muon collider stems from its potential to reach very high energies and the possibility that it could be constructed at an existing national laboratory.

An explanation of the matter–antimatter asymmetry in our universe requires first an understanding of the origin of mass. In addition, important supporting information can be obtained by studying rare interactions of a variety of known particles, including B mesons, K mesons, and muons. Such experiments are ongoing at several laboratories, and with the advent of B factories and intensity upgrades of proton synchrotrons at Brookhaven and Fermilab, more will be carried out in the future.

A unique feature of the symposium was the bringing together of many physicists who will have a major impact on the future direction of the field. Especially important was the set of presentations made by the Department of Energy Director of Energy Research M. Krebs, by B. Kayser of the National Science Foundation, and by the directors of the three U.S. High Energy Physics laboratories, J. Peoples (Fermilab), B. Richter (Stanford Linear Accelerator Center), and N. Samios (Brookhaven National Laboratory). Their perspectives, combined with presentations by internationally distinguished high energy and accelerator physicists, provided a comprehensive picture of the issues involved in formulating appropriate goals for the future. The difficult aspects and the far reaching consequences of the decisions that must be made were further clarified during a unique panel discussion designed to initiate the process of reaching accord in the high energy community as to what the future physics and accelerator priorities should be. Although there is not unanimity of opinion in all matters, there is a consensus that an NLC should be built somewhere in the world, and vigorous R & D should be pursued in promising new areas, such as the muon collider concept and new modes of particle acceleration.

Given the long lead times necessary for the design of new particle accelerators, it is important for the high energy physics community to decide soon what types and energy ranges of colliders are necessary to address the additional questions which the LEP and LHC may not be able to answer.

The perspectives that were presented at the symposium on the state and future of high energy physics are vital ingredients in the continuing discussion of how the U.S. High Energy Physics community should best marshal its national scientific resources while continuing a high level of international collaboration. They should provide valuable input for ongoing discussions and in making decisions regarding the future direction of the field.

I would like to thank all the authors for providing the write-ups of their talks. In addition, for one reason or another, several speakers were not able to provide the write-ups of their talks.

I nevertheless, thank them for their stimulating talks and participation in the symposium. In most cases, copies of the transparencies from their talks can be found in the report BNL-52524. These include presentations by Drs. D. Gross, J. Peoples, and B. Kayser on "Perspective on Future High Energy Colliders"; G. Jackson on "TeV2000, Accelerator Issues"; D. Burke and R. Toge on "Next Linear Colliders"; and R. Palmer on "Overview of Muon Collider Design and Simulations." For a complete list of all the presentations see the program schedule given in the Appendix.

I would like to thank Dr. W. J. Marciano and the advisory committee, all speakers, conveners, and participants for making the symposium a unique and stimulating experience. I also thank the ITP Director, Manager, and staff for providing a beautiful setting and making sure the meeting ran smoothly.

Zohreh Parsa
Chairperson, Symposium
Institute for Theoretical Physics,
UCSB, Santa Barbara, California
Brookhaven National Laboratory, Upton, New York

References

Z. Parsa, "Future High Energy Colliders Summary Report," BNL-52524.
Z. Parsa, "Collision Crossroads, CERN Courier," Vol. 37, **2**, March 1997, Ed. G. Fraser.

The *Future High Energy Colliders* Symposium Santa Barbara was organized and chaired by Zohreh Parsa, Brookhaven National Laboratory (2nd from left). Especially significant presentations were given by major US high energy physics laboratory directors (left to right): Nick Samios (Brookhaven National Laboratory), Burt Richter (Stanford Linear Accelerator Center, SLAC), and John Peoples (Fermilab).

Perspectives on Future High Energy Physics*

Nicholas P. Samios

Brookhaven National Laboratory
Upton, New York 11973

In preparing this talk on perspectives on the future high energy physics, I was again reminded that one should be wary of attempting to forecast the future, if at all, one should do it in general terms. If one examines previous attempts, one finds very few accurate predictions. Indeed, some are notoriously bad. Lord Kelvin was reported to have stated in 1895 that heavier than air machines are impossible, in 1897 that radio has no future and in 1950 that x-rays are a hoax. Of equal inaccuracy is the quote attributable to Michaelson in 1898.

"While it is safe to affirm that the future of physical science has no marvels in store even more astonishing than those of the past, it seems probable that most of the grand underlying principles have been firmly established and that further advances are to be sought chiefly in the rigorous applications of these principles to all phenomena which come under our notice. An eminent physicist has remarked that the future tasks of physical science are to be looked for in the 6th place of decimals."

To take just one example from the recent past, L.C.L. Yuan edited in 1964 a compendium of essays on the Nature of Matter by a group of pre-eminent scientists. Although erudite and thoughtful in expounding in what had been accomplished as well as outlining many of the short term extrapolation of scientific endeavors--they all missed the extraordinary evolution of the standard model with its family of quarks, leptons, gluons and bosons. In our own day, the recent book by John Hazen, the End of Science, is another example of poor prognostication. In this case, he forecasts the end of everything, physics, cosmology, etc., but doesn't make the case. Indeed the tone is reminiscent of the Michaelson quote. In sum, be wary of predictions.

If that is so, how can one proceed? The answer is with caution and humility and in two general ways. The first is to utilize the state of knowledge in the field and thereby provide theoretical and experimental guidance on future directions. Such a brief synopsis is outlined in Table I enumerating the ingredients of the standard model. In essence, all observables are described by the model, however, there are ~ 17 constants that are not presently calculable.

*Work performed under the auspices of the U.S. Department of Energy.

Table I. Standard Model

SU(3) x SU(2) x U(1)
Constituents -- 3 sets of doublets

Quarks

u	c	t
d	s	b	Gluons

Leptons

ν_e	ν_τ	ν_τ
e	μ	τ	γ, W, Z

Coupling α_i I = 1-3
K-M matrix

Electro-weak Symmetry Breaking
    Higgs - Introduce 1 or 2 complex scalar bosons.
    At least one Higgs boson at e-m scale mass
  Not fundamental
    Low Energy Approximation

As such it is not fundamental theory but possibly a low energy approximation to a more global theory. What are the expectations? Some people have hypothesized a desert, from $10^2 - 10^4$ GeV. If so, all possible next generation machines will produce nothing and this field will come to a hiatus, at least until some for future data when someone willcome up with a clever, clever idea for a super energy machine. A more likely scenario involves electro weak symmetry breaking by some mechanism, possibly one or more Higgs bosons. Predictions of masses for such particles range from 120 GeV up to 1,000 GeV with varying degrees of confidence. Even more exciting is the possibility of super-symmetry with multiple Higgs bosons and which is a natural consequence of string theory which unites all the forces with gravity. If this turns out to be a correct description of nature, then for every boson there is a fermion and vice versa. There should be many states to be found with masses less than 1,000 GeV. We can reasonably conclude that although we don't know which description is correct (although most theoreticians bet on super-symmetry), something should happen at a mass scale of 1,000 GeV. Therefore, exploration of the energy mass range of several TeV could be extremely desirable.

    The second approach is technical, namely, how well can one do in going to higher energies with present techniques or new accelerator principles. This was the argument for the BNL (AGS), a new principle allowing for higher energies at lower cost, as well as for the SLAC complex, a large extrapolation of the known linear accelerator technique. In both cases the physics justification were far off the mark, nowhere is there even a hint of the major discoveries at the AGS (2ν, J, CP) and

SLAC (Ψ, partons, τ). Excerpts from documents proposing the contribution of each of these projects are noted as follows.

AGS Justification

"...the Cosmotron has, during its relatively short operational use, yielded much new data on meson yields, and on the energy dependence of π meson and fast neutron cross sections and has even led to the observation of certain hitherto unobserved heavy meson phenomena. That extension of the available energy would yield many fruitful results seems unquestionable; indeed, it is already possible to visualize many useful experiments requiring considerably higher energies. Although many of these will be made possible by the 6 BeV soon to be available at the University of California Bevatron, still further extension seems highly desirable, for specific and predictable reasons as well as on the general grounds that past extensions of energy have always proved highly profitable."

Stanford Linear Electron Accelerator

"Any high-energy accelerator is useful in the following general categories: (1) as a controlled source of artificially-produced particles which are otherwise only produced in very low intensity from cosmic rays; (2) as a source of particles to explore nuclear structure and to study the interactions of these particles in nuclear matter; (3) as a tool to study the basic production process in the creation of particles. Added to these, in the case of electron accelerators, is: (4) as a tool to study the high-energy limits of quantum elecrodynamics."

For assistance in devising a strategy for technical accelerator possibilities, we can look at the historical record, Fig. 1, the Blewett-Livingston plot where accelerator energy is plotted as function of calendar time for various types of accelerators. One notes the well known pattern of exponential increase over decades of time driven by new ideas. The demise of the SSC which was the logical next machine and its partial replacement by the LHC in 2004 is certainly a setback. Our present high energy complement of machines is the Tevatron at 1.8 TeV, the SLC at 100 GeV and LEP at 200 GeV, and HERA the only CP collider. These are supplemented by the AGS and CESR at the lower energies and with high intensities. There are certainly ideas and projects for going to the multi-TeV range technically. Further along are designs for e^+e^- linear colliders in the 0.5 TeV to 1 TeV energy range. A novel idea for $\mu^+\mu^-$ colliders has recently emerged and a vigorous R&D program is clearly warranted. In addition, there are prospects for a 60-100 TeV pp collider. All should be pursued. What we should not do is duplicate machines. Recall the saga of PEP, PETRA, and TRISTAN all chasing the top quark which originally was expected at 15 GeV and emerged at 170 GeV.

Whatever emerges is going to have constraints, political and economic. Lessons learned from the demise of the SSC would serve us well. There are numerous reasons for the lack of success in this project among which were the lack of 100%

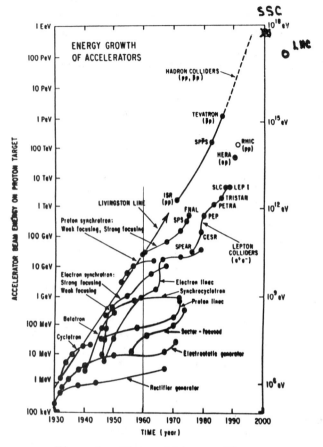

Figure 1. Blewett-Livingston Plot

support of the U.S. high energy community, similar reluctance on the part of the U.S. physics community, notably those in condensed matter, the confusing signals of the international community vis a vis the LHC, the competition of NASA projects especially in Houston, Texas, and finally, the luke-warm endorsement by the AAUP President contingent on no reductions in any other areas of science for support of the SSC. As a consequence, the project was lost, the logical game plan destroyed and the money, $600M per year, disappeared.

What is to be done. First one must stabilize U.S. science funding for all of science and in particular for high energy physics. There has been some recent success, in that the previous downward budget projections for DOE have been revised to flat. The high energy budget in FY'97 has two construction projects underway, the main injector at Fermilab and the B factory at SLAC totaling ≈

$100M which goes to zero in FY'99. We must retain this money in the high energy program. In order to improve the budget situation for science, it is imperative that all of science speak in one voice, chemists, biologists, physicists, medical people, etc. And in doing this, we must do a better job of communicating to the public and the Congress the importance and value of what we do.

The future strategy is straightforward. We must incrementally upgrade and run our present accelerator facilities producing exciting and forefront research. We are pressing on several frontiers with protons, electrons, in energy and intensity constantly testing the standard model. At the same time, we should sharpen our theoretical tools both extrapolating from lower energies (100 GeV) to high (multi TeV) and vice versa. The U.S. should be involved in the LHC, both in the accelerator and experimental areas as is presently proposed and planned. We should have an extensive R&D program on accelerators for a multi-TeV capability, emphasizing e^+e^- and $\mu^+\mu^-$ colliders. Finally, we should maintain and strengthen our international cooperative activities for it appears that scale and costs dictate that the next multi-TeV machine will be truly international.

I close by quoting an observation by Julian Schwinger on the future of fundamental physics.

"And one should not overlook how fruitful a decision to curtail the continued development of an essential element to the society can be. By the 15th century the Chinese had developed a mastery of ocean voyaging for beyond anything existing in Europe. Then in an abrupt change of intellectual climate, the insular party at court took control. The great ships were burnt and the crews disbanded. It was in these years that small Portugese ships rounded the Cape of Good Hope."

This should provide a lesson for all of us.

Appendix

For completeness I have added an appendix briefly commenting on the high energy experimental program at the AGS at BNL. This utilizes the high intense proton beam, 6×10^{13} protons per pulse allowing for the exploration of very rare processes in K decays and the anomolous magnetic moment of the muon. These are listed in Table II where Group A involves ongoing experiments and Group B proposed experiments. One notes present sensitivities of 10^{-10} to 10^{12}. Such measurements are relevant to parameters of K-M matrix and particle masses of 100 TeV coupling to μe.

The proposed experiments increase this sensitivity especially in the μ to e scattering, which is relevant to SU SY particles. The neutral K^o decay is a direct measurement of CP and the others either look for effects beyond the standard model or measure a K-M matrix parameters. This work complements that from other labs especially the B factories.

BNL has a long-range plan in high energy and nuclear physics and this is graphically displayed in Table III. The vertical axis is proportional to funding levels and the horizontal axis is calendar time. As is evident, there are many activities starting with the ongoing AGS program whose content was previously outlined. By

Table II

Group A

$K^+ \to \pi^+ \nu\nu$	BR ~ 1 x 10^{-10}
	$|V_{td}|$ ~ 15-25%
	$(K^+ \to \pi^+\gamma\gamma,\ K^+ \to \pi^+\mu^+\mu^-,\ K^+ \to \mu\nu\gamma)$
$K^+ \to \pi^+ \mu e$	3 x 10^{-12} goal.
	Also, $K^+ \to \pi^+\mu^+\mu^-,\ \pi^+ e\nu$
$K^+ \to \mu e$	8 x 10^{-13} goal. (Current 2 x 10^{-11})
g-2	$\Delta a_\mu \approx 20$ x 10^{-11}.

Group B

$K^\circ \to \pi^\circ \nu\nu$	10^{-12} sensitivity.
	Direct test of CP (η) ($V_{td}^* V_{ts}$)
$K^+ \to \pi^\circ \mu^+ \nu$	Transverse Pol. of μ. $P_\mu^t \approx 1.3$ x 10^{-4}.
$K^+ \to \pi^\circ e^+ \nu$	$|V_{us}|$, $\lambda \pm 0.35$ % Cabibbo.
$\mu^- Ae \to e^- Ae$	10^{-16} sensitivity. LOI

Above sensitivity to SUSY.

FY'00, RHIC is operational with a full heavy ion program and four detectors. The potential for discoveries with this new facility are great, centered around the creation of the quark gluon plasma, the restoration of chiral symmetry and the observation of the excited vacuum. The possibility of augmenting this program with polarized pp interactions is now assured with funding from Japan via RIKEN and should take place in the early years of operation. As also noted, there is a commensurate reduction in the AGS program in FY'00. Since RHIC is paying for full operation for the complex, the high energy physics program pays incremental costs for its implementation. As such, an AGS 2000 program has been formulated, a set of unique, crucial, and competitive experiments. The extent of this program will be reviewed by DOE in the next year.

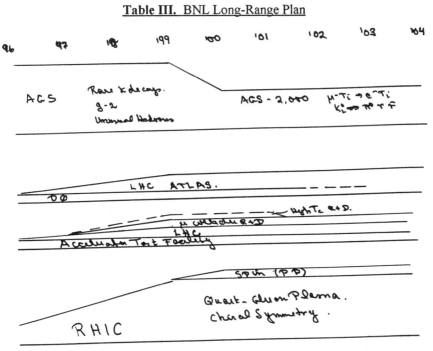

Table III. BNL Long-Range Plan

For completeness, we note experimental participation in both the D0 detector at the TEVATRON and the ATLAS detector at the LHC at CERN. The D0 participation has been longstanding and will continue with BNL being major contributors to the calorimetry and computing. With respect to ATLAS, BNL is again a major player and in this instance will be the center for the U.S. segment of the collaboration.

The rather extensive accelerator R&D activities are also noted. These involve a Center for Accelerator Physics with the Accelerator Test Facility which has been in existence for $\approx$ 10 years. This provides a facility for advanced accelerator concept experimentation with three beam lines and user groups from academia, university and other national labs. The BNL participation in the LHC accelerator is also noted with the main responsibility being the dipole magnets for the insertion areas as well as the short sample measurement of all the superconducting cable. The last entry is the muon collider R&D. This effort led by Bob Palmer addresses the issue of reaching multi TeV energies in the constituent reference frame. For a variety of reasons, the use of muons as the colliding particle is very attractive. The lack of beam strahlung on the part of the muon at high energies (> 1 TeV) in contrast to electrons is also an advantage. Much R&D effort has to be done to ascertain that such a collider is feasible and cost effective. BNL is actively pursuing this approach with Fermilab and LBL. This may supply the answer to Rabi's query on who ordered the muon--to make a muon collider.

View From Washington: Perspectives on Future High Energy Physics

Martha Krebs

Director of Energy Research, U.S. Department of Energy

The Office of Energy Research in the U.S. Department of Energy (DOE) is the only federal entity that concentrates on the science of energy. Whether we examine the fundamental nature of energy and matter, or investigate the effects of energy use, or develop new ways of maintaining energy security, DOE's mission in this area is one of the clearest science missions in the Federal government. However, the energy business and the science of energy is vast and multi-disciplinary by nature. And so DOE's programs, including basic energy sciences, fusion research, health and environmental research, nuclear science, and computational technology development, continue to push the boundaries of scientific disciplines — always keeping our focus, though, on our unique responsibilities and capabilities. And one of the most unique responsibilities and capabilities of our research efforts are those found in the science of high energy physics. That is where our competitive advantage resides.

DOE has consistently supported and advanced research and research infrastructure in high energy physics. DOE is far and away the major supporter if we look at the federal budget over time. The Department maintains 28 user facilities that provide access to almost 600 distinct university, industry, and government institutions catering to thousands of researchers nationwide. These facilities help researchers implement experiments across many energy-related disciplines, many of which are directed towards questions in high energy physics. One of the top priorities in the FY 1997 budget is sustaining at adequate levels our efforts in this field. There is, indeed, a commitment in DOE to encourage and enhance research in high energy physics. It is a unique and vital role that DOE plays in the advancement of science.

However, we must realize that the mood in Washington right now is volatile, and not always sensitive to long-term investments in the sight of short-term savings. Federal support of science research and development often falls on the cutting board of federal budget choppers. We know that and we understand the pressures Congress is under. But that does not mean we should discontinue our efforts to educate these decision makers. We need to make sure the country does not fall into the trap of being "penny wise and pound foolish." Investments in science, especially high energy physics, are designed for future-oriented perspectives. Short-term savings should not cripple, and won't in our proposed budget, the long-term benefits that purposeful support of science always generates.

One investment we plan to make is the U.S. involvement in the Large Hadron Collider. Formal negotiations began in January 1996 and final negotiations will take place in Washington this December. We fully intend to support this international project. We realize that international partnerships will help the scientific community develop the infrastructure it needs to answer the questions that point us to the future. Our proposed budget includes this integral part of high energy physics and it is one example of the long-term investments we must convince Congress are essential for continued U.S. leadership in science and technology.

The energy research effort in the federal government rests securely in the Department of Energy. Our mission has not changed -- knowledge generation, knowledge exchange, and knowledge applied for public benefit is still our focus. But we do need to revise our approach in the current climate. We must focus on productivity, outreach on the benefits of our science, and build an integrated research system through partnerships and collaborations. As one of the major science agencies of the government, DOE must balance the political realities of Washington, DC with the demands of successful science. Our goal is to do just that.

The Standard Model and Beyond (In Search of "New Physics")

William J. Marciano

Brookhaven National Laboratory
Upton, New York 11973

Abstract. The status of the Standard Model and guiding role of symmetry in its development are reviewed. Some outstanding problems are surveyed and possible solutions in the form of additional "Hidden Symmetries" are discussed. Experimental approaches to uncover and explore "New Physics" associated with those symmetries are described with emphasis on the utility of high energy colliders. An outlook for the future is given.

STANDARD MODEL OVERVIEW

A Symmetry - A Historical Perspective [1,2]

Since antiquity, symmetry has been synonymous with beauty, simplicity, and harmony. As such, it inspired advances in the art, architecture, science, etc. of ancient civilizations. Nowhere is that influence more apparent than in Greek philosophy and mathematics. The Greeks viewed the circle and sphere as manifestations of nature's perfect symmetries. Their fascination with those forms led to the development of Euclidean Geometry, a tremendous intellectual achievement. It also engendered an appreciation for the regularity of heavenly motion and the birth of astronomy. However, in that case symmetry became an obsession. The complex epicycle celestial model of Ptolemy with circles upon circles became accepted dogma. Failures of that model were perceived as observational distortions due to the imperfections of man and his methods. That viewpoint and the geocentric epicycle model lasted until the Renaissance years of 1500 A.D. Philosophical blindness had stifled the development of the scientific method and led to almost 2000 years of scientific stagnation. A lesson that we should always remember.

B The Age of Reason

The fall of the geocentric epicycle model and rise of the scientific method resulted from the observations and studies of men such as Copernicus, Brahe, Kepler, and Galileo. It culminated with Newton's "Universal Theory of Gravity". Physics overcame metaphysics. Dynamics and equations of motion replaced the aesthetics of pure thought and the idealized symmetry of fantasies. Calculus was invented and the algebraic approach to problem solving largely replaced geometry. An "Age of Reason" resulted in which any problem, scientific or social, was viewed as solvable. Along with that view, the experimental approach prospered and modern science was born. Man's ability to understand the laws of nature made fast steady progress and culminated in the 1860's with Maxwell's equations and the mastery of electromagnetism. Classical physics became so well understood that Michelson made his now famous pronouncement in 1894

> "The more important fundamental laws and facts of Physical Science have all been discovered, and these are now so firmly established that the possibility of their ever being supplanted in consequence of new discoveries is exceedingly remote... Our future discoveries must be looked for in the sixth place of decimals."

This insightful statement is often maligned as an end of physics message of despair (which was not the intention). It is then pointed out that just two years later in 1896 Becquerel's discovery of radioactivity ushered in the age of Modern Physics and its wonderous advances. Interestingly, Michelson's message is at least as appropriate today as it was 100 years ago. Will history repeat itself?

C Symmetry Strikes Back

What happened to symmetry as a guiding principle during the great scientific advances of Newton... Maxwell? During the 19th century, the mathematics of symmetry was formalized by the development of Group Theory (Galois, Lie, and others). Symmetries and their associated conservation laws (energy, momentum, angular momentum etc.) were certainly known and used, often quite elegantly, in physics problem solving, but they seemed to have little to do with fundamentals.

The importance of symmetry in physics was brought into prominence by the three great advances of the early twentieth century [2]: 1) Special Relativity (1905), 2) General Relativity (1916), and 3) Quantum Mechanics (1925). The last of these, Quantum Mechanics, was particularly instrumental for incorporating the language of group theory into the modern physics vocabulary.

In that case, global symmetries were found to be powerful aids in classifying eigenvalue solutions to quantum equations. The elegance and importance of global symmetries was emphasized in the classic textbook by Wigner [3]. However, the prevailing view in those endeavors was that symmetries were useful tools but would be unnecessary if we could exactly solve the equations of motion. Physics respected certain symmetries but was not governed by them.

The more revolutionary view of symmetry as playing a fundamental role in physics came about from the work and insight of Einstein. He first showed that space and time were symmetric, a radical realization. That exact symmetry of nature had been present but hidden in Maxwell's equations. Its unveiling required the genius of Einstein. The resulting symmetry of Poincaré invariance is now at the foundation of elementary particle physics and quantum field theory. The 10 generators of translations, rotations, and boosts provide a group structure for classifying elementary particles as irreducible representations labeled by their Casimir invariants of mass and spin.

Einstein's formulation of the equivalence principle and general relativity was even more fundamental and insightful than space-time symmetry. He showed that invariance under general coordinate transformations, a local symmetry, gave rise to gravitational field equations. Einstein's recognition that

"Symmetry Dictates Dynamics"

is a great legacy [2]. It gave us a profound understanding of how in the case of gravity local symmetry requirements give rise to fundamental interactions. Extensions of that insight are the bases for modern elementary particle physics and the standard model, as well as efforts to go beyond it. Indeed, Einstein's breakthrough was followed by: 1) Herman Weyl's formulation of electromagnetism as following from local U(1) gauge invariance and the introduction of the gauge field potential $A_\mu(x)$. 2) The Yang-Mills generalization of local gauge invariance from U(1) to non-abelian SU(N) symmetries. 3) The Weinberg-Salam [4] $SU(2)_L \times U(1)_Y$ local symmetry of electroweak unification. 4) The emergence of local $SU(3)_c$ quark color symmetry as a complete theory of strong interactions, Quantum Chromodynamics (QCD). Collectively, those advances constitute the "Standard Model" of strong and electroweak interactions. Let me next describe the status of that very successful theory.

D The Standard Model

The $SU(3)_c \times SU(2)_L \times U(1)_Y$ local gauge theory of strong and electroweak interactions accommodates all known elementary particles and elegantly incorporates the proven symmetries and successes of Poincaré invariance, quantum electrodynamics, the Four-Fermi V-A theory, quark model, etc. It correctly predicted weak neutral currents [4] as well as the observed properties of $W^\pm$, Z, and gluons. In addition, because that theory is renormalizable, its predictions can be scrutinized at the quantum loop level by high precision measurements. Remarkably, a wealth of experimental data has now been confronted at 1% or better without any signal of disagreement or inconsistency [5]. Those impressive successes have earned for the $SU(3)_c \times SU(2)_L \times U(1)_Y$ theory the title "Standard Model", a label that describes its acceptance as a proven standard or paradigm against which future experimental findings and alternative theories must be compared.

As a summary of the standard model content, I have illustrated in Table 1 its minimal spectrum of particles along with some of their basic properties. The fermions are grouped into three generations of spin 1/2 leptons and quarks which span an enormous mass range. Experiments are consistent with massless neutrinos as required by the minimal standard model (i.e. no right-handed neutrinos and only a Higgs scalar doublet). There are, however, some hints of very small neutrino masses (and mixing) from solar and atmospheric neutrino experiments. Should non-zero neutrino masses be established, they could be easily accommodated but would point indirectly to "new physics". For example, many grand unified theories (GUTS) naturally predict small neutrino masses. All of the particles in Table 1 have been observed (directly or indirectly) except for the spin 0 Higgs scalar, H.

TABLE 1. Elementary Particles and Their Properties

Particle	Symbol	Spin	Charge	Color	Mass (GeV)	
Electron neutrino	ν_e	1/2	0	0	$< 4.5 \times 10^{-9}$	
Electron	e	1/2	-1	0	0.51×10^{-3}	First
Up quark	u	1/2	2/3	3	5×10^{-3}	Generation
Down quark	d	1/2	-1/3	3	9×10^{-3}	
Muon neutrino	ν_μ	1/2	0	0	$< 1.6 \times 10^{-4}$	
Muon	μ	1/2	-1	0	0.106	Second
Charm quark	c	1/2	2/3	3	1.35	Generation
Strange quark	s	1/2	-1/3	3	0.175	
Tau neutrino	ν_τ	1/2	0	0	$< 2.4 \times 10^{-2}$	
Tau	τ	1/2	-1	0	1.777	Third
Top quark	t	1/2	2/3	3	174 ± 6	Generation
Bottom quark	b	1/2	-1/3	3	4.5	
Photon	γ	1	0	0	0	
W Boson	$W^\pm$	1	± 1	0	80.34 ± 0.12	
Z Boson	Z	1	0	0	91.188 ± 0.002	
Gluon	g	1	0	8	0	
Higgs	H	0	0	0	$70 \leq m_H < 800$	

Quarks and leptons interact by exchanging spin 1 gauge bosons as dictated by the local gauge symmetries. Eight massless gluons couple to the color $SU(3)_c$ charge and mediate strong interactions, while the $W^\pm$, Z, and γ of the $SU(2)_L \times U(1)_Y$ sector are responsible for weak and electromagnetic interactions. The $SU(3)_c$ gauge theory, quantum chromodynamics (QCD), taken on its own, has no arbitrary or free parameters. (It is a perfect theory.) QCD is scale invariant; so, even its dimensionless gauge coupling constant can be traded in for a mass scale [6] (dimensional transmutation), Λ, which merely serves as a unit of mass. All low hadronic masses are proportional to Λ, $m_i = C_i \Lambda$, with the C_i calculable predicted numbers. In principle, non-perturbative schemes such as lattice QCD should be capable of computing the C_i. All that is needed is a powerful computer and clever algorithm. QCD is a beautiful theory, a simple yet elegant field theory capable of explaining all strong interaction spectroscopy and dynamics. Nevertheless, exploring its rich dynamical consequences and subtleties of its non-linearity (confinement, exotic spectroscopy, proton structure, the quark-gluon plasma etc.) remains extremely interesting and may still reveal surprises. We must continue to study it.

In contrast with QCD, the electroweak sector has many arbitrary independent parameters. Most stem from the Higgs sector which is used to break the $SU(2)_L \times U(1)_Y$ symmetry and endow particles with masses. To accommodate observed phenomenology, a complex Higgs scalar isodoublet is appended to the electroweak theory via $\lambda \phi^4$ interactions (the linear sigma model). Remarkably, its vacuum expectation value $v \simeq 246$ GeV, the electroweak scale, is capable of generating all electroweak masses, mixing, and even CP violation. The disparity of particle masses in Table 1 is accommodated by the size of their coupling to the Higgs. Unfortunately, those coupling strengths are arbitrary and merely determined by observation rather than predicted.

It is generally believed that the simple Higgs mechanism is incomplete and "new physics" must emerge as shorter distances (higher energies) are probed. That conviction is based on shortcomings of the Higgs mechanism, e.g. $\lambda \phi^4$ is trivial (non-interacting) when considered alone and exhibits fine-tuning hierarchy problems when embedded in a grand unified theory or theory of gravity (i.e. Why is the electroweak scale so much smaller than the Planck scale, 10^{19} GeV, the natural scale of quantum gravity?). In addition, one hopes that the truly final fundamental theory, we seek, will be free of arbitrary parameters and will elucidate the origin of mass.

What are the parameters of the standard model? If we define our mass units by the electron volt (with $\hbar = c = 1$), then the scale of QCD in an effective 5 flavor scheme [7] using modified minimal subtraction ($\overline{MS}$) is

$$\Lambda_{\overline{MS}}^{(5)} \simeq 209^{+93}_{-72} \text{ MeV} \tag{1}$$

which corresponds to a gauge coupling at scale $\mu = m_Z$

$$\alpha_3(m_Z)_{\overline{MS}} = \frac{g_3^2(m_Z)}{4\pi} = 0.118 \pm 0.007 \qquad (2)$$

where the quoted error is probably overly conservative. In the $SU(2)_L \times U(1)_Y$ electroweak sector, one finds gauge couplings

$$\alpha_2(m_Z)_{\overline{MS}} = \frac{g_2^2(m_Z)}{4\pi} = 0.03382 \pm 0.00005$$
$$\alpha_1(m_Z)_{\overline{MS}} = \frac{g_1^2(m_Z)}{4\pi} = 0.01694 \pm 0.00002 \qquad (3)$$

Note, the values in Eq. (3) are not so much smaller than the QCD coupling in Eq. (2). Indeed, the values of all three couplings can be related via quantum loop renormalizations if we embed the standard model in a grand unified theory (GUT) such as SU(5), SO(10) etc. (broken at $\sim 10^{16}$ GeV), and introduce new physics such as supersymmetry at an intermediate mass scale ~ 1 TeV.

The Higgs mechanism appends a complex scalar doublet field ϕ and its potential

$$\lambda_0(\phi^2 - v_0^2/2)^2$$
$$\phi = \frac{1}{\sqrt{2}} \begin{pmatrix} \phi_1 + i\phi_2 \\ \phi_3 + i\phi_4 \end{pmatrix} \qquad (4)$$

to the theory with a minimum at $|\phi| = v_0/\sqrt{2}$. The gauge coupling of the scalar doublet to $W^\pm$ and Z bosons leads to natural bare masses and coupling relations

$$m_W^0 = m_Z^0 \cos\theta_W^0 = g_2^0 v_0/2$$
$$\tan\theta_W^0 = \sqrt{\frac{3}{5}} \frac{g_1^0}{g_2^0} \qquad (5)$$

The measured value of m_W then implies

$$v_0 \simeq 246 \text{ GeV} \qquad (6)$$

as the scale of electroweak symmetry breaking and source of all electroweak masses. A remnant of that mechanism is a single physical scalar particle H, the Higgs. Its mass is set by the arbitrary parameter λ_0.

$$m_H^0 = \sqrt{2\lambda_0} v_0 \qquad (7)$$

Determination of λ requires a measurement of the Higgs mass, m_H, or a study of longitudinal gauge boson scattering, e.g. $W_L W_L \to W_L W_L$.

The main source of arbitrary electroweak parameters is the Higgs-fermion Yukawa couplings. In the quark sector, there are 18 independent complex couplings of the form $G_{ij}\bar{q}_{i_L}\phi q_{j_R}$ which connect the 3 left-handed doublets and 6 right-handed singlets to the Higgs. That constitutes 36 independent real parameters. Most are, however, unobservable. They reside in undetectable right-handed mixing angles or relative quark phases. Left-over are 6 masses and 4 parameters of the CKM (Cabibbo-Kobayashi-Maskawa) mixing matrix. The quark masses are given in Table 1. The first two rows of CKM elements are (experimentally)

$$\begin{aligned} |V_{ud}| &= 0.9736 \pm 0.0007 \\ |V_{us}| &= 0.2196 \pm 0.0023 \\ |V_{ub}| &= 0.0036 \pm 0.0021 \\ |V_{cd}| &= 0.215 \pm 0.016 \\ |V_{cs}| &= 0.98 \pm 0.12 \\ |V_{cb}| &= 0.040 \pm 0.006 \end{aligned} \tag{8}$$

They are related by and consistent with unitarity. The third row involves the top quark and currently can be only indirectly inferred from loop effects and unitarity considerations. One finds

$$\begin{aligned} |V_{tb}| &= 0.9992 \pm 0.0003 \\ |V_{ts}| &\simeq |V_{cb}| \\ |V_{td}| &\simeq 0.01 \end{aligned} \tag{9}$$

with $|V_{td}|$ roughly determined by B_d^0-$\bar{B}_d^0$ oscillations. It is amusing that $|V_{tb}|$ is the best (indirectly) known CKM element.

Lepton masses are also determined by their Yukawa couplings to the Higgs doublet. If neutrinos have mass, then one expects mixing in the lepton sector analogous to the CKM matrix.

Given the central role of the elementary Higgs mechanism in electroweak symmetry breaking and mass generation, one would certainly like experimental confirmation or negation of the H's existence. In the simple Higgs doublet scenario, it is instructive to examine the $\lambda\phi^4$ sector of the theory after identifying

$$\begin{aligned} w^{\pm} &= (\phi_1 \mp i\phi_2)/\sqrt{2} \\ z &= \phi_4 \\ H &= \phi_3 - v_0 \end{aligned} \tag{10}$$

In terms of those fields, the Higgs potential becomes [8]

$$\mathcal{L}_{\text{int}} = -\lambda_0 \left(w^+ w + \frac{1}{2}z^2 + \frac{1}{2}H^2 + v_0 H \right)^2 \tag{11}$$

From the quadratic terms, one finds three massless Goldstone bosons $w^{\pm}$, z which become longitudinal components of the $W^{\pm}$, Z gauge fields and the physical Higgs scalar, H, with mass $\sqrt{2\lambda_0}\, v_0$. In a sense, the $w^\pm$ and z were discovered when massive $W^\pm$ and Z bosons were found and only the H remains to be uncovered. Finding that remnant of spontaneous symmetry breaking or whatever "new physics" replaces it, is a major goal of high energy physics.

How will the Higgs scalar be discovered? The most likely means of finding the Higgs depends on its mass, m_H; so, let me briefly discuss mass constraints. Searches at LEP via Z decays and more recently ZH production at LEP II have failed to find the Higgs and provide the lower bound

$$m_H \gtrsim 70 \text{ GeV} \quad \text{(LEP)} \tag{12}$$

LEP II will push the Higgs search to about $90 \sim 95$ GeV via $e^+e^- \to ZH$. Beyond that probably requires a new collider facility, although the Fermilab $p\bar{p}$ collider with its upgraded luminosity may be able to cover the 90–130 GeV region via $p\bar{p} \to W^\pm H$ or ZH with $H \to b\bar{b}$ and leptonic tagging of the $W^\pm$ or Z. The LHC should be capable of finding a Higgs in the entire mass range 90–800 GeV. At the lower end of that mass range, one searches for the (rare) loop induced decay $H \to \gamma\gamma$ or $W + H \to b\bar{b}$. That covers the 90–130 GeV region. From 130–182 GeV, the decay $H \to ZZ^*(Z^* = \text{virtual } Z) \to 4$ leptons provides the discovery. For $m_H \gtrsim 182$ GeV, $H \to ZZ \to 4$ leptons should be discernible up to about 800 GeV. Above 800 GeV, the Higgs width becomes rather broad and the signal fades. Indeed, for $m_H \gg m_W$ the Higgs width into gauge boson pairs

$$\Gamma(H \to W^+W^- \text{ or } ZZ) \simeq \frac{3g_2^2}{128\pi}\frac{m_H^3}{m_W^2} \tag{13}$$

grows like m_H^3. (At $m_H \simeq 1$ TeV, $\Gamma_H \simeq 500$ GeV.) The reason for the Γ_H growth is easily seen from Eq. (11). The HW^+W^- coupling for longitudinal (i.e. scalar) W's is given by

$$HWW \text{ coupling} = -2i\lambda_0 v_0 \simeq -ig_2\frac{m_H^2}{2m_W} \tag{14}$$

The Higgs mass grows like $\sqrt{\lambda}$; so, large Higgs mass corresponds to very strong self-coupling and probably indicates underlying new dynamics. If λ is very large, there are various pathologies in the model. For example, examining the S-matrix for $W_L^+W_L^- \to W_L^+W_L^-$ at large $\sqrt{s}$, one finds that perturbative partial wave unitarity in the $J = 0$, $I = 0$ channel requires [9]

$$\left|\frac{5\lambda}{16\pi}\right| < 1/2 \tag{15}$$

which implies $m_H \lesssim 780$ GeV. For larger λ, unitarization of the S-matrix suggests "new physics" such as ρ-like spin 1 mesons. Such resonances would manifest themselves in WW scattering (analogous to $\pi\pi$ scattering), but sorting out signal from background will be difficult and probably requires very high energy colliders with capabilities beyond the LHC.

Although the Higgs scalar is a focus of our quest, it is generally believed that the Higgs mechanism is only part of a larger underlying structure waiting to be uncovered. There may be a whole spectrum of new particles and interactions which provide a deeper understanding of mass generation, quark mixing, CP violation etc. Suggestions regarding what new physics might be expected are based on ideas about symmetry as well as responses to the outstanding Standard Model problems, some of which I briefly now recall.

OUTSTANDING PROBLEMS AND COMPELLING QUESTIONS

Although the standard model accommodates all known phenomenology and must be viewed as one of the great scientific triumphs of the twentieth century, it cannot be the final word. There remain too many open issues which must be resolved. The ad hoc description of mass generation via the Higgs mechanism and unexplained pattern of fermion masses and mixing (including CP violation) are the most unsatisfactory aspects. There are, in addition, many other problems and questions which must also be confronted before we can claim to understand the basic laws of nature. I mention below a few of the compelling questions

<u>Electroweak Symmetry Breaking</u>: Is there an elementary or dynamical Higgs scalar? What is its mass? What are its properties and origin?

<u>Top Quark Physics</u>: Why is top so heavy? What are its properties? Alternatively, why are the other fermions so light?

<u>Fermion Masses, Mixing, and CP Violation</u>: What is the underlying physics of fermion mass generation? How well can we test standard model predictions for quark mixing and CP violation?

<u>Neutrino Masses and Mixing</u>: Do neutrinos have non-zero masses? Are they part of dark matter? Do neutrinos oscillate?

<u>Generations</u>: Why are there 3 generations? Are there exotic heavy fermions?

<u>Parity</u>: Why is parity violated? We accommodate but do not understand the chiral structure of electroweak symmetry.

<u>Non-Standard CP Violation</u>: Is there CP violation beyond the Standard Model? Is it related to the matter-antimatter asymmetry of the universe?

QCD Dynamics: What is the structure of the proton? Can we better understand quark confinement? Are there exotic quark-gluon bound states? Is there a quark-gluon plasma?

Grand Unification: Can we confirm grand unification of strong and electroweak interactions? Is proton decay observable?

Gravity: What is the connection between gravity and the standard model? Why does a hierarchy of mass scales exist?

POSSIBLE ANSWERS - ADDITIONAL SYMMETRIES

Given the success of local gauge invariance in explaining strong and electroweak interactions, it is not surprising that we continue to seek guidance via possible additional symmetries. In fact, most conjectured solutions to the above problems entail local (short-distance) symmetry enlargements which remain hidden until new physics associated with them is uncovered. Let me mention a few leading possibilities.

i) Extra Gauge Bosons: Enlarging the Standard Model gauge group by appending an $SU(2)_R$ or $U(1)'$ symmetry would lead to additional $W_R^\pm$ or Z' gauge bosons. The $SU(2)_R$ appendage has the nice feature of providing Left-Right symmetry and elucidating the origin of parity violation. Additional $U(1)'$ symmetries could result from superstrings or GUTS. Currently, the Fermilab $p\bar{p}$ collider explores the gauge boson mass range ~ 500 GeV and has not seen evidence for such particles. With anticipated luminosity upgrades, they can reach ~ 1 TeV. The LHC should probe as high as 5 TeV.

ii) Grand Unification: Embedding the standard model in a simple gauge group such as $SU(5)$, $SO(10)$, E_6...has some very attractive features. It leads to strong-electroweak unification $\alpha_3^0 = \alpha_2^0 = \alpha_1^0 = \alpha_G^0$ at very short-distances. There is in fact some evidence for such unification in supersymmetric GUTS. Grand Unification also implies proton decay, which if observed would be a revolutionary discovery. The unification scale of $\sim 10^{16}$ GeV is too high for direct high energy probes. Instead, one will have to rely on precision measurements (remember Michelson's prophecy) and searches for forbidden reactions to uncover such very short-distance hidden symmetries.

iii) Technicolor Dynamics: Just as $SU(3)_c$ local gauge invariance leads to rich QCD dynamics, a conjectured much stronger local $SU(N)_{TC}$ symmetry called technicolor would dynamically break $SU(2)_L \times U(1)_Y$ and endow the $W^\pm$ and Z with masses. Such a scenario is attractive but looses appeal when one attempts to generate fermion masses. Very complicated extended technicolor models have been proposed to accomplish that task, but they lack simplicity and are problematic on several fronts. A generic prediction of such models is a plethora of new technicolor spectroscopy and rich dynamics at $\mathcal{O}(1\text{ TeV})$ as well as lower mass pseudo-goldstone bosons. So far, there is no experimental support for technicolor. Progress in that area will likely require experimental guidance. If a new strong dynamics symmetry like technicolor occurs at ~ 1 TeV, much work will be required to resolve its properties and new high energy colliders will be of central importance in that effort.

iv) <u>Supersymmetry</u>: The most radical, most appealing, most ambitious new symmetry idea is supersymmetry (SUSY). It is also the most likely possibility in the opinion of many theorists. The basic idea is to enlarge the Poincaré algebra with an additional spinor generator Q_α (or several such spinors). The resulting graded Lie algebra is the only known way to consistently expand the concept of space-time. That symmetry enlarges irreducible particle representations to include both bosons and fermions. If supersymmetry were exact, every known fermion (boson) would have a degenerate boson (fermion) partner. Since that is not the case, supersymmetry must be broken. But is the breaking at the Planck scale $\sim 10^{19}$ GeV or much closer to the electroweak scale ~ 250 GeV? Only experiment can answer that question.

Motivation for supersymmetry comes from various sources. Extending global supersymmetry to general coordinate transformations leads to supergravity which finds a natural origin in superstrings. Such a scenario can give a finite theory of quantum gravity and solve the hierarchy problem (why is $m_W \ll m_{\text{planck}}$?). It may also turn out to be unique and parameter free. Superstrings could revolutionize both physics and mathematics and radically alter our understanding of nature.

Is SUSY relevant for experimental particle physics? If the scale of SUSY particles is $\lesssim 1$ TeV, the answer is certainly yes. It would imply that every known elementary particle has a supersymmetric partner (sparticles) waiting to be discovered. There are also other exciting implications. Minimal SUSY predicts 5 spin 0 scalars with the lightest Higgs like particle $\lesssim 130$ GeV. The lightest supersymmetric particle (presumably a spin 1/2 neutralino) could be stable and weakly interacting if R-parity is respected. It is a leading cold dark matter candidate. Wouldn't it be amazing if most of the mass in our universe turns out to consist of supersymmetric partner particles.

If supersymmetry is manifested at low energies, then much will be discovered by the next generation of colliders. In fact, SUSY would be a much bigger prize than the Higgs scalar, since it dramatically alters our view of space-time. Currently, the only evidence for SUSY comes from the unification of couplings in a SUSY GUT framework. It will be interesting to see if that hint is in fact the first harbinger of SUSY or merely a misleading coincidence.

EXPERIMENTAL APPROACHES

Given the success of local symmetries and promise of superstrings, perhaps experiments are no longer needed. Instead, one might contemplate an all out theoretical blitz to find an elegant, aesthetically appealing, possibly unique superstring model which explains everything we currently know. Indeed, such a view is consistent with Einstein's famous quote from his 1933 Herbert Spencer Lecture:

> "I am convinced that we can discover by means of purely mathematical constructions the concepts and the laws connecting them with each other, which furnish the key to the understanding of natural phenomena."

It is not possible to find a counter quote from someone with anything close to Einstein's credentials. Instead, I borrow from the fictional super sleuth Sherlock Holmes who said:

> "It is a capital mistake to theorize before one has data. Insensibly one begins to twist facts to suit theories, instead of theories to suit facts."

Anyone who has a pet theory can recognize the wisdom in that statement.

Ultimately, I believe that Einstein will be correct, but it is much too premature to abandon experiments. We have the experimental capabilities to find the Higgs or whatever is responsible for electroweak symmetry breaking. Supersymmetry and other new particles may also be within reach. In addition, revolutionary discoveries may come from non-accelerator physics, e.g. proton decay, cosmic neutrinos etc. We have the technology to push further and that knowhow must be exploited. Indeed, pushing the frontiers of technology has the synergistic benefit of also advancing technology and stimulating new ideas.

To significantly advance our knowledge and address the many compelling questions before us, requires a broad diverse experimental program with lots of discovery potential. It must be capable of testing the standard model but at the same time be sensitive to "new physics". The program should utilize accelerators but also support non-accelerator physics initiatives. Roughly speaking, we must push as hard as possible on the High Energy, High Intensity, and High Precision frontiers.

High energy accelerators take us to new domains where top, Higgs, and "new physics" can be directly produced and studied. Right now, the Fermilab 1.8 TeV $p\bar{p}$ collider has the highest center-of-mass energy of any accelerator in the world and thus has unique discovery potential. Ongoing luminosity upgrades and an energy increase to 2 TeV will make it our premier high energy tool for most of the next decade. The LHC, scheduled for 2005 will take us to 14 TeV with very high luminosity $\simeq 10^{34}$ cm^{-2}s^{-1}. Besides finding the Higgs, it will be

TABLE 2. Existing and anticipated bounds (at 90% C.L.) on various muon-number violating reactions. The last column lists some speculations on how far the bounds might be pushed at upgraded existing or contemplated new facilities. The current bound on $\mu^- Ti \to e^- Ti$ is a preliminary result from SINDRUM II.

Reaction	Current Bound	Ongoing Exp.	Future (?)
$B(\mu^+ \to e^+e^-e^+)$	$< 1.0 \times 10^{-12}$	—	$\sim 10^{-13}$
$R(\mu^- Ti \to e^- Ti)$	$< 7 \times 10^{-13}$	$\sim 2 \times 10^{-14}$ (SINDRUM II)	$\sim 10^{-16}(10^{-18})$
$B(\mu^+ \to e^+\gamma)$	$< 4.9 \times 10^{-11}$	$\sim 4 \times 10^{-12}$ (MEGA)	$\sim 10^{-14}$
$B(K_L \to \mu e)$	$< 2.4 \times 10^{-11}$	$\sim 8 \times 10^{-13}$ (BNL 871)	$\sim 2 \times 10^{-14}$
$B(K^+ \to \pi^+ \mu e)$	$< 2.1 \times 10^{-10}$	$\sim 3 \times 10^{-12}$ (BNL 865)	$\sim 5 \times 10^{-14}$
$B(K_L \to \pi^0 \mu e)$	$< 3.2 \times 10^{-9}$	$\sim 10^{-11}$ (FNAL)	$\sim 10^{-13}$

capable of uncovering supersymmetry, Z' bosons, technicolor or many other scenarios with "new physics" $\lesssim 1$ TeV. Beyond those facilities, new ideas and technologies are required. The Next Linear Collider (e^+e^-) offers an exciting viable possibility. Recently, there has also been growing enthusiasm for a $\mu^+\mu^-$ collider with high energy $\gtrsim 4$ TeV and luminosity $> 10^{35}$ cm^{-2}s^{-1}. Such a facility, if feasible, would be a significant technological leap forward.

A different approach to finding "new physics" involves studies of very rare, or even forbidden processes, including CP violation. Searches for rare μ, K, B, and τ decays, proton decay, neutrino oscillations, electric dipole moments, etc. are all well motivated and could provide big payoffs. Indeed, a discovery in any of those areas would revolutionize our thinking and open up new areas of research. To illustrate the state of affairs, I have given in Table 2 some current bounds on muon number violating μ and K reactions along with projected capabilities of ongoing experiments and suggested future possibilities. I am particularly excited by the prospect of developing a new intense muon source at the AGS with currents of 10^{11}–$10^{13}\mu$/sec. With such a facility, the coherent conversion of a muon into an electron could be searched for at the extraordinary level of 10^{-16}–10^{-18}. Hopefully, searches for rare B and τ decays, such as $\tau \to \mu^+\mu^-\mu^+$ can also make significant advances.

A third means of testing the standard and searching for "new physics" relies on high precision measurements of fundamental parameters such as m_W, m_Z, Γ_Z, $\sin^2 \theta_W$, CKM parameters, etc. Those experiments probe predicted quantum loop effects. A deviation from expectations would signal the presence of physics beyond the standard model. Ultimately, this approach may provide our best test of GUT and superstring structure. (Remember Michelson's quote.)

FUTURE COLLIDERS

High energy experiments have been in somewhat of a lull. We are anxiously waiting for the next dramatic experimental discovery which will rekindle our imaginations. Fortunately, anticipated future collider facilities offer broad discovery potential. Asymmetric B factories will provide new ways to explore CP violation. LEPII will push its e^+e^- center-of-mass energy to $\sqrt{s} \simeq 190$ GeV. If a standard model or SUSY Higgs with mass $\lesssim 90$ GeV exists, it should be found. I think there is a reasonable chance. Perhaps, they will also get a first glimpse of SUSY. On the bread and butter side, the $W^\pm$ mass will be measured to about ± 50 MeV at LEPII. That will provide an interesting constraint on the Higgs mass via quantum loop relations.

On the hadronic collider front, the Fermilab main injector upgrade will allow the $p\bar{p}$ Tevatron to operate at $\sqrt{s} = 2$ TeV and luminosity $10^{32} \sim 10^{33}$. Those improvements broaden the discovery potential while allowing precision measurements and searches for rare B and τ decays. The Higgs mass region of $80 \sim 130$ GeV may be explored via $W^\pm H$ and ZH associated production if the $H \to b\bar{b}$ mode is resolvable [10]. We might also get a glimpse of SUSY.

In the longer term (~ 2005), the LHC pp collider with $\sqrt{s} = 14$ TeV should find the Higgs scalar or tell us it doesn't exist. If SUSY exists $\lesssim 1$ TeV, it will be discovered. Hopefully, completely unexpected revelations will also be made.

Beyond the LHC, various collider options are possible. The Next Linear Collider (NLC) would start e^+e^- collisions at $\sqrt{s} = 500$ GeV and be upgradable to 1–1.5 TeV. It would have high luminosity $> 5 \times 10^{33}$ and e^- polarization. The NLC also offers $\gamma\gamma$, e^-e^-, and $e^-\gamma$ collider options which expand its physics potential. Recently, there has been discussion of possible future e^+e^- colliders with $\sqrt{s} \simeq 5$ TeV, a major step, if achievable. The NLC is a superb tool for studying the Higgs, SUSY, Technicolor etc. [11].

Less advanced possibilities are the $\mu^+\mu^-$ collider and Very Large Hadron Collider (pp with $\sqrt{s} \simeq 100$ TeV). The muon collider concept is extremely interesting, but how can one demonstrate the technology? An effort at BNL will aim to produce very intense muon beams and use them to do physics (such as $\mu^- N \to e^- N$). Such hands on efforts combined with a vigorous R&D program could lead to the First Muon Collider, but at what energy, 91 GeV, 500 GeV, 4 TeV? In my view, the 4 TeV facility is most complementary to the LHC and currently best motivated.

The Very Large Hadron Collider with $\sqrt{s} \simeq 100$ TeV and $\mathcal{L} \simeq 10^{35}$ looks technically feasible but is very expensive. People are working on new ideas to significantly reduce the cost. An interesting study would be a comparison of pp vs. $\mu^+\mu^-$ physics potential at very high energy and luminosity. Such information will be extremely important for determining our communities future course.

OUTLOOK AND COMMENTARY

Given the demise of the SSC, "The future isn't what it used to be." However, we do have the Standard Model. It represents a tremendous scientific achievement and guide to future exploration. Many compelling questions remain. The primary issue, the source of electroweak symmetry breaking and mass generation is nearly within grasp and will be addressed by the next generation of colliders, particularly the LHC.

Where do we go from here? In my view the NLC physics case is extremely compelling [11]. Such a facility must be built, but where and at what cost? Whatever country rises to that challenge is likely to be the leader in high energy physics during the first half of the next century. Upgrades of such a facility offer decades of forefront physics.

The muon collider concept [12] is an idea whose time may be near. Now it requires serious study and R&D. It has the attractive feature of fitting on an existing laboratory site and using the existing infrastructure. If it can work, it should be built.

Does a Very Large Hadron Collider with $\sqrt{s} \simeq 100$ TeV have viability? Our SSC experience suggests a prohibitive cost and difficult construction issues because of its size. However, interesting new ideas about inexpensive magnets and tunnels offer hope.

Perhaps it is appropriate to recall the words of the great experimentalist Ernest Rutherford

"We haven't got the money, so we have to think"

Even if we get some money we better think carefully about our next big initiative. We must find the source of electroweak symmetry breaking and mass generation, open new frontiers, find new symmetries, and continue Einstein's legacy. We don't want to be responsible for another 2000 years (or even 20 years) of scientific stagnation.

REFERENCES

1. W.J. Marciano, Keynote Address in *Proc. of Snowmass 1996*. The write-up in that Proceedings and this talk are very similar.
2. C-N. Yang, Oskar Klein Memorial Lecture; *Phys. Today* **33**,42 (1980); A. Zee, "Fearful Symmetry", Macmillan 1986.
3. E.P. Wigner, *Group Theory and its Application to the Quantum Mechanics of Atomic Spectra*, Academic Press, New York, 1959.
4. S. Weinberg, *Phys. Rev. Lett.* **19** (1967) 1264; A. Salam, in *Elementary Particle Theory*, ed/ N. Svartholm (Almquist & Wiksells, 1968) p. 367.
5. See A. Sirlin, *Comments on Nucl. and Part. Phys.*, **21**, 287 (1994).
6. D. Gross and F. Wilczek, *Phys. Rev. Lett.* **30**, 1323 (1973); H. D. Politzer, *Phys. Rev. Lett.* **30**, 1346 (1973).

7. W. Marciano, *Phys. Rev.* **D29**, 580 (1984).
8. W. Marciano and S. Willenbrock, *Phys. Rev.* **D37**, 2509 (1988).
9. W. Marciano, G. Valencia, and S. Willenbrock, *Phys. Rev.* **D40**, 1725 (1989).
10. A. Stange, W. Marciano, and S. Willenbrock, Phys. Rev. **D50**, 4491 (1994).
11. S. Kuhlman *et al.*, *Physics Goals of the Next Linear Collider*, BNL report 63158.
12. *The $\mu^+\mu^-$ Collider Feasibility Study*, BNL preprint 52503.

Higgs Physics:
An Historical Perspective

Scott Willenbrock

*Department of Physics, University of Illinois at Urbana-Champaign,
1110 West Green Street, Urbana, IL 61801-3080*

Abstract. "Weakly-coupled" and "strongly-coupled" models of electroweak symmetry breaking are introduced by analogy with the Fermi theory of the weak interaction and the low-energy interaction of pions, respectively. The implications of these two classes of models for colliders beyond the LHC and NLC are discussed.

INTRODUCTION

As we have heard repeatedly at this meeting, uncovering the mechanism which breaks the electroweak symmetry will be a crucial step forward in our quest to understand nature at a deeper level. Electroweak symmetry breaking is the target of future accelerators, and we will hear a great deal about the phenomenology of this physics in talks on the LHC, NLC,[a] and $\mu^+\mu^-$ colliders. I have therefore chosen topics with an eye towards minimizing overlap with other talks.

As my title indicates, I discuss electroweak symmetry breaking from an historical perspective. Since we have not yet discovered the physics of electroweak symmetry breaking, this requires some imagination. I begin with a very brief history of the weak interaction. I then discuss the two categories of electroweak symmetry breaking, usually referred to as "weak coupling" and "strong coupling." I next discuss the phenomenological implications of these two categories of electroweak symmetry breaking for accelerators beyond the LHC and NLC. I conclude with a few historical remarks.

---

[a] NLC is used to generically denote an e^+e^- collider with $\sqrt{s} = 0.5 - 1.5$ TeV.

A BRIEF "HISTORY" OF THE WEAK INTERACTION

This section is intended to recall some of the milestones in the development of the standard model of the electroweak interaction. It is too sketchy to be a proper history. Furthermore, it is more a history of the way things could have gone, rather than the way they actually went. For a proper historical account, see Ref. [1].

The study of the weak interaction began 100 years ago, with the discovery of beta decay by H. Becquerel. It was not recognized at the time that this was due to a new force, however. The first "modern" theory of the weak interaction was due to Fermi (1933) [2], and involved a four-fermion interaction Lagrangian

$$\mathcal{L} = G_F \bar\psi \gamma^\mu \psi \bar\psi \gamma_\mu \psi \qquad (1)$$

where G_F is the Fermi constant and the fermion fields are those of the proton, neutron, electron, and electron neutrino.[b] The discovery of the muon in 1947 allowed the study of the weak interaction in another context, and it was soon realized that the weak interaction is the same for electrons and muons. The universality of the weak interaction suggested that it is mediated by a gauge boson, in analogy with quantum electrodynamics, and that the electron and muon have the same weak "charge." However, gauge bosons are exactly massless, while the hypothetical weak gauge boson is necessarily massive, since it yields the Fermi theory at energies much less than the weak-boson mass. This obstacle was surmounted in 1967 by Weinberg and Salam, who argued that the weak interaction is indeed a gauge theory, but with the gauge symmetry spontaneously broken, such that the weak gauge boson acquires a mass [3].

We now jump to 1996, where we have a beautiful theory of the electroweak (and strong) interactions acting on three generations of quarks and leptons. It is fair to say that the gauge interactions are understood, and that the masses of the weak gauge bosons, the W and Z, are understood to be a consequence of electroweak symmetry breaking. However, the mechanism responsible for electroweak symmetry breaking is unknown. This mechanism is also responsible, at least in part, for the fermion masses and the Cabibbo-Kobayashi-Maskawa matrix (including CP violation), so its complete elucidation is essential to our quest to understand nature at a deeper level.

Although we do not know what the electroweak-symmetry-breaking mechanism is, we know that it involves new particles and new forces. It is possible that it will take another 100 years before we completely understand the nature of these particles and forces.

---

[b] Today we know that it is only the left-handed components of the fields that are involved in the interaction. The normalization is also different from that given in Eq. (1).

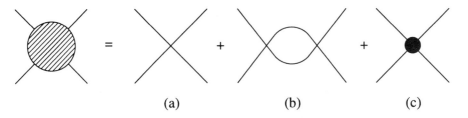

FIGURE 1. Fermion scattering in the Fermi theory of the weak interaction: (a) leading order, (b) next-to-leading-order correction, with the loop-momentum integration cut off at Λ, (c) next-to-leading-order correction from loop momenta greater than Λ.

ELECTROWEAK SYMMETRY BREAKING

Weak Coupling

To explain what is meant by weakly-coupled electroweak symmetry breaking, I draw an analogy with the Fermi theory, which is the low-energy limit of a weakly-coupled theory, namely the electroweak gauge theory.

Fermi Theory

Consider the calculation of a four-fermion amplitude in the Fermi theory of the weak interaction, as shown in Fig. 1. We use the interaction Lagrangian, Eq. (1), to calculate the amplitude perturbatively. We can use dimensional analysis to derive the dependence of the amplitude on the typical energy in the process, E. The Fermi constant, G_F, has units of inverse energy squared. Since the amplitude is dimensionless, the leading term in the amplitude, from the diagram in Fig. 1(a), is proportional to $G_F E^2$.

At next order in perturbation theory, one has the one-loop diagram in Fig. 1(b). Using dimensional analysis, we see that this diagram is proportional to $G_F^2 E^4$. Thus we are performing a perturbative expansion in powers of the dimensionless quantity $G_F E^2$.

This statement is in fact correct, but the argument is somewhat more subtle, because the loop integration is ultraviolet divergent. The modern attitude towards this divergence is encompassed by the idea of an "effective field theory" [4,5]. Let's say we only believe the Fermi theory up to some energy $\Lambda \gg E$. We restrict the loop integration to momenta less than Λ, and profess ignorance about what happens for momenta greater than Λ. However, we cannot simply throw away the contribution from momenta greater than Λ.[c] Instead,

---

[c] This is reminiscent of a quote from the early days of renormalization theory: "Just because something is infinite does not mean it is zero!" [6].

we parameterize it by adding additional terms to the interaction Lagrangian of the form

$$\mathcal{L} = c\, G_F^2 \partial^\nu(\bar\psi\gamma^\mu\psi)\partial_\nu(\bar\psi\gamma_\mu\psi) + ... \qquad (2)$$

which are characterized by having two derivatives, and an unknown dimensionless coefficient c. These terms yield a tree-level contribution to the amplitude proportional to $c\, G_F^2 E^4$, indicated in Fig. 1(c). Thus the next-to-leading-order amplitude is given schematically by

$$A = G_F E^2 + G_F^2 E^4 \ln\Lambda + c\, G_F^2 E^4 \qquad (3)$$

where the ultraviolet divergence of the loop diagram is evidenced by the dependence of the amplitude on $\ln\Lambda$. Combining the last two terms using $c' \equiv c + \ln\Lambda$ yields

$$A = G_F E^2 + c'\, G_F^2 E^4 \qquad (4)$$

where c' is to be taken from experiment. Thus the amplitude is indeed an expansion in powers of the dimensionless quantity $G_F E^2$, despite the ultraviolet divergence.

At low energies, $E \ll G_F^{-1/2}$, only the first few terms in the expansion are numerically important, and the amplitude depends on just a few coefficients which must be extracted from experiment. However, for $E \sim G_F^{-1/2}$, every term in the expansion is equally important, and the amplitude depends on an infinite number of unknown coefficients. The theory therefore loses all predictive power. It is sometimes said that the theory "breaks down" or becomes "strongly-interacting," but in fact the theory simply becomes useless.

There is a quick way to estimate the critical energy at which the theory becomes unpredictive, which also allows us to get the numerical factors straight. Unitarity implies that the partial waves of a two-particle scattering amplitude have a real part which does not exceed 1/2 in magnitude. If the leading term in the expansion were to greatly exceed this bound, the higher-order terms in the expansion would have to be just as large as the leading term in order to restore unitarity. But if that were the case, then every term in the expansion is equally important, and the theory is unpredictive. So we can estimate the critical energy by imposing the unitarity bound on the leading-order amplitude. This yields [7]

$$E_{critical} \approx \left(\frac{\sqrt{2}\pi}{G_F}\right)^{1/2} \approx 600 \text{ GeV}. \qquad (5)$$

Thus the Fermi theory loses predictive power for $E \sim 600$ GeV.

It is a logical possibility that physics simply becomes very complicated at energies in excess of the critical energy. However, history has taught us to

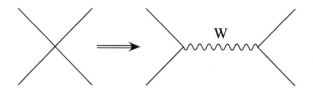

FIGURE 2. The Fermi theory of the weak interaction is the low-energy approximation to a gauge theory.

expect the opposite; the theory becomes simpler at higher energy. In the case of the Fermi theory, the four-fermion interaction is replaced by the exchange of a weak gauge boson, as shown in Fig. 2. The amplitude is proportional to

$$A \sim g^2 \frac{E^2}{E^2 - M_W^2} \qquad (6)$$

where g is the gauge coupling, and the denominator is from the W propagator. Since

$$G_F \sim g^2/M_W^2 \qquad (7)$$

the amplitude reduces to that of the Fermi theory for $E \ll M_W$. However, for $E \gg M_W$, the amplitude is proportional to g^2. Hence the expansion parameter is a (small) constant, and the theory is predictive for all energies.

In the case of the Fermi theory, new physics, in the form of the W boson, enters before the critical energy. Formally,

$$M_W^2 \sim \frac{g^2}{4\pi} E_{critical}^2 < E_{critical}^2, \qquad (8)$$

using Eqs. (5) and (7).[d] The fact that the new physics enters before the critical energy can thus be related to the fact that $g^2/4\pi < 1$, i.e., that the theory is weakly coupled.

Electroweak Theory

In the previous section, the W boson was responsible for regulating the growth of the four-fermion amplitude at high energy. Now consider the scattering of the W bosons themselves, in particular longitudinal (helicity-zero) W bosons, as shown in Fig. 3. The top row of Feynman diagrams depicts the gauge interactions responsible for the scattering. The resulting amplitude is proportional to $G_F E^2$, just as in the case of the Fermi theory. Using unitarity, we estimate the critical energy to be [8]

---

[d] More precisely, using $G_F/\sqrt{2} = g^2/8M_W^2$.

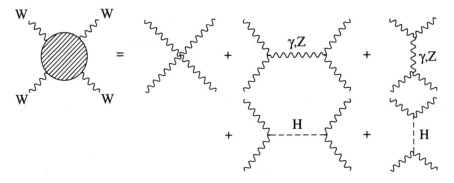

FIGURE 3. Weak-vector-boson scattering in the standard model of the electroweak interaction.

$$E_{critical} \approx \left(\frac{4\sqrt{2}\pi}{G_F}\right)^{1/2} \approx 1.2 \text{ TeV} . \qquad (9)$$

Thus the electroweak theory loses predictive power for $E \sim 1.2$ TeV.

By analogy with the Fermi theory, we might expect new physics, in the form of a new particle, to regulate the growth of the amplitude. This is exactly what happens in the standard Higgs model. The Higgs boson gives rise to additional Feynman diagrams, shown in the second row in Fig. 3. These diagrams cancel the terms proportional to $G_F E^2$, leaving behind an amplitude proportional to $G_F m_H^2 \sim \lambda$, where λ is the Higgs self-coupling. Thus the expansion parameter is a (small) constant, and the theory is predictive for all energies.[e] Higgs-higgs scattering is also proportional to λ, so the theory is complete.

The formal relation between the Higgs mass and the critical energy is

$$m_H^2 \sim \frac{\lambda}{4\pi} E_{critical}^2 < E_{critical}^2 \qquad (10)$$

where the last relation relies on $\lambda/4\pi < 1$, known from non-perturbative studies of the standard Higgs model [9]. Thus the new physics, namely the Higgs boson, enters before the critical energy because the theory is weakly coupled. In fact, it has been shown that the standard Higgs model, and variations of it (such as two Higgs doublets, as used in supersymmetric models), are the only weakly-coupled models of electroweak symmetry breaking [10].

Strong Coupling

To explain what is meant by strongly-coupled electroweak symmetry breaking, I draw an analogy with low-energy pion physics, which is the low-energy

---

[e] Ignoring the running of λ, to be discussed in a later footnote.

limit of a strongly-coupled theory, namely QCD.

Pion theory

The low-energy interaction of pions is dictated by the fact that they are the (approximate) Goldstone bosons of spontaneously-broken chiral symmetry. This is embodied by an effective field theory of pions called "chiral perturbation theory" [4]. The leading interaction of pions is

$$\mathcal{L} = G_\pi \pi \cdot \partial^\mu \pi \pi \cdot \partial_\mu \pi \qquad (11)$$

where $G_\pi = 1/(2f_\pi^2)$, with f_π the pion decay constant. The coupling G_π has dimensions of inverse energy squared, in analogy with the Fermi constant, G_F.

The leading-order amplitude for $\pi\pi$ scattering is proportional to $G_\pi E^2$. Using unitarity to estimate the critical energy at which the theory loses predictive power yields

$$E_{critical} \sim \left(\frac{4\pi}{G_\pi}\right)^{1/2} \approx 450 \text{ MeV}. \qquad (12)$$

What actually happens in nature in $\pi\pi$ scattering at 450 MeV? One begins to encounter a plethora of new particles, beginning with the σ meson.[f] The theory of pion interactions becomes complicated above 450 MeV, and loses all predictive power. Unlike the case of a weakly-coupled theory, there is no new particle (the analogue of a W or a Higgs boson) that restores predictivity to the theory.[g]

Although the pion theory becomes complicated above 450 MeV, we have learned that nature is nevertheless simple: there is a new description of physics in terms of quarks and gluons, interacting via QCD. At low energies this theory is strongly-interacting, and gives rise to all the complications of hadron physics. But even at low energies the theory itself is simple, in the sense that it is described by a Lagrangian with just a few terms in it, and the only parameters are the strong coupling constant and the quark masses.

Electroweak Theory

Now consider WW scattering near the critical energy of 1.2 TeV, and imagine that nature does not provide a Higgs boson to regulate the growth of the

---

[f] After a twenty-year absence, the σ meson has once again been recognized by the Particle Data Group, but they are hedging on its mass: $f_0(400 - 1400)$ [11].
[g] One might be tempted to interpret the σ as the Higgs boson of the pion theory. This interpretation is invalidated by the $\rho(770)$, which is comparable in mass to the σ. A true Higgs theory would have no such particle.

amplitude with energy. What will happen? There is no answer to this question within the electroweak theory; it simply becomes unpredictive. However, based on our experience with pion physics, we might expect that there is a deeper, simpler theory, which is hidden from view by virtue of the fact that it is strongly interacting, and therefore manifests itself in a complicated way. This would mean that there are new particles in nature, experiencing a new strong interaction.

For example, one can imagine that this simpler theory is analogous to QCD. This is the idea behind the so-called Technicolor theory [12,13]. In that case, one would encounter the analogues of the σ, $\rho(770)$, etc., at energies above 1.2 TeV. These "Technimesons" would be made from strongly-interacting "Techniquarks," bound together by "Technigluons." This class of models is reviewed by Appelquist at this symposium.

PHENOMENOLOGY BEYOND THE LHC/NLC

In this section I make some observations about the phenomenological implications of electroweak symmetry breaking for colliders with energy in excess of the LHC ($\sqrt{s} = 14$ TeV) and the NLC ($\sqrt{s} = 0.5 - 1.5$ TeV). I begin where I left off, with the case of strongly-coupled electroweak symmetry breaking.

Strong Coupling

Let's continue our analogy with pion physics. To study the strong interaction at low energy, one performs scattering experiments involving pions. For example, one can perform $\pi\pi$ scattering experiments, or study the coupling of the pion to virtual photons, as depicted in Fig. 4(a). The analogues of these processes for the electroweak interaction are shown in Fig. 4(b), where the longitudinal W bosons replace the pions. The incident fermions could be either quarks and antiquarks (LHC) or electrons and positrons (NLC). These colliders will be capable of probing these processes at energies of about 1 TeV, comparable to the critical energy of 1.2 TeV at which the theory loses predictive power.

Recall that the critical energy for $\pi\pi$ scattering is about 450 MeV. How much would we know about the strong interaction if we had data from $\pi\pi$ scattering and $\gamma^* \to \pi\pi$ at energies only up to 450 MeV? The answer is very little. We might know about the σ, but we would see only the low-energy tail of the ρ, and the heavier mesons would be completely out of sight. More importantly, we would not know that the mesons are composed of strongly-interacting quarks, interacting via QCD.

The moral is that if the mechanism of electroweak symmetry breaking is indeed strongly-coupled, it will likely require energies greatly in excess of 1 TeV for the complete elucidation of this physics. This implies the need for

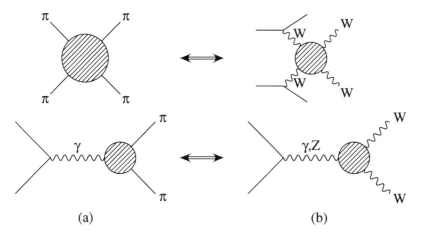

FIGURE 4. (a) Experimental probes of the pion and (b) their counterparts for weak vector bosons.

colliders beyond the LHC and NLC. Although these machines are likely to tell us *something* about strongly-coupled electroweak symmetry-breaking, it is hard to imagine they will be able to tell us *everything* about it.

Weak Coupling

The case of weakly-coupled electroweak symmetry breaking is apparently in stark contrast with that of the strongly-coupled case. For example, the standard Higgs model has a Higgs boson with mass less than 700 GeV [9], and nothing else. The LHC and NLC are both capable of discovering this particle. Once discovered, and its coupling to itself and other particles measured, we have learned everything there is to learn about electroweak symmetry breaking. Can it really be this simple?

Probably not. Although the standard Higgs model is predictive at all energies, it suffers from another disease - it is "unnatural" [14]. We know that the Higgs field acquires a vacuum-expectation value of $v = (\sqrt{2}G_F)^{-1/2} \approx 250$ GeV. The diagram in Fig. 5(a) is one of the one-loop corrections to the vacuum-expectation value, from a Higgs loop. There are similar one-loop corrections from loops of weak bosons and fermions. All these one-loop diagrams share the feature that they are quadratically divergent [13]. Let us regard the standard Higgs model as being valid up to some energy Λ. The relation between the bare vacuum-expectation value, v_0, and the actual vacuum-expectation value, v, may be approximated by cutting off the loop integration at momenta of order Λ:

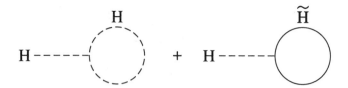

FIGURE 5. (a) Quadratically-divergent one-loop correction to the Higgs vacuum-expectation value from a Higgs loop; (b) the quadratic divergence is cancelled by a Higgsino loop in a supersymmetric theory.

$$v^2 = v_0^2 + O\left(\frac{\Lambda^2}{(4\pi)^2}\right) \tag{13}$$

where the factor $(4\pi)^2$ is the usual factor which arises from loop diagrams. There is also an (unknown) contribution from momenta greater than Λ, but we will assume that it does not conspire to cancel the contribution from momenta less than Λ.[h] If $\Lambda \gg v$, then v_0 must be tuned to almost exactly cancel the one-loop contribution to the vacuum-expectation value. Such a fine tuning is unnatural. Instead, it is natural to expect $\Lambda/(4\pi) \leq v$. Thus if nature makes use of the standard Higgs model, we anticipate that it is replaced by a more fundamental theory at energy $\Lambda \leq 4\pi v \approx 3$ TeV, regardless of the Higgs mass.[i]

Thus we see that weakly-coupled electroweak symmetry breaking may not be so different from strongly-coupled electroweak symmetry breaking. The strongly-coupled approach involves new physics at the TeV scale, while the weakly-coupled approach suggests new physics (beyond the standard Higgs model) by at least the TeV scale. It could be that this new physics lies well below the TeV scale, in which case it may be possible to discover all of it with the LHC/NLC. But if it really lies at the TeV scale, it will likely require colliders beyond the LHC/NLC for its complete elucidation.

---

[h] It has been argued that just such a cancellation occurs if the underlying theory, to which the standard Higgs model is a low-energy approximation, is conformally invariant [15]. However, a realistic example of such a theory has yet to be constructed.

[i] A separate argument can be given for the incompleteness of the standard Higgs model. The Higgs self-coupling, λ, is a running coupling, and increases with energy (for sufficiently-large Higgs mass), eventually blowing up. New physics must intervene before this occurs. This can be used to place an upper bound on the Higgs mass for a given energy scale of the new physics [16]. For example, if the new physics is at the Planck scale, then there is an upper bound on the Higgs mass of about 200 GeV [17]. However, naturalness implies that there is new physics by at least 3 TeV, so it is not realistic to imagine that there is no new physics until the Planck scale. Imposing the condition that the Higgs coupling not blow up below 3 TeV yields an upper bound on the Higgs mass of only about 700 GeV.

A well-known example of new physics in the weakly-coupled scenario is supersymmetry. If nature is supersymmetric, every particle is accompanied by a superpartner. The one-loop correction to the Higgs vacuum-expectation value in Fig. 5(a) is accompanied by a second diagram, shown in Fig. 5(b), in which the Higgs loop is replaced by a Higgsino loop. The quadratic divergences of each diagram cancel, and the theory becomes natural. The same sort of cancellation occurs for loops of weak bosons and fermions and their superpartners.

If nature is supersymmetric, then supersymmetry must be broken, since we don't observe the superpartners of the known particles. Let's refer to the mass scale of the superpartners as M_{SUSY}. It M_{SUSY} were much larger than v, the quadratic divergence would reappear, cut off only when the momenta reach M_{SUSY}:

$$v^2 = v_0^2 + O\left(\frac{M_{SUSY}^2}{(4\pi)^2}\right) \tag{14}$$

Thus the model is only natural if $M_{SUSY} < 4\pi v \approx 3$ TeV. More careful estimates yield a somewhat lower scale [18].

If M_{SUSY} is at the TeV scale, it may require colliders beyond the LHC/NLC to discover all the superpartners and measure their couplings. However, if M_{SUSY} is a few hundred GeV, then all the physics of supersymmetry may be within reach of the LHC/NLC. This is the only natural scenario I know of in which the entire physics of electroweak symmetry breaking is elucidated by these colliders. However, even in this scenario, there may be motivation for higher-energy colliders. The physics of supersymmetry breaking may lie at the 10 TeV scale, as in the class of models in which dynamical supersymmetry breaking is communicated to the standard model via new gauge interactions [19]. It is possible that the elucidation of the physics of electroweak symmetry breaking will reveal to us the scale of supersymmetry breaking.

It is also possible that the new physics which subsumes the standard Higgs model at the TeV scale is strongly-interacting, and has nothing to do with supersymmetry [20]. This class of models is similar to the strongly-coupled electroweak-symmetry-breaking scenario, with the exception that there is a Higgs boson. The complete elucidation of these models would likely require colliders beyond the LHC/NLC, as in the strongly-coupled models.

CONCLUDING HISTORICAL REMARKS

The history of the weak interaction was punctuated by periods of confusion, followed by clarification, which ultimately led to a beautiful theory. For example, the early measurements of beta decay indicated that the emitted electron was monoenergetic. It took twenty years to establish that the electron is emitted with a continuous spectrum of energies. This was the first evidence for the existence of the neutrino, which was a vital ingredient in Fermi's theory of the weak interaction [21].

The same sequence of events is likely to occur for the electroweak-symmetry-breaking mechanism. In fact, we are already in the first stage, the period of confusion. There are already several experimental results which have been interpreted as a Higgs boson [22–28]. These interpretations have been made either in the context of a multi-Higgs model, or with additional new physics – none has had an interpretation in terms of the standard Higgs model with no new physics.

A particularly noteworthy example is Ref. [26], which was an attempt to interpret the $\zeta(8.3)$ as a Higgs boson with enhanced coupling to b quarks.[j] This would imply an enhanced coupling to muons, which led the authors to conclude that if their interpretation were correct, "the case for construction of a dedicated muon collider for Higgs boson studies will become as compelling as it is technically feasible." The idea of using a muon collider for resonant Higgs production (even the standard Higgs) has lately resurfaced and been considerably refined [29].

My final historical remark concerns our ability to foresee the physics of the future. There are many examples of incorrect theories and prognostications, many more than correct ones, and these are often held up as examples of our inability to predict what will be found in the next generation of colliders. While I have some sympathy for this point, I also feel it can be overstated. A well-known example is the top-quark mass; predictions ranged from 15 GeV (just above the reach of PEP) to 230 GeV, and everything in between. But perhaps this makes us lose sight of the real achievement: we knew the top quark existed long before it was discovered, based entirely on indirect evidence.

Today many feel that there is indirect evidence for weak-scale supersymmetry. Although the evidence is not as compelling as it was for the top quark, an advocate might make the following statement:

> "There can be no two opinions about the practical utility of supersymmetry, but until clear experimental evidence for the existence of supersymmetry can be obtained, supersymmetry must remain purely hypothetical. Failure to detect any evidence of supersymmetry is no evidence against its existence."

In fact, this quote is paraphrased from a 1937 meeting of the Royal Society, except I have substituted "supersymmetry" for "the neutrino" throughout [30].

---

[j] The $\zeta(8.3)$ was conjectured to be responsible for a monoenergetic photon signal in Υ decay, via $\Upsilon \to \zeta\gamma$. The experiment turned out to be erroneous.

ACKNOWLEDGEMENTS

I am grateful for conversations with D. Dicus, R. Leigh, and W. Marciano, and for assistance from Z. Sullivan. This work was supported in part by the Department of Energy under Grant No. DE-FG02-91ER40677 and by the National Science Foundation under Grant No. PHY94-07194.

REFERENCES

1. A. Pais, *Inward Bound* (Oxford University Press, Oxford, 1986).
2. E. Fermi, Z. Phys. **88** 161 (1934).
3. S. Weinberg, Phys. Rev. Lett. **19**, 1264 (1967); A. Salam, in *Elementary Particle Physics*, edited by N. Svartholm (Almqvist and Wiksell, Stockholm, 1968), p. 367.
4. S. Weinberg, Physica **96A**, 327 (1979).
5. S. Weinberg, *The Quantum Theory of Fields* (Cambridge University Press, Cambridge, 1995), Vol. I.
6. Ref. [5], p. 35.
7. E. Abers and B. Lee, Phys. Rep. **9**, 1 (1973).
8. M. Chanowitz and M. K. Gaillard, Nucl. Phys. **B261**, 379 (1985); W. Marciano, G. Valencia, and S. Willenbrock, Phys. Rev. D **40**, 1725 (1989).
9. M. Lüscher and P. Weisz, Phys. Lett. **B212**, 472 (1988).
10. J. M. Cornwall, D. Levin, and G. Tiktopoulos, Phys. Rev. Lett. **30**, 1268 (1973); Phys. Rev. D **10**, 1145 (1974); C. Llewellyn-Smith, Phys. Lett. **B46**, 233 (1973).
11. Review of Particle Properties, Phys. Rev. D **54**, 1 (1996).
12. S. Weinberg, Phys. Rev. D **19**, 1277 (1979).
13. L. Susskind, Phys. Rev. D **20**, 2619 (1979).
14. G. 't Hooft, in *Recent Developments in Gauge Theories*, Proceedings of the NATO Advanced Study Institute, Cargèse, 1979, eds. G. 't Hooft *et al.* (Plenum, New York, 1980), p. 135.
15. W. Bardeen, FERMILAB-CONF-95-391-T, to appear in the *Proceedings of the 1995 Ontake Summer Institute*, Ontake Mountain, Japan, Aug. 27 - Sept. 2, 1995.
16. M. Lindner, Z. Phys. C **31**, 295 (1986).
17. L. Maini, G. Parisi, and R. Petronzio, Nucl. Phys. **B136**, 115 (1978); N. Cabbibo, L. Maini, G. Parisi, and R. Petronzio, Nucl. Phys. **B158**, 295 (1979).
18. G. Anderson and D. Castano, Phys. Rev. D **52**, 1693 (1995).
19. M. Dine and A. Nelson, Phys. Rev. D **48**, 1277 (1993).
20. D. Kaplan and H. Georgi, Phys. Lett. **B136**, 183 (1984); D. Kaplan, S. Dimopoulos, and H. Georgi, Phys. Lett. **B136**, 187 (1984).
21. Ref. [1], p. 11.
22. D. McKay and H. Munczek, Phys. Rev. Lett. **34**, 432 (1975).
23. E. Ma, S. Pakvasa, and S. Tuan, Phys. Rev. D **16**, 568 (1977).

24. R. Willey, Phys. Rev. Lett. **52**, 585 (1984).
25. H. Georgi and S. Glashow, Phys. Lett. **B143**, 155 (1984).
26. S. Glashow and M. Machacek, Phys. Lett. **B145**, 302 (1984).
27. H. Haber and G. Kane, Nucl. Phys. **B250**, 716 (1985).
28. M. Shin, H. Georgi, and M. Axenides, Nucl. Phys. **B253**, 205 (1985).
29. V. Barger, M. Berger, J. Gunion, and T. Han, Phys. Rev. Lett. **75**, 1462 (1995;)hep-ph/9602415.
30. Ref. [1], p. 18.

A Simplified Summary of Supersymmetry

John F. Gunion*

*Davis Institute for High Energy Physics, Department of Physics,
University of California at Davis, Davis CA 95616*

Abstract. I given an overview of the motivations for and theory/phenomenology of supersymmetry.

INTRODUCTION

The overview consists of three parts. Namely,

- WHY do theorists find supersymmetry so attractive?
- WHAT do we expect to see experimentally?
- HOW do we go about observing what is predicted?

In each part, I will give only the barest outline of the relevant issues and discussions. In order to keep the presentation simple, I will often be less than precise in the technicalities. Well-known results will not be referenced in detail. Further discussion and references can be found, for example, in [1–4].

WHY SUPERSYMMETRY IS ATTRACTIVE

First, there are some very general aesthetic considerations.

- SUSY is the only non-trivial extension of the Lorentz group which lies at the heart of quantum field theory; the simplest such extension is referred to as $N = 1$ supersymmetry and requires the introduction of a single (two-component) spinorial (anti-commuting) dimension to space, the extra dimension(s) being denoted θ. Taylor expansions of a superfield $\hat{\Phi}(x,\theta)$ then take the form $\hat{\Phi}(x,\theta) = \phi(x) + \sqrt{2}(\theta\psi(x)) + (\theta\theta)F(x)$ (higher orders in θ being zero), implying an automatic association of a spin-1/2 field $\psi(x)$ with every spin-0 field $\phi(x)$ and vice-versa ($F(x)$ is an auxiliary field, *i.e.* it does not represent a dynamically independent field degree of freedom).

- If SUSY is formulated as a *local* symmetry, then a spin-2 (graviton) field must be introduced, thereby leading automatically to (SUGRA) models in which gravity is unified with the other interactions. Further, SUGRA reduces to general relativity in the appropriate limit.

- SUSY appears in superstrings.

- Historically, adding the Lorentz group to quantum mechanics required introducing an antiparticle for every particle; why should history not repeat — SUSY + quantum field theory requires a 'sparticle' for every particle.

Of course, since we have not detected any of the superpartners of the known Standard Model (SM) particles, it is clear that supersymmetry is a broken symmetry. Thus, it is possible that supersymmetry could be irrelevant at the energy scales where experiments can currently be performed. However, there are many reasons to suppose that the superpartners have masses below a TeV and, therefore, could be discovered anytime now.

- String theory solutions with a non-supersymmetric *ground* state are quite problematic, implying that all the physical states of the theory should have similar masses aside from the effects of supersymmetry breaking (which should be a perturbation on the basic string theory solution).

- SUSY solves the hierarchy problem, *e.g.* $m_{\text{Higgs}}^2 < (1 \text{ TeV})^2$ (as required to avoid a non-perturbative WW sector), via spin-1/2 loop cancellation of spin-0 loop quadratic divergences:

$$m_H^2 \sim [m_H^0]^2 + \lambda^2(m_{\text{boson}}^2 - m_{\text{fermion}}^2) \ln \frac{\Lambda^2}{\langle m_b^2, m_f^2 \rangle}, \quad (m_f \sim m_{\text{SUSY}}); \quad (1)$$

m_H^2 can be small if $[m_H^0]^2$ is, *provided* $m_f^2 \lesssim (1 \text{ TeV})^2$.

- SUSY implies gauge coupling unification (at $M_U \sim$ few $\times 10^{16}$ GeV) if $m_{\text{SUSY}} \lesssim 1 - 10$ TeV and there is nothing (other than complete SU(5) representations) between m_{SUSY} and M_U. Of course there are some qualifications to this statement.

  a) Perturbative unification requires that the number of families (only complete families lead to unification) must be ≤ 4 [5].

  b) There is a possible difficulty in the string theory context in that $M_U < M_S$ if we accept the perturbative result that $M_S = M_{\text{Planck}}/\sqrt{8\pi} \sim 2 \times 10^{18}$ GeV. However, it has been emphasized [6] that in non-perturbative approaches it may be possible for M_S to be substantially lower, in principle even as low as a TeV.

c) Unification takes place only if there are exactly 2 Higgs doublets (+ singlets) below M_U. This is also the minimal Higgs field content required to give both up and down quarks masses and to guarantee anomaly cancellation.

d) The precise scale $m_{\rm SUSY}$ preferred for exact unification depends upon the precise value of $\alpha_s(m_Z)$. For currently accepted values of $\alpha_s(m_Z) \lesssim 0.118$, an effective $m_{\rm SUSY}$ value above 1 TeV is seemingly preferred although other subtle issues could alter this preference [7].

- Electroweak symmetry breaking (EWSB) occurs automatically for the simplest universal boundary conditions at M_U by virtue of the fact that the mass-squared parameter for the Higgs field coupled to the top-quark ($m_{H_2}^2$) is driven negative (during evolution from M_U down to m_Z) by the large top-quark Yukawa contribution to its renormalization group equation (RGE). The associated symmetry breaking occurs very naturally at an energy scale in the vicinity of m_Z if $m_{H_2}^2 < (1-2 \text{ TeV})^2$ at M_U.

- If the supersymmetric partner of the QCD gluon has mass below ~ 1 TeV, then most GUT boundary conditions imply that the lightest supersymmetric particle (LSP) will have mass on the 100 GeV scale and would interact weakly and have other properties that make it a natural candidate for the cold dark matter of the universe. However, this is only true if this LSP is essentially stable. In certain variants of supersymmetry, to be discussed later, this is not the case.

WHAT WE EXPECT TO SEE

A) Sparticles: in the minimal supersymmetric model (MSSM) every normal SM particle has its supersymmetric counterpart.

$$[u,d,c,s,t,b]_{L,R} \quad [e,\mu,\tau]_{L,R} \quad [\nu_{e,\mu,\tau}]_L \quad g \quad \underbrace{W^\pm, H^\pm}_{\tilde{\chi}^\pm_{1,2}} \quad \underbrace{\gamma, Z, H_1^0, H_2^0}_{\tilde{\chi}^0_{1,2,3,4}}$$

$$[\tilde{u},\tilde{d},\tilde{c},\tilde{s},\tilde{t},\tilde{b}]_{L,R} \quad [\tilde{e},\tilde{\mu},\tilde{\tau}]_{L,R} \quad [\tilde{\nu}_{e,\mu,\tau}]_L \quad \tilde{g}$$

The quark, lepton and neutrino partners are the spin-0 squarks, sleptons and sneutrinos; the partner of the gluon is the spin-1/2 gluino; and the partners of the charged (neutral) vector bosons and Higgs bosons are the spin-1/2 charginos (neutralinos). Often the latter can be approximately separated into the spin-1/2 bino and wino gaugino partners of the U(1) B and SU(2) W gauge fields and the higgsino partners of the Higgs fields. In other cases, these states are strongly intermixed.

There is a possibly exact discrete symmetry of the theory, called R-parity, such that SM particles have $R = +$ while the sparticle partners have $R = -$. If R-parity is an exact symmetry then any physical process must always involve an even number of sparticles, and the LSP, normally the $\tilde\chi_1^0$, will be stable against decay to SM particles.

B) A **very** special Higgs sector [8].

- For the minimal two-doublet Higgs sector, the physical eigenstates comprise two CP-even scalars (h^0, H^0), a CP-odd scalar (A^0), and a charged Higgs pair ($H^\pm$).
- SUSY implies that the Higgs self couplings have strength of order g, the SU(2) SM coupling, which has the consequence that $m_{h^0} \leq$ 130 GeV ($\leq$ 150 GeV, if singlet Higgs fields are added).
- For boundary conditions such that EWSB is an automatic consequence of the RGE's, $m_{H^0} \sim m_{A^0} \sim m_{H^\pm} > 200$ GeV is very probable, in which case the h^0 will have properties very much like those predicted for the SM Higgs (h_{SM}) in the minimal one-doublet SM, while the H^0, A^0 decouple from WW, ZZ.

Thus, there is little question as to what we should see, but the very uncertain nature of supersymmetry breaking implies that there is a great deal of uncertainty as to the exact mass scale at which we should see the new sparticles and as to the new experimental signatures that will appear when sparticles are produced.

There is one important general point. If R-parity is exact, the sparticles must be produced in pairs. (Limits on R-parity violation suggest that single sparticle production is at best very weak in any case.) Thus, in order to observe supersymmetric particles at hadron machines we must have significant gg and/or $q\bar q$ luminosity at $\sqrt{s}_{q\bar q, gg}$ above (very roughly) $2m_{\rm SUSY}$; at an e^+e^- or $\mu^+\mu^-$ collider the $\sqrt{s}$ must exceed the sum of the masses of the the two sparticles one hopes to observe. Large masses ($\gtrsim 1$ TeV) for some sparticles are certainly possible, in which case a lepton collider with $\sqrt{s} \sim 3-4$ TeV will be needed.

With regard to the Higgs sector, the h^0 is guaranteed to be light and should be easily detected, perhaps even at LEP2 or the Tevatron. If a light h^0 is not found, then we must abandon the possibility of low-energy supersymmetry as we now understand it. In the $m_{A^0} \sim m_{H^0} \sim m_{H^\pm} > 200$ GeV decoupling limit, the heavier Higgs must be pair produced, e.g. $e^+e^-, \mu^+\mu^- \to H^0 A^0, H^+ H^-$. Energy reach could again prove to be crucial.

Supersymmetry breaking

The couplings of the complete complex of sparticles and particles are simultaneously fixed by the superpotential, denoted W, which involves products of superfields (with their particle and sparticle component fields). As a result, almost all couplings involving sparticles are related by 'Clebsch-Gordon' factors to the couplings of their SM counterparts. The only exception is the possible presence in W of R-parity violating ($\not R$) couplings. I shall temporarily ignore such couplings, but will return later to this subject.

Most of the uncertainty in phenomenology is related to the many possible scenarios for supersymmetry breaking, which lead to many different predictions for the masses (especially relative masses) of the sparticles and for the detailed experimental signatures that will be present when they are produced. The main constraint on supersymmetry breaking is that it should be 'soft' in the sense that it should not destroy the very attractive SUSY solution to the naturalness and hierarchy problems. The possible supersymmetry breaking terms in the Lagrangian can then be enumerated.

- gaugino masses: $M_i \lambda_i \lambda_i$ where $i = 3, 2, 1$ for SU(3), SU(2), U(1) and λ_i denotes the spin-1/2 partner of the corresponding gauge field.

- scalar masses: e.g.

$$m_{H_1}^2 |H_1|^2 + m_{H_2}^2 |H_2|^2 + m_{(\tilde{t},\tilde{b})_L}^2 (\tilde{t}_L^\star \tilde{t}_L + \tilde{b}_L^\star \tilde{b}_L) + m_{\tilde{t}_R}^2 \tilde{t}_R^\star \tilde{t}_R + m_{\tilde{b}_R}^2 \tilde{b}_R^\star \tilde{b}_R \quad (2)$$

- 'A' terms: e.g. $A_t \lambda_t (\tilde{t}_L H_2^0 - \tilde{b}_L H_2^+) \tilde{t}_R^\star$.

- 'B': $B\mu(H_1^0 H_2^0 - H_1^+ H_2^-)$, where μ is the parameter appearing in the superpotential term $W \ni \mu \widehat{H}_1 \widehat{H}_2$.

Altogether, including CP-violating phases, the above comprise 105 independent and unknown (although limited in magnitude if we are to maintain the naturalness of the model and avoid a charge and/or color breaking ground state) parameters beyond the SM. The obvious question is whether the possibility of so many a priori unknown parameters is a good or bad thing.

- Bad: If we want to know ahead of time exactly what to look for, then it is bad in that the uncertainties associated with so many parameters imply a very large range of phenomenological possibilities.

- Good: If we want to be confident that we will learn something from what we observe, then the existence of so many a priori unknown parameters is good. In particular, by evolving the low-energy parameters up to M_U one can hope to uncover M_U-scale boundary conditions that imply an underlying organization for the 105 parameters that can be associated with an attractive and theoretically compelling GUT/string model.

The lesson for machine builders and experimentalists is that being prepared for a wide range of possibilities as to HOW? we see and fully explore SUSY is a necessary evil. The rest of the talk reviews some of the many possibilities discussed to date.

HOW TO LOOK FOR SUPERSYMMETRY

It will be important to check that the Higg sector fits within the supersymmetric model constraints and to learn whether it contains more than the minimal two doublet fields, in particular whether or not there are additional singlet Higgs fields. Direct discovery of sparticles will be even more important. We would hope to eventually observe all the sparticles of the theory. I give a short description of Higgs phenomenology and then turn to sparticle phenomenology.

Detecting the SUSY Higgs bosons

It has been clearly established that for a SUSY Higgs sector consisting of only the minimal two-doublets at least one of the SUSY Higgs bosons will be detectable at the large hadron collider (LHC) and at the planned $\sqrt{s} \sim$ 500 GeV next linear lepton collider (NLC). In the $m_{A^0} > 2m_Z$ decoupling limit, it is always the h^0 whose observability is guaranteed. Detection of the $H^0, A^0, H^\pm$ is not guaranteed. In particular, at the NLC $H^0 A^0$ and $H^+ H^-$ pair production is not kinematically allowed if $m_{A^0} \gtrsim \sqrt{s}/2$. Also, at the LHC there is a region, see Fig. 1, in the standard $(m_{A^0}, \tan\beta)$ parameter space (which specifies the tree-level properties of the Higgs bosons) characterized by moderate $\tan\beta \gtrsim 3$ and $m_{A^0} > 200$ GeV such that only the h^0 will be detectable.

The two most urgent and best-motivated questions are:

- Is discovery of one Higgs boson guaranteed if singlet Higgs fields are added to the minimal two doublet fields? This question is particularly important in light of the fact that string theories typically lead to a Higgs sector with extra singlets.

- What is required in order to discover the $H^0, A^0, H^\pm$ of the MSSM? Here, the region of concern is the $m_{A^0} > 200$ GeV parameter region that is natural when EWSB is automatically broken by virtue of the RGE's.

Both questions have been explored in the literature and I summarize the conclusions to date.

- It can be demonstrated [10-13] that at least one of the light CP-even Higgs bosons ($h_{1,2,3}$) of a two-doublet plus one singlet Higgs sector will be observed at the NLC. This result follows from the fact that if the lightest (h_1) has weak ZZ coupling then (one of) the heavier ones must actually be almost as light and have substantial ZZ coupling (and thus be discoverable in the $e^+e^- \to Z^\star \to Zh$ production mode). This result probably extends to the inclusion of several singlet Higgs fields.

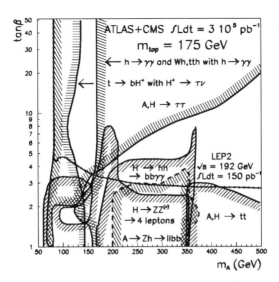

FIGURE 1. Higgs discovery contours (5σ) in the parameter space of the minimal supersymmetric model for ATLAS+CMS at the LHC: $L = 300$ fb^{-1}. Two-loop/RGE-improved radiative corrections to the MSSM Higgs sector are included assuming $m_{\tilde{t}} = 1$ TeV and no squark mixing. From Ref. [9].

- At a $\mu^+\mu^-$ collider, direct s-channel $\mu^+\mu^- \to h_1$ production is guaranteed to be visible [14]. Only the (predicted) ($\mu^+\mu^- h_1$) coupling is needed; the h_1 can be decoupled from ZZ,WW without affecting its detectability. A scan search in the ≤ 150 GeV region is required with $\Delta E_{\text{beam}} \lesssim 0.01\%$.

- At the LHC, discovery of at least one Higgs boson is no longer guaranteed if a singlet Higgs field is present in addition to the minimal two doublet fields [15]. The no-discovery "holes" in parameter space are never terribly large, but are certainly not insignificant when $\tan\beta$ is in the moderate $\tan\beta \sim 3 - 10$ range.

- At an e^+e^- collider, the only means for detecting any of the $H^0, A^0, H^\pm$ when $m_{A^0} \gtrsim \sqrt{s}/2$ (so that $H^0 A^0, H^+ H^-$ pair production is forbidden) is to run the collider in the $\gamma\gamma$ collider mode [16,17]. In this way, single $\gamma\gamma \to H^0, A^0$ production can be observed for $m_{A^0} \lesssim 0.8\sqrt{s}$ provided high integrated luminosity (e.g. $L \sim 200$ fb^{-1}) is accumulated. There is still no certainty that a $\gamma\gamma$ collider facility will be included in the final NLC plans, nor regarding the luminosity that would be available.

- At a $\mu^+\mu^-$ collider, direct $\mu^+\mu^- \to H^0, A^0$ production will be observable for any m_{A^0} between $\sqrt{s}/2$ (the pair production limit) and $\sqrt{s}$ provided that $\tan\beta \gtrsim 3$ (below this the LHC will find the H^0, A^0, see Fig. 1) and that $L = 100$ fb^{-1} (for $\sqrt{s} \sim 500$ GeV) is available for the $\sqrt{s}/2 \to \sqrt{s}$ scan [18]. It must be possible to maintain high instantaneous luminosity throughout the scan region (perhaps requiring two cheap storage rings designed for optimal luminosity in different $\sqrt{s}$ ranges).

- At both the e^+e^- and $\mu^+\mu^-$ collider, it is envisioned that the machine will be upgraded to increasingly large $\sqrt{s}$. Once $\sqrt{s} \gtrsim 2m_{A^0}$, $H^0 A^0$ and H^+H^- pair production will be observable. It has been shown that this will be true even if these Higgs bosons have substantial decays to sparticles [19,20]. The hierarchy motivations for supersymmetry suggest that m_{A^0} will certainly lie below $1-1.5$ TeV. A strong form of naturalness suggests $m_{A^0} \lesssim 500$ GeV [21].

If a SM-like h^0 is observed, the most precise determination of all its properties requires *both* $\sqrt{s} = 500$ GeV running at the NLC *and* an s-channel scan of the resonance peak in $\mu^+\mu^- \to h^0$ collisions at $\sqrt{s} \sim m_{h^0}$ [22]. This argues strongly for having *both* types of machine, especially since very substantial $L = 200$ fb^{-1} accumulated luminosity is needed at *both* machines in order to achieve good precision for a model-independent determination of all the Higgs couplings and its total width. Once the H^0, A^0 and/or $H^\pm$ have been detected (whether in pair production or single production at the muon collider), the relative branching ratios for $H^0, A^0, H^\pm$ decays to different types of channels can be measured and very strong constraints will be placed on a host of the important parameters of the SUSY model, and thus on the GUT boundary conditions [19,20]. At a muon collider, $\mu^+\mu^- \to H^0, A^0$ studies would require substantial L at a $\sqrt{s}$ that is almost certain to be significantly different than that employed for $\mu^+\mu^- \to h^0$ studies. Perhaps more than one muon collider will turn out to be needed.

Detecting the sparticles

The phenomenology of sparticles is determined by the nature and source of supersymmetry breaking and whether or not R-parity is broken. A selection of possibilities is discussed below. A useful recent review covering some of the following theoretical material is [23].

1 Minimal Supergravity (mSUGRA)

This is the simplest and most fully investigated model. It assumes that the boundary conditions are set at M_U, that there is a desert in between M_U and $m_{\rm SUSY}$, and that R-parity is exact. Further, at M_U the supersymmetry breaking parameters listed earlier are taken to be universal and to have no

CP-violating relative phases. Universal boundary conditions are natural[1] in the picture where SUSY-breaking (S$\cancel{U}$SY) arises in a hidden sector and is only communicated to the visible sector at the GUT/string scale via interactions involving gravity (which of course knows nothing about quantum numbers). The result is a model specified by just five parameters and one sign. The GUT boundary conditions are:

$$m_{H_1} = m_{H_2} = m_{\tilde{q}_i} = m_{\tilde{\ell}_i} \ldots = m_0, \quad M_3 = M_2 = M_1 = m_{1/2},$$

$$A_{ijk} = A_0, \quad |\mu|, \quad \text{sign}(\mu), \quad B.$$

Beginning with these boundary conditions at M_U, evolution to low energies $\lesssim m_{\text{SUSY}}$ yields some simple and important results.

- $M_3 : M_2 : M_1 \sim \alpha_3 : \alpha_2 : \alpha_3 \sim 7 : 2 : 1$, leading to a similar ratio of the ($\overline{\text{MS}}$) gluino to wino to bino masses.[2]

- Approximate squark degeneracy is maintained for the first two generations, implying acceptably small FCNC.

- The third family stops are normally significantly mixed and split, and the lightest squark will be the lighter $\tilde{t}_1$ eigenstate.

- EWSB is automatic, as described earlier — it is convenient to trade the parameters $|\mu|$ and B for $\tan\beta$ and the (known) value of m_Z.

- $|\mu|$ and $m_{A^0} \sim m_{H^0} \sim m_{H^\pm}$ are typically large (> 200 GeV) unless $\tan\beta \gg 1$;

- In the usual case where $\mu > M_2$, one finds the following: the LSP is the (stable) $\tilde{\chi}_1^0 \simeq \tilde{B}$ with $m_{\tilde{\chi}_1^0} \sim M_1$ and is a good (cold) dark matter candidate; the lightest chargino ($\tilde{\chi}_1^\pm$) and 2nd lightest neutralino ($\tilde{\chi}_2^0$) are approximately SU(2) charged and neutral winos and have similar mass $\sim M_2$; the heavier chargino ($\tilde{\chi}_2^\pm$) and neutralinos ($\tilde{\chi}_{3,4}^0$) are charged and neutral higgsinos with masses $\sim |\mu|$.

- If $m_0^2 \gg m_{1/2}^2$, then squarks, sleptons, *etc.* are heavier than the lighter (gaugino) $\tilde{\chi}$'s and are approximately degenerate;

- If $m_0^2 \ll m_{1/2}^2$, then $m_{\tilde{\ell}} < m_{\tilde{q}}$ and $m_{\tilde{\ell}_i}$ is distinctly smaller than the other $m_{\tilde{q}_i}$; charged sleptons can be lighter than most gauginos and it is possible (not included in later discussions) that the LSP could be the $\tilde{\nu}_\tau$.

The experimental signatures are determined by how the sparticles decay. In mSUGRA, the above-outlined mass hierarchy $m_{\tilde{\chi}_{3,4}^0} \sim m_{\tilde{\chi}_2^\pm} > m_{\tilde{\chi}_2^0, \tilde{\chi}_1^\pm} > m_{\tilde{\chi}_1^0}$

---

[1] This assumes a sufficiently simple form for the Kahler potential.
[2] The gluino pole mass is substantially larger than its $\overline{\text{MS}}$ mass.

implies that most signatures will result from chain decays — *e.g.*

$$\tilde{g} \to q\bar{q}\tilde{\chi}_1^\pm \to q\bar{q} + \begin{cases} \ell^\pm \nu \tilde{\chi}_1^0 \\ q\bar{q}\tilde{\chi}_1^0 \end{cases}$$

(It is important to note that $\tilde{g}$ decay leads with equal probability to either ℓ^+ or ℓ^-.) Since masses and mass differences are both substantial, events will be characterized by large $\not{E}_T$ from the final $\tilde{\chi}_1^0$'s and fairly energetic jets and leptons.

◇ Tevatron and LHC

Extensive Monte Carlo studies have determined the region of mSUGRA parameter space for which direct discovery of sparticles will be possible. Cascade decays lead to events with jets, missing energy, and various numbers of leptons. The Tev33 option at Fermilab will cover [24] the most natural [21] portion of parameter space. The maximum reach at the LHC is in the 1ℓ+jets+$\not{E}_T$ channel; one will be able to discover squarks and gluinos with masses up to several TeV [25]. Some particularly important types of events are the following.

- $pp \to \tilde{g}\tilde{g} \to$ jets + $\not{E}_T$ and $\ell^\pm \ell^\pm$+jets+$\not{E}_T$, the latter being the like-sign dilepton signal [26]. The mass difference $m_{\tilde{g}} - m_{\tilde{\chi}_1^\pm}$ can be determined from jet spectra end points [26–28], while $m_{\tilde{\chi}_1^\pm} - m_{\tilde{\chi}_1^0}$ can be determined from ℓ spectra end points in the like-sign channel [26–28] — an absolute scale for $m_{\tilde{g}}$ can be estimated ($\pm 15\%$) by separating the like-sign events into two hemispheres corresponding to the two $\tilde{g}$'s [26], by a similar separation in the jets+$\not{E}_T$ channel [25] or variations thereof [27,28].

- $pp \to \tilde{\chi}_1^\pm \tilde{\chi}_2^0 \to (\ell^\pm \tilde{\chi}_1^0)(Z^* \tilde{\chi}_1^0)$, which yields a trilepton + $\not{E}_T$ final state when $Z^* \to \ell^+ \ell^-$; $m_{\tilde{\chi}_2^0} - m_{\tilde{\chi}_1^0}$ is easily determined if enough events are available [29].

- $pp \to \tilde{\ell}\tilde{\ell} \to 2\ell + \not{E}_T$, detectable at the LHC for $m_{\tilde{\ell}} \lesssim 300$ GeV [25], *i.e.* the region of parameter space favored by mixed dark matter cosmology [30].

- Squarks will be pair produced and, for $m_0 \gg m_{1/2}$, would lead to $\tilde{g}\tilde{g}$ events with two extra jets emerging from the primary $\tilde{q} \to q\tilde{g}$ decays.

◇ NLC

Important discovery modes include the following [31].

- $e^+e^- \to \tilde{\chi}_1^+ \tilde{\chi}_1^- \to (q\bar{q}\tilde{\chi}_1^0$ or $\ell\nu\tilde{\chi}_1^0) + (q\bar{q}\tilde{\chi}_1^0$ or $\ell\nu\tilde{\chi}_1^0)$; $m_{\tilde{\chi}_1^\pm}$ and $m_{\tilde{\chi}_1^0}$ will be well-measured using spectra end points and beam energy constraints.

- $e^+e^- \to \tilde{\ell}\tilde{\ell} \to (\ell\tilde{\chi}_1^0)(\ell\tilde{\chi}_1^0), (\nu\tilde{\chi}_1^-)(\bar{\nu}\tilde{\chi}_1^+), \ldots$; masses will be well-measured.

The $\tilde{q}$'s and (if m_0 is big) $\tilde{\ell}$'s can be too heavy for pair production at $\sqrt{s} = 500$ GeV. If $m_0 \sim 1-1.5$ TeV (the upper limit allowed by naturalness), then

$\sqrt{s} \gtrsim 2-3$ TeV is required. It could be that such energies will be more easily achieved in $\mu^+\mu^-$ collisions than in e^+e^- collisions.

For certain choices of the mSUGRA parameters, in can happen that the phenomenology is much more 'peculiar' than the canonical situation described above. One such choice of boundary conditions is that termed the **'Standard' Snowmass 96 Point** [28]:

$$m_0 = 200 \text{ GeV}, \quad m_{1/2} = 100 \text{ GeV}, \quad A_0 = 0, \quad \tan\beta = 2, \quad \mu < 0.$$

These boundary conditions predict that the gluino mass is approximately the same as the average squark mass; only the $\tilde{t}_1$ and $\tilde{b}_L$ are lighter than the $\tilde{g}$. Consequently, $\tilde{g} \to \tilde{b}_L b$ almost 100% of the time. Also, masses are not very large: e.g. $m_{\tilde{g}} = 285$ GeV, $m_{\tilde{b}_L} = 266$ GeV. At the LHC, there will be millions of spectacular events with $\tilde{g}\tilde{g} \to b\bar{b}\tilde{b}_L\tilde{b}_L \to \ldots$. At the NLC, all $\tilde{q}\tilde{q}$ thresholds are $> \sqrt{s} = 500$ GeV, but $\tilde{\ell}\tilde{\ell}$ and $\tilde{\chi}_1^+\tilde{\chi}_1^-, \tilde{\chi}_{1,2}^0\tilde{\chi}_{1,2}^0$ pair processes are all kinematically allowed.

There are particular choices for mSUGRA boundary conditions that have strong theoretical motivation in the context of strings and/or supergravity. These include:

♣ No-Scale [32]: $m_{1/2} \neq 0$, $m_0 = A_0 = 0$;

♣ Dilaton or Dilaton-Like [33]: $m_{1/2} = -A_0 = \sqrt{3}m_0$

Here the dilaton is the string 'modulus' field associated with coupling constant strength in string models. To many, the dilaton-like boundary conditions appear to be particularly worthy of being taken seriously. To deviate significantly from dilaton-like boundary conditions in Calabi-Yau and Orbifold models requires going to an extreme in which the dilaton has nothing, or almost nothing, to do with supersymmetry breaking. The dilaton and no-scale models have many common features. Most importantly, the small (or zero) m_0 compared to $m_{1/2}$ implies that sleptons are light, which in turn leads to excellent LEP2, TeV33, LHC, and NLC leptonic signals for supersymmetry [34], the only uncertainty being the overall rates as determined by the overall mass scale, $m_{1/2}$. ($\tan\beta$ and sign(μ) are the only other free parameters in these models.)

A quite different set of mSUGRA boundary conditions is that motivated by the assumption that there is hidden-sector dynamical SUSY breaking without gauge singlets. In this case it is natural for all dimension-3 SUSY-breaking operators in the low-energy theory to be very small, i.e. $m_{1/2}, A \sim 0$ (and possibly $B \sim 0$ as well) [35]. Such boundary conditions, in combination with existing experimental constraints, imply that the gluino would be lighter than the lightest neutralino [36], and would tend to emerge as part of a relatively long-lived gluon-gluino bound state (denoted R^0) with mass ~ 1.4 GeV (according to lattice calculations). Remarkably, this scenario cannot yet be absolutely excluded by accelerator experiments [37]. In addition, it is possible for the photino in such a model to have the necessary properties to be a dark matter candidate [38]. Ongoing experiments and analyses will be able to exclude this scenario in the near future. For example, for such boundary conditions the lightest chargino must have $m_{\tilde{\chi}_1^\pm} < m_W$ and should be discovered at LEP2 once substantial luminosity is accumulated at $\sqrt{s} \sim 190$ GeV, despite the non-canonical nature of predicted signals. Searches for hadrons containing gluinos could also provide strong constraints [39].

2 Beyond mSUGRA: Non-universality

Despite the very attractive nature of the mSUGRA boundary conditions, it is certainly possible to find motivation for many possible sources of non-universality. The two general classes of non-universality are gaugino mass non-universality and scalar mass non-universality. Ref. [40] provides a useful review and references. Due to lack of space I discuss only (and very briefly) gaugino mass non-universality. Models characterized by such non-universality include:[3]

- The O-II orbifold string model in which all matter fields lie in the $n = -1$ untwisted sector and SUSY-breaking is dominated by the overall size modulus (not the dilaton).

- F-term supersymmetry breaking with $F \neq$ SU(5) singlet, leading to $\mathcal{L} \sim \frac{\langle F \rangle_{ab}}{M_{\text{Planck}}} \lambda_a \lambda_b$, where $\lambda_{a,b}$ ($a, b = 1, 2, 3$) are the gaugino fields. If F is an SU(5) singlet then $\langle F \rangle_{ab} \propto c \delta_{ab}$ and we get standard universality, $M_1 = M_2 = M_3$ (at M_U). But, more generally F can belong to a non-singlet SU(5) representation: $F \in (\mathbf{24 \times 24})_{\text{symmetric}} = \mathbf{1} \oplus \mathbf{24} \oplus \mathbf{75} \oplus \mathbf{200}$. which implies that $\langle F \rangle_{ab} = c_a \delta_{ab}$, with c_a depending on the representation (an arbitrary superposition of representations is also possible).

The $M_{3,2,1}$ for these different cases are compared in Table 1.

I have room for only a few remarks regarding how the phenomenology changes as a function of boundary condition scenario. I focus on the most extreme changes. See Ref. [40] for more details and references.

- The general relations that $m_{\tilde{\chi}_1^0} \sim \min(M_1, M_2)$ and $m_{\tilde{\chi}_1^\pm} \sim M_2$ imply that $m_{\tilde{\chi}_1^\pm} \simeq m_{\tilde{\chi}_1^0}$ (both are winos) in the $\mathbf{200}$ and O-II scenarios, where $M_2 \lesssim M_1$. Some important consequences of this result are [41]:

    - At the NLC, $e^+e^- \to \tilde{\chi}_1^+ \tilde{\chi}_1^-$ would be very hard to see since the (invisible) $\tilde{\chi}_1^0$ would take all of the $\tilde{\chi}_1^\pm$ energy in the $\tilde{\chi}_1^\pm \to \ell^\pm \nu \tilde{\chi}_1^0, q'\bar{q}\tilde{\chi}_1^0$ decays. One must employ $e^+e^- \to \gamma \tilde{\chi}_1^+ \tilde{\chi}_1^-$.

---

[3] Readers not familiar with string terminology can simply take these specifications as names.

TABLE 1. M_a at M_U and m_Z for the four F_Φ irr. reps. and in the $\delta_{GS} \sim -4$ O-II model. From Ref. [40].

F	M_U			m_Z		
	M_3	M_2	M_1	M_3	M_2	M_1
1	1	1	1	~ 7	~ 2	~ 1
24	2	-3	-1	~ 14	~ -6	~ -1
75	1	3	-5	~ 7	~ 6	~ -5
200	1	2	10	~ 7	~ 4	~ 10
$O-II$, $\delta_{GS}=-4$	1	5	$\frac{53}{5}$	~ 6	~ 10	$\sim \frac{53}{5}$

- At the LHC, the like-sign dilepton signal (coming again from $\tilde{\chi}_1^\pm \to \ell^\pm \nu \tilde{\chi}_1^0$ decays) would be very weak. In the O-II model, $m_{\tilde{g}} \sim m_{\tilde{\chi}_1^\pm} \simeq m_{\tilde{\chi}_1^0}$ means that the jets from $\tilde{g} \to q\bar{q}\tilde{\chi}_1^0$ decay would also be soft; the standard jets+$\not{E}_T$ signature would be much weaker than normal and the maximum $m_{\tilde{g}}$ for which SUSY could be discovered would be smaller than in the usual case.

- Dark matter phenomenology would be substantially altered. In particular, the very close degeneracy $m_{\tilde{\chi}_1^\pm} \simeq m_{\tilde{\chi}_1^0}$ implies very similar $\tilde{\chi}_1^\pm, \tilde{\chi}_1^0$ densities (due to very similar Boltzmann factors) at freeze-out which in turn leads to $\tilde{\chi}_1^0 \tilde{\chi}_1^\pm$ 'co-annihilation' that greatly reduces relic dark matter ($\tilde{\chi}_1^0$) density.

• Generally speaking, if $m_{\tilde{g}}$ is not large then distinguishing between the five scenarios would be quite easy. For example, if we keep m_0 and $m_{\tilde{g}}$ the same as at the Snowmass 96 point described earlier, there would be many millions of $\tilde{g}\tilde{g}$ pair events at the LHC, and the different scenarios would lead to the following very different event characteristics [40]:

  - **1**: $\not{E}_T$, $\ell^+\ell^-$, $4b$'s;
  - **24**: $\not{E}_T$, no $\ell^+\ell^-$, $8b$'s, with 2 pairs having mass m_{h^0}.
  - **75**: traditional cascade chain decay signals;
  - **200**: $\not{E}_T$, $4b$'s, no leptons;
  - **O-II**: $\tilde{g} \to \tilde{\chi}_1^\pm, \tilde{\chi}_1^0$ + soft and $\tilde{\chi}_1^\pm \to \tilde{\chi}_1^0$ + very soft; soft jet cuts required to observe.

3 R-parity violating models

I next consider the possibility that R-parity is violated. R-parity violation can come in two different forms:

- Hard: that is explicit terms in the superpotential of the form $W_{\not R} = \lambda_{ijk}(\hat{L}_i\hat{L}_j)\widehat{E}_k + \lambda'_{ijk}(\hat{L}_i\hat{Q}_j)\widehat{D}_k + \lambda''_{ijk}\widehat{U}_i\widehat{D}_j\widehat{D}_k$.

- Soft: $W_{\not R} = \mu_i\hat{L}_i\widehat{H}_1$

I make a few brief remarks regarding the former. (Phenomenology for the latter is reviewed in Ref. [42].) First, $\lambda, \lambda' \neq 0$ implies lepton-number violation, while $\lambda'' \neq 0$ implies baryon-number violation. Both cannot be present; if they were, the proton lifetime would be short. Otherwise, current constraints are not strong [43]: λ's $\lesssim 0.1 - 0.01$. Phenomenologically, the crucial point is that unless the λ's are *very, very* small, the LSP $\tilde{\chi}_1^0$ will decay inside the detector. Thus, the signals for supersymmetry will no longer involve missing energy associated with the $\tilde{\chi}_1^0$. Impacts on phenomenology are substantial.

At the LHC [44]:

- If $\lambda'' \neq 0$, then $\tilde{\chi}_1^0 \to 3j$. The large jet backgrounds imply that we would need to rely on the like-sign dilepton signal. For universal boundary conditions, this signal turns out to be sufficient for supersymmetry discovery out to $m_{\tilde{g}}$ values somewhat above 1 TeV. However, if the leptons are very soft, as occurs if $m_{\tilde{\chi}_1^\pm} \sim m_{\tilde{\chi}_1^0}$ (as in the **200** and O-II models) then the discovery reach would be much reduced — the combination of $\not R$ and $M_2 \lesssim M_1$ would be a bad scenario for the LHC.

- If $\lambda \neq 0$, $\tilde{\chi}_1^0 \to \mu^\pm e^\mp \nu, e^\pm e^\mp \nu$, and there would be many very distinctive multi-lepton signals.

- If $\lambda' \neq 0$, $\tilde{\chi}_1^0 \to \ell 2j$ and again there would be distinctive multi-lepton signals.

At the NLC:

- Even $e^+e^- \to \tilde{\chi}_1^0\tilde{\chi}_1^0 \to \underbrace{(3j)(3j)}_{\lambda''\neq 0}, \underbrace{(2\ell\nu)(2\ell\nu)}_{\lambda\neq 0}, \underbrace{(\ell 2j)(\ell 2j)}_{\lambda'\neq 0}$ could yield an observable SUSY signal.

At HERA:

- Squark production via R-parity violating couplings could be an explanation for the HERA anomaly at high x and large Q^2 [45]. For example, if λ'_{111} is near its current upper limit, a lepto-quark like signal of the type $e^+d \to \tilde{u} \to e^+d$ would be detected if $m_{\tilde{u}} \lesssim 220$ GeV.

Finally, if R-parity is violated then supersymmetry will no longer provide a source for dark matter, the LSP no longer being stable. One would have to turn to neutrinos (which, however, only provide a source of hot dark matter, whereas some cold dark matter also seems to be needed).

4 Event-Motivated Models

A number of specific supersymmetry breaking parameter choices have been proposed in order to explain particular anomalous events seen at the Tevatron and at LEP. Here, I only have room to list some of these and make a few remarks. In all cases the M_U-scale boundary conditions would be required to be non-universal (assuming the usual desert between low energy and M_U).

CDF $ee\gamma\gamma$ event = $\tilde{e}\tilde{e}$, $\tilde{\chi}_1^+\tilde{\chi}_1^-$, ... production?

Let us focus on the $\tilde{e}\tilde{e}$ explanation. One proposed explanation of this event [46] as $\tilde{e}\tilde{e}$ production requires that the one-loop decay $\tilde{\chi}_2^0 \to \gamma\tilde{\chi}_1^0$ dominate over all tree level decays *and* that $BF(\tilde{e} \to e\tilde{\chi}_2^0) \gg BF(\tilde{e} \to e\tilde{\chi}_1^0)$. If both are true, then the process

$$\tilde{e}\tilde{e} \to (e\tilde{\chi}_2^0)(e\tilde{\chi}_2^0) \to (e\gamma\tilde{\chi}_1^0)(e\gamma\tilde{\chi}_1^0)$$

would explain the observed event provided masses are also appropriate. For the above decays to be dominant requires $\tilde{\chi}_2^0 = \tilde{\gamma}$ and $\tilde{\chi}_1^0 =$ higgsino, which in turn is only the case if $M_2 \sim M_1$ $\tan\beta \sim 1$ $|\mu| < M_{1,2}$. This, in combination with the observed kinematics of the event and a cross section consistent with its having been detected, implies masses for all the neutralinos and charginos that are mostly in the $50 - 150$ GeV range. For such masses there should be a large number of other equally distinctive events in the $L \sim 100$ pb^{-1} of Tevatron data currently being analyzed and clear signals should also emerge at LEP when run at $\sqrt{s} \sim 190$ GeV. At the Tevatron, $L \sim 100$ pb^{-1} implies $N(2\ell + X + \not{E}_T) \geq 30$, $N(\gamma\gamma + X + \not{E}_T) \geq 2$, $N(\ell\gamma + X + \not{E}_T) \geq 15$, $N(2\ell\gamma + X + \not{E}_T) \geq 4$, $N(\ell 2\gamma + X + \not{E}_T) \geq 2$, $N(3\ell + X + \not{E}_T) \geq 2$. At LEP190 with $L = 500$ pb^{-1}, one finds $N(2\ell + X + \not{E}_T) \geq 50$, $N(2\gamma + X + \not{E}_T) \geq 3$. In the above, $X =$ additional leptons, photons, jets. Finally, I note that the small μ value needed implies that we must have soft mass non-universality in the form $m_{H_1}^2 \neq m_{\tilde{q},\tilde{\ell}}^2$.

CDF and D0 'di-lepton top events' that don't look like top events

There are two events in the CDF di-lepton top sample and one event in the D0 sample that have very low probability, in terms of their kinematics, to be top events. It has been proposed [47] that these could be from $\tilde{q} \to q\tilde{\chi} \to q(\nu\tilde{\ell})$ with $\tilde{\ell} \to \ell\tilde{\chi}_1^0$, where $\tilde{\chi} = \widetilde{W}_3, \widetilde{W}^\pm$. For this explanation to work in terms of kinematics and cross section requires: $m_{\tilde{g}} \simeq 330$, $m_{\tilde{q}} \simeq 310$, $m_{\tilde{\ell}_L} \simeq 220$, $m_{\tilde{\nu}_\ell} \sim 220$, $m_{\tilde{\ell}_R} \simeq 130$, $\mu \sim -400$, $M_1 \simeq 50$, and $M_2 \simeq 260$. Note that these values are inconsistent with the previous $ee\gamma\gamma$ event explanation. An abundance of other signals should be hidden in the full Tevatron data set for such parameter choices.

The Four-Jet 'Signal' at ALEPH

The 4-jet signal at ALEPH is well-known. It has $M + M' \sim 106$ GeV and $M - M' \sim 10$ GeV (i.e. $M \sim 58, M' \sim 48$ GeV), significant 'rapidity-weighted' jet charge and no b jets. One interpretation [48] is as $\tilde{e}_L \tilde{e}_R$ production with $\tilde{e}_L, \tilde{e}_R \to 2j$ via $\slashed{R}$ coupling $W_{\slashed{R}} \ni \lambda'_{ijk}(\hat{L}^i \hat{Q}^j)_L \overline{\hat{D}^k_R}$. $M_1 \lesssim 100$ GeV is required for the needed $\sigma \sim 1 - 2$ pb cross section level, and is also helpful in suppressing $\tilde{e}_L \tilde{e}_L, \tilde{e}_R \tilde{e}_R$ production. To suppress (unobserved) $\tilde{e} \tilde{\nu}_e$ production we need the largest $m_{\tilde{\nu}_e}$ possible, which occurs for $\tan\beta \sim 1$. For $\tilde{e}_L \to \overline{u}_j d_k$ to dominate over the more standard $\tilde{e}_L \to e e \tilde{e}_R$ decays via $\tilde{\chi}^0_1$ exchange requires $\lambda_{1jk} \gtrsim$ few $\times 10^{-4}$, which is entirely consistent with known bounds. In this picture, the $\tilde{e}_R$ must decay by mixing, $\tilde{e}_R \to \tilde{e}_L \to \overline{u}_j d_k$ rather than via $\tilde{e}_R \to e e \overline{u} d$ from virtual $\tilde{\chi}^0_1, \tilde{e}_L$ exchange; the required mixing angle, $\sin\phi \gtrsim 10^{-4}$, is easily accommodated. One finds that a displaced vertex might be observable. The absence of $\tilde{\mu}$ and $\tilde{\tau}$ pair signals requires that these states have substantially heavier masses than the $\tilde{e}$, as would only be possible if slepton masses are non-universal. Finally, lots of signals would be expected at LEP2: for $M_1 = 100$ GeV, $m_{\tilde{\nu}_e} = 58$ GeV and $\sqrt{s} = 186$ GeV,

$$\sigma(\tilde{e}_L \tilde{e}_R, \tilde{e}_L \tilde{e}_L, \tilde{e}_R \tilde{e}_R, \tilde{\nu}_{eL} \tilde{\nu}_{eL}, \tilde{\chi}^0_1 \tilde{\chi}^0_1) = 1.33, 0.23, 0.2, 0.29, 0.62 \text{ pb}$$

5 Models with gauge-mediated supersymmetry breaking

Underlying all our previous discussions has been the implicit assumption that SUSY is almost certainly an exact local symmetry (as in strings) which, if not accidental, should be spontaneously broken in a 'hidden sector'. The SUSY-breaking is then fed to the ordinary superfields in the form of an effective Lagrangian, $\mathcal{L}_{\text{eff}}$. The mass expansion parameter for $\mathcal{L}_{\text{eff}}$ is the mass scale of the sector responsible for the communication between the hidden sector and the ordinary superfields. For gravity-mediated communication (as appropriate in SUGRA/superstring theories), this mass scale is most naturally $\sim M_{\text{Planck}}$ and one arrives at the $\mathcal{L}_{\text{eff}} \sim \frac{\langle F \rangle}{M_{\text{Planck}}} \lambda_a \lambda_a$ expression given earlier. For gaugino masses of order m_W, this requires $\langle F \rangle \equiv m^2_{\text{SUSY}} \sim m_W M_{\text{Planck}} \sim (10^{11} \text{ GeV})^2$.

The gravity-mediated scenario is certainly very attractive but has its doubters, primarily based on the fact that keeping FCNC phenomena at a sufficiently small level is not guaranteed. Indeed, although gravity being 'flavor blind' suggests universal soft scalar masses, terms that violate universality are certainly possible in the Kahler potential and (since not forbidden by symmetry) are generated at some level by radiative corrections, even if not present at tree-level. Further, a certain amount of FCNC and lepton-number violation can arise during evolution from $M_S \to M_U$, although, as noted earlier, $M_S \sim M_U$ is also possible non-perturbatively. A final point of concern is that EWSB via the RGE's is not guaranteed for all possible boundary condition choices. All these problems can be resolved if the scale of supersymmetry breaking, m_{SUSY}, is much lower.

The specific and popular models that incorporate this idea are the gauge-mediated supersymmetry breaking (GMSB) models. In these models, M_{Planck} above is replaced by M, the mass scale of a new messenger sector, and $m^2_{SUSY} \sim m_W M$ can be much lower — $m_{SUSY} \sim 10 - 100$ TeV is often discussed. The GMSB models have rather few parameter, at least in current incarnations. They lead to dramatic signals, but the signals could well be somewhat different than at first anticipated because of issues related to the mass scale of supersymmetry breaking. I briefly outline the basics [49,4].

In the standard GMSB models there are three sectors.

I: First, there is the SUSY-breaking sector, sometimes called the 'secluded' sector, containing hidden particles that interact via strong gauge interactions which cause supersymmetry breaking characterized by a scale $\sqrt{F}$. The SUSY-breaking is then fed via *two-loops* in the strong interaction into

II: the 'messenger' sector, which contains a SU(3)×SU(2)×U(1) (but not necessarily SU(5)) singlet superfield $\hat{S}$ with non-zero vacuum expectation values for both its scalar component and its F-term component ($\langle S \rangle \neq 0$ and $\langle F_S \rangle \neq 0$). The two-loop communication between the two sectors implies that $\langle F_S \rangle \sim (\alpha_m^2/16\pi^2)F$, where α_m characterizes the strength of the gauge interactions that are responsible for the two-loop communication between the SUSY-breaking sector and the messenger sector. In addition to $\hat{S}$, the messenger sector must contain some messengers that transform under SU(3), SU(2) and U(1). In order to maintain actual gauge-coupling unification, these messengers should form a complete anomaly free GUT representation (*e.g.* **5** + **5̄** with messengers $\hat{q}, \hat{\bar{q}}$ and $\hat{\ell}, \hat{\bar{\ell}}$ superfield triplets and doublets in the SU(5) case). These messenger superfields must communicate with $\hat{S}$; a typical superpotential might be $W = \lambda_1 \hat{S}\hat{q}\hat{\bar{q}} + \lambda_2 \hat{S}\hat{\ell}\hat{\bar{\ell}}$. The mass scales of the component boson (b) and fermion (f) fields of these messenger superfields are given very roughly by

$$M \equiv m_{f_{mess.}} \sim \lambda \langle S \rangle; \quad m^2_{b_{mess.}} \sim \begin{pmatrix} \lambda^2 \langle S \rangle^2 & \lambda \langle F_S \rangle \\ \lambda \langle F_S \rangle & \lambda^2 \langle S \rangle^2 \end{pmatrix} \quad (3)$$

(where λ is the typical $\lambda_{1,2}$), implying that $\lambda \langle S \rangle^2 > \langle F_S \rangle$ is required to avoid a negative mass-squared eigenvalue. Ratios of m_b^2 eigenvalues < 30 (no fine tuning) suggests $\lambda \langle S \rangle^2 / \langle F_S \rangle \gtrsim 1.05$.

III: These messengers communicate SUSY-breaking to the normal superfield sector according to a scale characterized by $\Lambda \equiv \langle F_S \rangle / \langle S \rangle$ (with $M/\Lambda = \lambda \langle S \rangle^2 / \langle F_S \rangle > 1$ required, as noted above). After integrating out the heavy messengers, and assuming that $\hat{S}$ is an SU(5) singlet, the masses of the particles important to low-energy phenomenology are as follows.

- The gauginos acquire masses at one-loop given by

$$M_i(M) = k_i N_{5,10} g\left(\frac{\Lambda}{M}\right) \frac{\alpha_i(M)}{4\pi} \Lambda, \qquad (4)$$

  where $k_2 = k_3 = 1, k_1 = 5/3$, and $N_{5,10}$ is the number of $\mathbf{5 + \bar{5}}$ plus three times the number of $\mathbf{10 + \overline{10}}$ messenger representations. $N_{5,10} \leq 4$ is required to avoid Landau poles. The gaugino mass ratios are the same as found for universal boundary conditions at M_U in mSUGRA models.

- The squarks/sleptons acquire masses-squared at two-loops: $m_i^2(M) =$

$$2\Lambda^2 N_{5,10} f\left(\frac{\Lambda}{M}\right)\left[c_3\left(\frac{\alpha_3(M)}{4\pi}\right)^2 + c_2\left(\frac{\alpha_2(M)}{4\pi}\right)^2 + \frac{5}{3}\left(\frac{Y}{2}\right)^2\left(\frac{\alpha_1(M)}{4\pi}\right)^2\right], \qquad (5)$$

  with $c_3 = 4/3$ (triplets) $c_2 = 3/4$ (weak doublets), $Y/2 = Q - T_3$. Note that the degeneracy among families is broken only by effects of order quark or lepton Yukawa couplings, implying no FCNC problems in the first two generations.

In the above, $g(\Lambda/M), f(\Lambda/M) \to 1$ for M/Λ not too near 1. For $\Lambda = M$, $g(1) = 1.4, f(1) = 0.7$.

- Spontaneous SUSY-breaking leads to a goldstone fermion, the goldstino, G. (In local SUSY, G is the longitudinal component of the gravitino.) G acquires mass determined by F, the goldstino decay constant, where F is the largest F-term vev (here, the F of the secluded true supersymmetry breaking sector):

$$m_G = \frac{F}{\sqrt{3} M_{\text{Planck}}} \sim 2.5\left(\frac{\sqrt{F}}{100 \text{ TeV}}\right)^2 \text{ eV}; \qquad (6)$$

  As sketched earlier, $\sqrt{F} \sim 10^8$ TeV is appropriate in the usual mSUGRA/string-motivated models; the G is then fairly massive and, since it is also very weakly interacting, it is irrelevant to low-energy physics. In GMSB models, the much smaller $\sqrt{F} \sim 100 - 1000$ TeV values envisioned imply that the G will be the lightest supersymmetric particle. In this case, a very crucial constraint is that m_G be $\lesssim 1$ keV to avoid overclosing the universe; i.e. $\sqrt{F} \lesssim 2000$ TeV is required for consistency with cosmology.

Some useful observations affecting the phenomenology of a GMSB model are the following.

- FCNC etc. problems are solved since the gauge interactions are flavor-blind and, thus, so are the soft terms.

- As already noted, if $\hat{S}$ is an SU(5) singlet, then we obtain the usual mSUGRA prediction: $M_3 : M_2 : M_1 = 7 : 2 : 1$ at scale $\sim m_Z$.

- Eq. (4) implies

$$\Lambda = \frac{\langle F_S \rangle}{\langle S \rangle} \sim \frac{80 \text{ TeV}}{N_{5,10}} \left(\frac{M_1}{100 \text{ GeV}} \right). \qquad (7)$$

- Eqs. (4) and (5) imply (for $g = f = 1$)

$$m_{\tilde{q}} : m_{\tilde{\ell}_L} : m_{\tilde{\ell}_R} : M_1 = 11.6 : 2.5 : 1.1 : \sqrt{N_{5,10}}, \qquad (8)$$

which in turns implies that the lightest of the standard sparticles (referred to as the next-to-lightest supersymmetric particle, denoted NLSP — the goldstino being the LSP) is the $\tilde{B}$ for $N_{5,10} = 1$ or the $\tilde{\ell}_R$ for $N_{5,10} \geq 2$.

- μ and B do not arise from the GMSB ansatz and are 'model-dependent', but EWSB driven by negative $m^2_{H^0_2}$ turns out to be completely automatic.

- The couplings of the G are fixed by a supersymmetric Goldberger-Treiman relation:

$$\mathcal{L} = -\frac{1}{F} j^{\mu\alpha} \partial_\mu G_\alpha + h.c.$$

where j is the supercurrent connecting a SM particle to its superpartner. For $\sqrt{F}$ values in the range of interest, this coupling is very weak, implying that all the superparticles other than the NLSP undergo chain decay down to the NLSP. The NLSP finally decays to the G: e.g. $\tilde{B} \to \gamma G$ (and ZG if $m_{\tilde{B}} > m_Z$) or $\tilde{\ell}_R \to \ell G$. The $c\tau$ for NLSP decay depends on $\sqrt{F}$; e.g. for $N_{5,10} = 1$

$$(c\tau)_{\tilde{\chi}^0_1 = \tilde{B} \to \gamma G} \sim 130 \left(\frac{100 \text{ GeV}}{M_{\tilde{B}}} \right)^5 \left(\frac{\sqrt{F}}{100 \text{ TeV}} \right)^4 \mu\text{m} \qquad (9)$$

If $\sqrt{F} \sim 2000$ TeV (the upper limit from cosmology), then $c\tau \sim 21$m for $M_{\tilde{B}} = 100$ GeV; $\sqrt{F} \sim 100$ TeV implies a short but vertexable decay length.

If the final decay of the NLSP occurs rapidly ($\sqrt{F} \sim 100$ TeV), then there are many highly observable signatures for a GMSB model [50]. For example, GMSB with the $\tilde{\chi}^0_1$ being the NLSP could explain the Tevatron $ee\gamma\gamma$ event as $\tilde{e}\tilde{e} \to e\tilde{\chi}^0_1 e\tilde{\chi}^0_1 \to ee\gamma\gamma GG$ (the G's yielding $\not{E}_T$). However, I will now argue that $\sqrt{F}$ is most naturally very near the $\sqrt{F} \sim 2000$ TeV upper bound, in which case rather few of the NLSP's decay inside the detector. In fact, in their current incarnation the GMSB models have a significant problem of scale related to the value of $f \equiv F/\langle F_S \rangle$. As noted earlier, the communication between the true supersymmetry breaking sector and the messenger sector occurs at two-loops, implying $f \sim \left(\frac{g_m^2}{16\pi^2}\right)^{-2} \sim 2.5 \times 10^4/g_m^4$ where g_m refers to the gauge group responsible for SUSY-breaking, whereas Eqs. (3) and (7) imply

$$f = \frac{F}{\langle F_S \rangle} = \left(\frac{\sqrt{F}}{2000 \text{ TeV}}\right)^2 \frac{625 \lambda N_{5,10}^2}{(M/\Lambda)} \left(\frac{100 \text{ GeV}}{M_1}\right)^2. \quad (10)$$

The problem is that a value as large as $f \sim 2.5 \times 10^4/g_m^4$ is generally rather inconsistent with basic phenomenological constraints if (as hoped) $g_m \lesssim 1$.

- To illustrate, consider $\lambda \sim 1$, $g_m = 1$ (i.e. $f = 2.5 \times 10^4$) and $M/\Lambda \sim 1$. (Recall $M/\Lambda > 1$ is required; the choice $M/\Lambda \sim 1$ minimizes the scale problem.) If $\sqrt{F} \sim 2000$ TeV, i.e. as large as possible consistent with $m_G \leq 1$ keV, then Eq. (10) implies: $M_1 \sim 16$ GeV (63 GeV) for $N_{5,10} = 1$ ($N_{5,10} = 4$). The former is experimentally ruled out. The latter might be acceptable, but for $N_{5,10} = 4$ the NLSP is the $\tilde{\ell}_R$ with $m_{\tilde{\ell}_R} \sim 35$ GeV, see Eq. (8), which is ruled out by Z data. Although the inconsistency in this latter case is not very bad and could be resolved by modest increases in λ and/or g_m, or corrections to our simple two-loop estimate for f, the $\tilde{\ell}_R$ NLSP would appear as a heavily ionizing track of substantial length in the detector when $\sqrt{F}$ is as large as assumed. Most probably such events would have been observed at the Tevatron in the $m_{\tilde{\ell}_R} \lesssim 100$ GeV range roughly consistent with $f \sim 2.5 \cdot 10^4$.

- A value of $\sqrt{F} \sim 100$ TeV, as taken in many phenomenological discussions and studies [50], is highly inconsistent with the two-loop expectations for f. (This has also been noted in Ref. [51], where it was used to motivate attempts to construct models in which the supersymmetry breaking sector and the messenger sector are one and the same.) For example, taking $\sqrt{F} = 100$ TeV, $M_1 = 100$ GeV, $M/\Lambda \sim 1$, $\lambda \sim 1$ and $N_{5,10} = 1$ ($N_{5,10} = 4$) in Eq. (10) results in $f = 1.56$ ($f = 25$).

- If an acceptable model with 1-loop communication between the SUSY-breaking sector and the messenger sector can be constructed, we would predict $f \sim 16\pi^2/g_m^2 \sim 160/g_m^2$. For $g_m = 1$ (i.e. $f = 16\pi^2$), $M_1 = 100$ GeV, $M/\Lambda \sim 1$, $\lambda \sim 1$ and $N_{5,10} = 1$ ($N_{5,10} = 4$), Eq. (10) yields $\sqrt{F} \sim 1000$ TeV ($\sqrt{F} \sim 250$ TeV). The associated NLSP lifetimes, see e.g. Eq. (9), would lead to easily detected vertices in events where sparticles are produced. Analysis of existing Tevatron data and forthcoming LEP2 data should readily uncover supersymmetric signals.

In short, for models in which the SUSY-breaking sector communicates at two (or even one) loops with the messenger sector, it seems to be inconsistent to take $\sqrt{F} \lesssim 100$ TeV. This implies that the simple phenomenology in which the NLSP decays ($\tilde{\chi}_1^0 \to \gamma G$ or $\tilde{\ell}_R \to \ell G$) almost immediately is not relevant, and the GMSB explanation of the Tevatron $ee\gamma\gamma$ event would be very unlikely. The large $\sqrt{F}$ values required for even a modicum of consistency imply that experimentalists should be looking for events with vertices a substantial distance from the interaction point, quite possibly in association with heavily ionizing tracks. In the most natural case, where $\sqrt{F} \sim 1000 - 2000$ TeV and $m_{\tilde{\chi}_1^0}$ and $m_{\tilde{\ell}_R}$ are below 100 GeV, $c\tau$ — see e.g. Eq. (9) — is typically many tens to several hundreds of meters and only a fraction $\sim 2R/[\langle\gamma\rangle c\tau]$ of the events will have at least one vertex inside a detector of radius R.

- If the NLSP is the $\tilde{\ell}_R$, the bulk of SUSY events will have several heavily ionizing tracks in the detector, a small fraction of which will suddenly terminate with the emission of a ℓ. Since there is surely no background to such events, this possibility should be excludable using existing Tevatron data (given the significant production rates associated with the low sparticle masses required for scale consistency).

- If the NLSP is the $\tilde{B}$, the bulk of events will not have a vertex inside the detector and will appear as typical $\not{E}_T$ supersymmetry signal events; high rates would be required to uncover such events. However, the small, but significant (given the low sparticle masses), number of very distinctive events in which a photon (or Z) suddenly emerges in the middle of the electromagnetic calorimeter should have a small background, in which case only a few events would be required for discovery. It would seem that this GMSB scenario might also be ruled out or confirmed with proper analysis of Tevatron, if not LEP2, data.

If the above kinds of events, consistent with the most natural $\sqrt{F} \sim 1000 - 2000$ TeV values are found, it will be useful to construct 'far-out' additions to the CDF and D0 detectors designed to reveal more of the decay vertices.

The cosmological implications/consistency of GMSB constitute an important issue [52,53], but one that is far too complex to elaborate on significantly here. I only note that if $\sqrt{F} \sim 100$ TeV, as disfavored by the scale consistency discussed above, then $m_G \sim 2.5$ eV implies that the goldstino cannot make a significant contribution to the dark matter of the universe; however, the messenger sector might contain an appropriate dark matter candidate [52]. In a model with $\sqrt{F} \sim 1000$ TeV, as preferred by the scale arguments, the G gives rise to a cosmologically significant abundance of warm dark matter; unfortunately, it would be invisible in halo detection experiments [52].

CONCLUSIONS

If supersymmetry is discovered, it will be a dream-come-true for both theorists and experimentalists. For theorists, it would a a triumph of aesthetic principles, naturalness, *etc.* For experimentalists, it would be a gold mine of experimental signals and analyses. As has always been the case in the past, the next step in theory will require experimental guidance and input. The many phenomenological manifestations and parameters of supersymmetry imply that many years of experimental work will be required before it will be possible to determine the precise nature of supersymmetry breaking and the associated boundary conditions. Our ultimate dream is that, armed with this information, we will be able to construct the 'final' theory.

ACKNOWLEDGEMENTS

This work was supported in part by the Department of Energy, by the Davis Institute for High Energy Physics and by the Institute for Theoretical Physics. I am grateful to B. Dobrescu and H. Murayama for helpful discussions.

REFERENCES

1. J. Wess and J. Bagger, *Supersymmetry and Supergravity* (Princeton University Press, Princeton, 1983), and references therein.
2. G. Ross, *Grand Unified Theories* (Frontiers in Physics, Benjamin Cummings, 1984)
3. H.E. Haber and G.L. Kane, Phys. Rep. **117**, 75 (1985).
4. M. Dine, hep-ph/9612389, and references therein.
5. See, for example, J.F. Gunion, D.W. McKay and H. Pois, Phys. Rev. **D53**, 1616 (1996).
6. J. Lykken, Phys. Rev. **D54**, 3693 (1996).
7. D. Pierce, J.A. Bagger, K. Matchev and R.-J. Zhang, hep-ph/9606211 and references therein.
8. Pedagogical treatments and detailed references can be found in: J.F. Gunion, H.E. Haber, G. Kane and S. Dawson, *The Higgs Hunters Guide* (Frontiers in Physics, Addison Wesley, 1990); *Perspectives on Higgs Physics*, ed. G. Kane (World Scientific Publishing, 1993); J.F. Gunion, A. Stange, and S. Willenbrock, *Weakly-Coupled Higgs Bosons*, preprint UCD-95-28 (1995), to be published in *Electroweak Physics and Beyond the Standard Model* (World Scientific Publishing), eds. T. Barklow, S. Dawson, H. Haber, and J. Siegrist.
9. D. Froidevaux, F. Gianotti, L. Poggioli, E. Richter-Was, D. Cavalli, and S. Resconi, ATLAS Internal Note, PHYS-No-74 (1995).
10. B.R. Kim, S.K. Oh and A. Stephan, *Proceedings of the 2nd International Workshop on "Physics and Experiments with Linear e^+e^- Colliders"*, eds. F. Harris, S. Olsen, S. Pakvasa and X. Tata, Waikoloa, HI (1993), World Scientific Publishing, p. 860.
11. J. Kamoshita, Y. Okada and M. Tanaka, Phys. Lett. **B328**, 67 (1994).
12. S.F. King and P.L. White, preprint SHEP-95-27 (1995).
13. U. Ellwanger, M.R. de Traubenberg and C.A. Savoy, Z. Phys. **C67**, 665 (1995).

14. V. Barger, M. Berger, J.F. Gunion and T. Han, unpublished.
15. J.F. Gunion, H.E. Haber and T. Moroi, hep-ph/9610337, to be published in the proceedings of the 1996 DPF/DPB Summer Study on *New Directions for High-Energy Physics* (hereafter referred to as Snowmass 96).
16. J.F. Gunion and H.E. Haber, Proceedings of the 1990 DPF Summer Study on *High Energy Physics: Research Directions for the Decade*, editor E. Berger, Snowmass (1990), p. 206; and Phys. Rev. **D48**, 5109 (1993).
17. D. Borden, D. Bauer, and D. Caldwell, Phys. Rev. **D48**, 4018 (1993).
18. V. Barger, M. Berger, J. Gunion and T. Han, Phys. Rev. Lett. **75**, 1462 (1995); hep-ph/9602415, to appear in Physics Reports.
19. J.F. Gunion and J. Kelly, hep-ph/9610495; a summary will appear in Snowmass 96.
20. J. Feng and T. Moroi, hep/ph-9612333.
21. G. Anderson and D. Castano, Phys. Lett. **B347**, 300 (1995); Phys. Rev. **D52**, 1693 (1995); Phys. Rev. **D53**, 2403 (1996).
22. J.F. Gunion, L. Poggioli and R. Van Kooten, to appear in Snowmass 96.
23. J. Amundson *et al.*, hep-ph/9609374, to appear in Snowmass 96.
24. See, for example, H. Baer. C.-H. Chen, F. Paige and X. Tata, Phys. Rev. **D54**, 5866 (1996); D. Amidei *et al.*, FERMILAB-PUB-96-082; and references therein.
25. H. Baer, C.-H. Chen, F. Paige and X. Tata, Phys. Rev. **D52**, 2746 (1995); Phys. Rev. **D53**, 6241 (1996).
26. R.M. Barnett, J.F. Gunion, and H.E. Haber, Phys. Lett. **B315**, 349 (1993);
27. I. Hinchliffe, F. Paige, M. Shapiro, J. Soderqvist and W. Yao, hep-ph/9610544.
28. A. Bartl *et al.*, to appear in Snowmass 96.
29. H. Baer, C.H. Chen, F. Paige and X. Tata, Phys. Rev. **D50**, 4508 (1994).
30. H. Baer and M. Brhlik, Phys. Rev. **D53**, 597 (1996).
31. T. Tsukamoto, K. Fujii, H. Murayama, M. Yamaguchi and Y. Okada, Phys. Rev. **D51**, 3153 (1995); H. Baer, R. Munroe and X. Tata, Phys. Rev. **D54**, 6735 (1996). For a review, see *Physics and Technology of the Next Linear Collider: a Report Submitted to Snowmass 1996*, SLAC Report 485.
32. See, for example, A. Lahanas and D. Nanopoulos, Phys. Rep. **145**, 1 (1987).
33. A. Brignole, L. Ibanez and C. Munoz, Nucl. Phys. **B422**, 125 (1994); hep-ph/9508258. See also V. Kaplunovsky and J. Louis, Phys. Lett. **B306**, 269 (1993).
34. H. Baer, J. Gunion, C. Kao and H. Pois, Phys. Rev. **D51**, 2159 (1995); J. Lopez, D. Nanopoulos, X. Wang, and A. Zichichi, Phys. Rev. **D52**, 142 (1995); J. Lopez, D. Nanopoulos and Z. Zichichi, Phys. Rev. **D52**, 4178 (1995).
35. T. Banks, D. Kaplan and A. Nelson, Phys. Rev. **D49**, 779 (1994).
36. G. Farrar and A. Masiero, hep-ph/9410401.
37. G.R. Farrar, Phys. Rev. **D51**, 3904 (1995).
38. G.R. Farrar and E.W. Kolb, Phys. Rev. **D53**, 2990 (1996).
39. G.R. Farrar, Phys. Rev. Lett. **76**, 4111 (1996).
40. G. Anderson, C.H. Chen, J.F. Gunion, J. Lykken, T. Moroi, Y. Yamada, hep-ph/9609457, to appear in Snowmass 96.

41. C.-H. Chen, M. Drees and J.F. Gunion, Phys. Rev. Lett. **76**, 2002 (1996); Phys. Rev. **D55**, 330 (1997).
42. R. Hempfling, hep-ph/9609528, hep-ph/9702412.
43. See, for example, V. Barger, G.F. Giudice, and T. Han, Phys. Rev. **D40**, 2987 (1989); S. Davidson, D. Bailey and B. Campbell, Z. Phys. **C61**, 613 (1994).
44. An incomplete set of references is: P. Binetruy and J.F. Gunion, proceedings of the *Workshop on Novel Features of High Energy Hadronic Collisions*, Erice, Italy, 1988, published in *Eloisatron: Heavy Flavors 1988*, p. 489; H. Dreiner, M. Guchait and D.P. Roy, Phys. Rev. **D49**, 3270 (1994); H. Baer, C.-H. Chen and X. Tata, Phys. Rev. **D55**, 1466 (1997); A. Bartl *et al.*, hep-ph/9612436.
45. An incomplete list is: D. Choudhury and S. Raychaudhuri, hep-ph/9702392; G. Altarelli, J. Ellis, G.F. Giudice, S. Lola and M. Mangano, hep-ph/9703276; H. Dreiner and P. Morawitz, hep-ph/9703279; J. Kalinowksi, R. Ruckl, H. Spiesberger and P.M. Zerwas, hep-ph/9703288; K.S. Babu, C. Kolda, J. March-Russell and F. Wilczek, hep-ph/9703299.
46. S. Ambrosanio, G.L. Kane, G.D. Kribs, S.P. Martin and S. Mrenna, Phys. Rev. **D55**, 1372 (1997).
47. R.M. Barnett and L.J. Hall, Phys. Rev. Lett. **77**, 3506 (1996); and hep-ph/9609313.
48. M. Carena, G.F. Giudice, S. Lola and C. Wagner, hep-ph/9612334.
49. See, for example, M. Dine, A. Nelson and Y. Shirman, Phys. Rev. **51**, 1362 (1995); M. Dine, A. Nelson, Y. Nir and Y. Shirman, Phys. Rev. **D53**, 2658 (1996); and references to earlier work therein.
50. See, for example, S. Dimopoulos, M. Dine, S. Raby and S. Thomas, Phys. Rev. Lett. **76**, 3494 (1996); S. Ambrosanio, G. Kane, G. Kribs, S. Martin and S. Mrenna, Phys. Rev. Lett. **76**, 3498 (1996); S. Dimopoulos, S. Thomas and J.D. Wells, Phys. Rev. **D54**, 3283 (1996); H. Baer, M. Brhlik, C.-H. Chen and X. Tata, Phys. Rev. **D55**, 4463 (1997); J. Bagger, K. Matchev, D. Pierce and R.-J. Zhang, Phys. Rev. Lett. **78**, 1002 (1997).
51. N. Arkani-Hamed, J. March-Russell and H. Murayama, hep-ph/9701286. See also, G. Dvali and M. Shifman, hep-ph/9612490.
52. S. Dimopoulos, G. Giudice and A. Pomarol, Phys. Lett. **B389**, 37 (1996).
53. A. de Gouvea, T. Moroi and H. Murayama, hep-ph/9701244.

Strongly Interacting New Physics

Thomas Appelquist[1]

Abstract.

In this talk, I will describe two aspects of the breaking of electroweak symmetry by new, strong interactions. First I will review the model independent approach to the low energy form of such new interactions – the electroweak chiral Lagrangian. Next I will summarize some of the phenomenological challenges facing technicolor theories, in particular those associated with the generation of the top and bottom masses.

I THE STANDARD MODEL AND THE ELECTROWEAK CHIRAL LAGRANGIAN

The electroweak interactions are described by a spontaneously broken $SU(2)_2 \times U(1)$ gauge theory, but the symmetry breaking mechanism remains a mystery. None of the symmetry breaking physics, other than the three eaten Goldstone bosons, have been seen. The new physics that must eventually emerge is heavy enough so that it has not been seen up through the energies that have been explored so far.

This being the case, the most economical description of observed electroweak

---

[1] Electronic address: twa@genesis3.physics.yale.edu

physics is a (gauged) chiral Lagrangian [1–4]. The only degrees of freedom are those that have been seen: the quarks, the leptons, the transversely polarized gauge bosons, and the three eaten Goldstone bosons. The electroweak chiral Lagrangian describes a non-renormalizable, effective low energy theory. It must break down at or below energies of order 1 Tev. In this lecture, I will discuss the possibility that the breaking takes place at TeV energies and is due to new, strong interactions that must set in there.

To construct the electroweak chiral Lagrangian, it is useful to start from the "minimal" (renormalizable) standard model with a single complex doublet field

$$\Phi = \begin{bmatrix} \phi^+ \\ \phi^0 \end{bmatrix} \tag{1}$$

with electroweak hypercharge $Y = 1$. The scalar potential $V = \lambda(\Phi^\dagger \Phi - (1/2)f^2)^2$, where λ is the dimensionless scalar self-coupling, leads to the the vacuum value $<\phi^0>_0 = f = 246 GeV$. The Higgs mass is $M_H = \lambda f^2$.

The scalar potential along with the scalar kinetic term respects a larger symmetry than $SU(2)_2 \times U(1)$. It can be exhibited by introducing the hypercharge conjugate scalar field $\tilde{\Phi} = i\tau_2 \Phi^*$, and then the matrix field $M(x) = \sqrt{2}(\tilde{\Phi}, \Phi) = \sigma(x) + i\vec{\tau}\cdot\vec{\pi}$. The vacuum value of $M(x)$ is $<M(x)> = f\mathbf{1}$, where $\mathbf{1}$ is the two-by-two unit matrix. $M(x)$ transforms from the left as a 2 of $SU(2)_L$ and from the right as a 2 of another $SU(2)$, which I will call $SU(2)_R$: $M(x) \to exp(-i\vec{\tau}\cdot\vec{\epsilon}_L/2)M(x)exp(i\vec{\tau}\cdot\vec{\epsilon}_R/2)$. This $SU(2)_L \times SU(2)_R$ symmetry is the larger symmetry of the scalar sector. It leads to the famous prediction $\rho \equiv M_W^2/M_Z^2 cos^2\theta_W = 1$, which is valid to lowest order in the electroweak interactions, and which agrees with experiment to less than $1/2\%$.

Since the Higgs particle has not been seen, the least theoretically biased way to proceed is to remove it from the theory. This can be done formally by taking the limit $M_H \to \infty$ ($\lambda \to \infty$), with f fixed, leading to the nonlinear constraint $M^\dagger(x)M(x) = M(x)M^\dagger(x) = f^2$. When imposed on the minimal standard model, this constraint leads to the lowest dimensional, and most familiar, terms in the electroweak chiral Lagangian. They must respect the (spontaneously broken) $SU(2)_L \times U(1)$ gauge symmetry. The chiral lagrangian is thus constructed using the dimensionless unitary unimodular matrix field $U(x) \equiv M(x)/f$. The covariant derivative of $U(x)$ is:

$$D_\mu U = \partial_\mu U + ig\frac{\vec{\tau}}{2} \cdot \vec{W}_\mu U - ig'U\frac{\tau_3}{2}B_\mu. \tag{2}$$

In constructing the most general chiral $SU(2)_L \times U(1)_Y$ invariant effective lagrangian order by order in the energy expansion, it is convenient to define the basic building blocks which are $SU(2)_L$ covariant and $U(1)_Y$ invariant as follows:

$$T \equiv U\tau_3 U^\dagger, \qquad V_\mu \equiv (D_\mu U)U^\dagger \tag{3}$$

$$F_{\mu\nu} \equiv \partial_\mu W_\nu - \partial_\nu W_\mu + ig[W_\mu, W_\nu] \tag{4}$$

where T, V_μ and $F_{\mu\nu}$ have dimensions zero, one, and two respectively.

The familiar pieces of the chiral lagrangian, that emerge for example from the $M_H \to \infty$ limit of the linear theory at tree level, are:

$$\mathcal{L}_0 \equiv \frac{1}{4}f^2 Tr[(D_\mu U)^\dagger(D^\mu U)] - \frac{1}{4}B_{\mu\nu}B^{\mu\nu} - \frac{1}{2}TrF_{\mu\nu}F^{\mu\nu}, \tag{5}$$

where $f \simeq 250 \text{GeV}$ is the symmetry breaking scale, and $B_{\mu\nu} \equiv \partial_\mu B_\nu - \partial_\nu B_\mu$. The first term has dimension two, while the second two (kinetic energy) terms have dimension four. The gauge couplings to the quarks and leptons must also be added to Eq. 5.

There is one, additional dimension-two operator allowed by the $SU(2)_L \times U(1)$ symmetry [2]:

$$\mathcal{L}_1' \equiv \frac{1}{4}\beta_1 f^2 [Tr(TV_\mu)]^2. \tag{6}$$

This term, which does not emerge from the $M_H \to \infty$ limit of the renormalizable theory at tree level, violates the $SU(2)_R$ (custodial) symmetry even in the absence of the gauge couplings. It is the low energy description of whatever custodial-symmetry violating physics exists, and has been integrated out, at energies above roughly $\Lambda_\chi \equiv 4\pi f \simeq 2$ TeV. Used at tree level, it contributes directly to the deviation of the ρ parameter from unity.

At the dimension-four level, there are a variety of new operators that can be written down. Making use of the equations of motion, and first restricting attention to CP-invariant operators, the list can be reduced to eleven independent terms:

$$\mathcal{L}_1 \equiv \frac{1}{2}\alpha_1 gg' B_{\mu\nu} Tr(TF^{\mu\nu}) \qquad \mathcal{L}_2 \equiv \frac{1}{2}i\alpha_2 g' B_{\mu\nu} Tr(T[V^\mu, V^\nu])$$

$$\mathcal{L}_3 \equiv i\alpha_3 g Tr(F_{\mu\nu}[V^\mu, V^\nu]) \qquad \mathcal{L}_4 \equiv \alpha_4 [Tr(V_\mu V_\nu)]^2$$

$$\mathcal{L}_5 \equiv \alpha_5 [Tr(V_\mu V^\mu)]^2 \qquad \mathcal{L}_6 \equiv \alpha_6 Tr(V_\mu V_\nu) Tr(TV^\mu) Tr(TV^\nu)$$

$$\mathcal{L}_7 \equiv \alpha_7 Tr(V_\mu V^\mu) Tr(TV_\nu) Tr(TV^\nu) \qquad \mathcal{L}_8 \equiv \frac{1}{4}\alpha_8 g^2 [Tr(TF_{\mu\nu})]^2$$

$$\mathcal{L}_9 \equiv \frac{1}{2}i\alpha_9 g Tr(TF_{\mu\nu}) Tr(T[V^\mu, V^\nu]) \qquad \mathcal{L}_{10} \equiv \frac{1}{2}\alpha_{10}[Tr(TV_\mu) Tr(TV_\nu)]^2$$

$$\mathcal{L}_{11} \equiv \alpha_{11} g \epsilon^{\mu\nu\rho\lambda} Tr(TV_\mu) Tr(V_\nu F_{\rho\lambda}) \tag{7}$$

The first ten terms were written down by Longhitano [2]. The operator $\mathcal{L}_{11}$ is new [3,11] and it completes the list of all CP invariant operators up to dimension four. $\mathcal{L}_{11}$ corresponds to a CP-conserving, but C and P violating, term in the general parameterization of the triple gauge boson vertex. In addition, there are several CP-violating operators at the dimension-four level [3]. I will not discuss them in is lecture.

Since experimental work is so far restricted to energies below the W-pair threshold, the only operators in the above list that have been directly constrained experimentally are those that contribute to the gauge boson two-point functions. The ones that correspond to deviations from the standard model are $\mathcal{L}'_1$, $\mathcal{L}_1$ and $\mathcal{L}_8$. They are sometimes referred to as oblique corrections and can be directly related to the S, T and U parameters introduced by Peskin and Takeuchi [5]. By setting the Goldstone boson fields to zero in these operators ("going to unitary gauge"), one finds

$$S \equiv -16\pi \frac{d}{dq^2} \Pi_{3B}(q^2)|_{q^2=0} = -16\pi\alpha_1, \tag{8}$$

$$\alpha T \equiv \frac{e^2}{c^2 s^2 m_Z^2}(\Pi_{11}(0) - \Pi_{33}(0)) = 2\beta_1, \tag{9}$$

$$U \equiv 16\pi \frac{d}{dq^2}[\Pi_{11}(q^2) - \Pi_{33}(q^2)]|_{q^2=0} = -16\pi\alpha_8. \tag{10}$$

The $\Delta\rho(\equiv \rho - 1)$ parameter is related to T by $\Delta\rho_{new} = \Delta\rho - \Delta\rho_{SM} = \alpha T$, where $\Delta\rho_{SM}$ is the contribution arising from standard model corrections.

The new generation of e^+e^- colliders will operate above the W pair production threshold, and will therefore be able to measure carefully the triple gauge vertices (TGV's). The most general polynomial structure of the TGV has been derived [6] by imposing Lorentz invariance and on-shell conditions for the W^+ and W^-. If it is restricted to terms that correspond only to operators of dimension four in the chiral Lagrangian, it is given by $ig_{WWV}\Gamma_V^{\mu\nu\rho}(p,q,k)$, where

$$\Gamma_V^{\mu\nu\rho}(p,q,k) = g_1^V(p-q)^\rho g^{\mu\nu} + (g_1^V + \kappa_V)(k^\mu g^{\rho\nu} - k^\nu g^{\rho\mu})$$
$$+ ig_4^V(k^\mu g^{\rho\nu} + k^\nu g^{\rho\mu}) + ig_5^V \epsilon^{\rho\mu\nu\lambda}(p-q)_\lambda$$

$$-\tilde{\kappa}_V \epsilon^{\rho\mu\nu\lambda} k_\lambda, \tag{11}$$

and where $k = p+q$ is the momentum of the neutral vector boson V. The first two terms in the above exression are C and P invariant (thus CP invariant), the g_5^V term is CP invariant but C and P violating, and the other two terms are CP violating.

Restricting attention to the CP invariant terms and using the convention of defining the parameters of the TGV's to include the effects of corrections to the W, Z and γ propagators and effects of the γZ mixing coming from physics above Λ_χ, these parameters are related to those in the chiral lagrangian through the latter's contribution to both three-point and two-point functions [3]:

$$g_1^Z - 1 = \frac{1}{(c^2-s^2)}\beta_1 + \frac{1}{c^2(c^2-s^2)}e^2\alpha_1 + \frac{1}{s^2c^2}e^2\alpha_3$$

$$g_1^\gamma - 1 = 0$$

$$\kappa_Z - 1 = \frac{1}{(c^2-s^2)}\beta_1 + \frac{1}{c^2(c^2-s^2)}e^2\alpha_1 + \frac{1}{c^2}e^2(\alpha_1 - \alpha_2) + \frac{1}{s^2}e^2(\alpha_3 - \alpha_8 + \alpha_9)$$

$$\kappa_\gamma - 1 = \frac{1}{s^2}e^2(-\alpha_1 + \alpha_2 + \alpha_3 - \alpha_8 + \alpha_9)$$

$$g_5^Z = \frac{1}{s^2c^2}e^2\alpha_{11}$$

$$g_5^\gamma = 0, \tag{12}$$

where $s \equiv \sin\theta_Z$, $c \equiv \cos\theta_Z$. The first two terms in the expressions for $g_1^Z - 1$ and $\kappa_Z - 1$ are the contributions from the two-point functions of the chiral Lagrangian. The other terms in $g_1^Z - 1$ and $\kappa_Z - 1$ come from the three-point function contribution of the chiral lagrangian. The parameters $\kappa_\gamma - 1$ and g_5^Z have contributions only from the three-point functions of the chiral lagrangian. Note that g_1^γ measures the electric charge of the W in unit of e and that the vanishing of g_5^γ is a consequence of $U(1)_{em}$ gauge invariance.

The right hand side of Eq. (12) measures deviations from the standard-model tree-level predictions, coming from new high energy physics. The α and β parameters are expected, on the basis of naturality, to be of order $\frac{1}{16\pi^2}$. In order to isolate these effects experimentally, one loop radiative corrections within the standard model (arising from momentum scales less than Λ_χ) must also be included. This program has been pursued by many researchers in recent years [4].

To conclude this brief discussion, I will estimate the size of the S, T and U parameters arising from physics above Λ_χ in a simple model consisting of one flavour-doublet of heavy fermions U and D. By assigning both U and D to the fundamental representation of an $SU(N)$ group, this model can be viewed as a simplified version of a technicolor theory with the technicolor interactions neglected. A small weak-isospin asymmetry arising from the mass splitting in the fermion doublet is also included. The masses of U- and D-type fermions are denoted by m_U and m_D respectively, and the small weak-isospin asymmetry parameter is defined by $\delta \equiv \frac{m_U - m_D}{m_U + m_D}$. The electric charges of the U- and D-type fermions are set to be $+\frac{1}{2}$ and $-\frac{1}{2}$ respectively by anomaly cancelation condition for the new fermionic sector.

The computation yields [5]:

$$S = \frac{N}{6\pi}(1 - Y \ln \frac{m_U^2}{m_D^2}) = \frac{N}{6\pi} \tag{13}$$

$$\alpha T \simeq \frac{Ne^2}{48\pi^2 s^2 c^2} \frac{\Delta m^2}{m_Z^2} \tag{14}$$

$$U \simeq \frac{8N}{15\pi} \delta^2 \tag{15}$$

where $\Delta m \equiv m_U - m_D$, and the hypercharge Y for each left-handed U and D doublet is zero for anomaly cancellation. These are useful starting-point expressions for estimating the S, T and U parameters. In reality, of course, the detailed computation of these parameters, and the others in Eq. 7, are difficult, involving uncertainties associated with the strong interactions that are expected at 1 TeV and above, and with the explicit form of the fermion mass generating interactions.

II TECHNICOLOR AND ELECTROWEAK TECHNICOLOR

I will turn next to a description of technicolor theories, focusing on the problem of t and b mass generation and precision electroweak studies. Technicolor models break electroweak symmetry with fermion condensates, generated by a strongly interacting gauge theory patterned after QCD [12]. The masses and mixing angles of quarks and leptons arise from an extended technicolor (ETC) sector [13] that communicates the spontaneous breaking of electroweak symmetry to the quark-lepton sector through broken gauge interactions. To accommodate a top mass $m_t \simeq 175$ GeV with perturbative ETC interactions requires a corresponding ETC scale less than or on the order of 1 TeV.

Such a low scale for new physics raises the possibility that ETC dynamics may visibly affect low energy precision data. For example, the weak interaction ρ parameter can receive important corrections from ETC interactions [17] which exhibit enough weak-isospin-breaking at the lowest ETC scale to explain the mass splitting between the top quark t and the bottom quark b. Such a large splitting requires realistic ETC models to be chiral [13,18], treating the t_R and b_R differently yet allowing the t_L and b_L to transform together.

I will describe chiral models with separate ETC groups acting on the t_R and b_R, allowing separate ETC couplings and breaking scales to be associated with each [19]. I will assume that the ETC gauge group commutes with $SU(2)_L$ and that the fermions transform in the fundamental representation of ETC.

The massive ETC gauge bosons in such models have couplings which violate weak-isospin symmetry and hence contribute to $\Delta \rho \equiv \rho - \rho^{(SM)}$, or αT in the notation of Ref. [22]. The contributions from ETC bosons associated with the broken *diagonal* generators of the ETC gauge groups have been calculated [15] to be near experimental limits. In a recent paper [19] it was noted that these models also contain massive ETC bosons in the *adjoint* representation of technicolor. Their exchange gives rise to $\Delta \rho$ contributions exceeding the experimental limits by at least an order of magnitude, if the corresponding M_{ETC} scales are of order 1 TeV (small enough to generate m_t perturbatively). Alternatively, if ETC bosons are an order of magnitude heavier, to suppress adequately the adjoint contribution to $\Delta \rho$, then generating m_t requires strong, and tuned, ETC interactions [23].

The t and b masses must be generated by ETC interactions connecting t and b quarks to their respective technifermion partners. As a specific example, I will describe the ETC breaking patterns responsible for the t and b masses, in a one doublet technicolor model. I will concentrate on the symmetry breaking patterns rather than the breaking mechanisms, simply noting that among other possibilities an underlying QCD-like model [25] can trigger the necessary breaking.

In a one doublet model, the ETC fermion multiplets above the lowest ETC scale are $\mathcal{U}_R$, $\mathcal{D}_R$ and $(\mathcal{U}, \mathcal{D})_L$, which contain the technicolor multiplets (U, D) and the QCD triplets of t and b quarks [27]. A particularly simple example of a chiral gauge structure puts both $(\mathcal{U}, \mathcal{D})_L$ and $\mathcal{U}_R$ into the same (fundamental) representation of a single ETC subgroup $SU(N+3)_L$, while $\mathcal{D}_R$ transforms under a separate group $SU(N+3)_{\mathcal{D}_R}$. At the ETC-breaking scale, the group structure then breaks in the pattern

$$SU(N+3)_L \times SU(N+3)_{\mathcal{D}_R} \to SU(N)_{TC} \times SU(3)_{QCD} \,. \tag{16}$$

One may wish to distinguish between possibly separate breaking scales: F_L, of order 1 TeV, for $SU(N+3)_L \to SU(N)_L \times SU(3)_L$; $F_{\mathcal{D}_R}$, of order 1 TeV or larger, for $SU(N+3)_{\mathcal{D}_R}$ similarly; and F_{mix}, less than or of order 1 TeV, for the "vector subgroup" mixing.

Each ETC group contains bosons transforming in the adjoint, the fundamental, and the singlet representations of technicolor. Since only one adjoint remains massless below the ETC scales to form the technicolor gauge bosons, chiral ETC models generate many massive ETC bosons. The group structure could be more elaborate than in the simple example above: each right-handed ETC multiplet might transform under a separate ETC group, and a sufficiently grandiose model could also distinguish different left-handed multiplets. Enlarging the number of distinct ETC groups creates more sets of massive gauge bosons, exacerbating the phenomenological difficulties I will describe.

The t and b masses are generated by the ETC bosons transforming in the fundamental representation of technicolor ("sideways" ETC bosons). There is one set of such bosons associated with each ETC gauge group, and they generate m_t and m_b as shown in Fig. 1. The ETC couplings cancel in the four-fermion approximation, which is applicable if the ETC boson masses ($\simeq gF/2$, with g the appropriate ETC coupling) are at least of order 1 TeV, larger than the dominant internal momentum (the technicolor scale) in the diagram. Then from the first diagram, $m_t \simeq \langle \bar{U}U \rangle / F_L^2$, where F_L is the decay constant of the Goldstone bosons formed and eaten at the breaking scale, and where the techni-up condensate $\langle \bar{U} U \rangle$ is roughly of order $4\pi v^3$. Generating $m_t \simeq 175$ GeV requires $F_L \lesssim 1$ TeV.

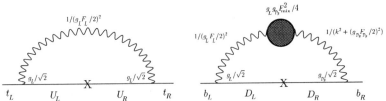

FIGURE 1. Perturbative generation of top and bottom masses, schematically indicating some coupling constant factors and heavy boson propagators. The heavy X represents a technifermion condensate, the blob represents ETC boson mixing induced by breaking to the "vector subgroup". The $\mathcal{D}_R$ boson requires a full propagator if it is light.

The second diagram similarly generates m_b, with additional factors from the ETC boson mixing and extra ETC ($\mathcal{D}_R$) boson propagator. If the latter boson's mass, approximately $g_{\mathcal{D}_R} F_{\mathcal{D}_R}/2$, is large enough to allow the four-fermion approximation, then $F_{\mathcal{D}_R}$ must exceed the other breaking scales so that $F^2_{\text{mix}}/F^2_{\mathcal{D}_R}$ suppresses m_b relative to m_t. Alternatively, if $F_L \simeq F_{\text{mix}} \simeq F_{\mathcal{D}_R}$, then obtaining the necessary $t - b$ mass splitting requires $g_{\mathcal{D}_R}$ to be much less than unity. The associated ETC boson is in that case lighter than the momentum scale in the diagram, invalidating the four-fermion approximation; $g^2_{\mathcal{D}_R}$ then remains uncancelled in the numerator, suppressing m_b.

I will next discuss the implications of chiral ETC for low energy precision measurements, focusing on the contributions to $\Delta \rho$. It is also important to try to estimate the S parameter in models such as these. The rough estimate (Eq. 13) indicates that if N (technicolor) is small, the S parameter may be within the experimental bound. For a more detailed discussion of S, I refer you to Ref. 11.

The dominant contribution to $\Delta \rho$ is from a loop of technifermions corrected by exchange of ETC bosons transforming in the adjoint representation of technicolor. Remember that with at least two ETC groups at the lowest ETC scale, there are at least two sets of gauge bosons transforming in the adjoint representation of technicolor. One set, the technigluons of unbroken technicolor, remains massless, with gauge coupling smaller than the smallest ETC coupling. The other set (with coupling of order the largest ETC coupling) acquires a mass of order at least 1 TeV set by the scale F_{mix}, independently of whether g_{D_R} is small or F_{D_R} large. Within each ETC group, the gauge bosons in different representations of technicolor share the same ETC coupling and breaking scale, which we have related to the $t-b$ mass splitting. The couplings of the massive adjoints, therefore, explicitly violate weak-isospin symmetry.

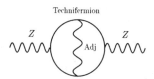

FIGURE 2. Contribution to m_Z from an ETC boson in the technicolor adjoint representation, exchanged across a technifermion loop.

I will first consider the lowest order contribution shown in Fig. 2. Since the adjoint representation ETC gauge boson mass, M_{adj}, exceeds the technicolor scale it generates an effective four-fermion interaction, and after a Fierz transformation the diagram becomes the product of two axial-axial current correlators. Neglecting strong technicolor interactions between the two technifermion loops, the lowest order contribution to $\Delta\rho \equiv M_W^2/c_\theta^2 M_Z^2 - 1$ can be estimated to be

$$\Delta\rho_{(1)} \simeq \frac{v^2}{N_D^2 M_{\text{adj}}^2} \left(g_U^{\text{adj}} - g_D^{\text{adj}}\right)^2 , \tag{17}$$

where g_U^{adj} and g_D^{adj} are the couplings of the adjoint ETC boson to the U and D respectively.

In the model considered here, the ETC axial coupling to the U vanishes, but g_D^{adj} is of order the largest of g_L and $g_{\mathcal{D}_R}$, independent of small $g_{\mathcal{D}_R}$ or large $F_{\mathcal{D}_R}$. (In more complicated models, $(g_U^{\text{adj}} - g_D^{\text{adj}})$ remains non-zero, reflecting the isospin splitting that must be present to generate $t - b$ mass splitting). Taking $(g_D^{\text{adj}})^2/M_{\text{adj}}^2 \simeq (1 \text{ TeV})^{-2}$ we find $\Delta \rho_{(1)} \gtrsim 6\%$. Even if trusted only in order of magnitude because of the approximate treatment of strong technigluon interactions, this is completely at odds with experiment.

Raising the lowest ETC scale above about 10 TeV quadratically suppresses the first order contribution $\Delta \rho_{(1)}$ to below current experimental bounds. Simultaneously maintaining $m_t \simeq 175$ GeV despite this decoupling, however, may then require tuning the ETC coupling g_L close to its critical value. The degree of tuning is at least of order 10% for a 10 TeV ETC scale [23].

Maintaining $m_t = 175$ GeV, though, prevents decoupling of some higher order contributions to $\Delta \rho$, due to technifermion $U - D$ mass splitting arising at second order in ETC boson exchange. When the splitting is small relative to the technifermion masses this "indirect" contribution [22] yields approximately

$$\Delta \rho_{(2)} \simeq \frac{N_{TC}}{12\pi^2 v^2} [\Delta \Sigma(0)]^2 \simeq 0.4\% \times N_{TC} \left(\frac{\Delta \Sigma}{m_t}\right)^2 \tag{18}$$

We expect $\Delta \Sigma(M_{ETC})$ to be of order m_t, since the t unifies with the U (and similarly the b with the D) at the lowest ETC scale, where their masses must be equal. That equality is preserved as the ETC scale increases.

Eq. (18) shows that when the ETC couplings are perturbative, $\Delta \rho_{(2)}$ is much smaller than $\Delta \rho_{(1)}$; but that when the ETC breaking scale is raised with the ETC coupling tuned to maintain m_t, $\Delta \rho_{(2)}$ remains close to experimental bounds instead of decoupling like $\Delta\rho_{(1)}$. The derivation treats strongly interacting physics naively, but should give trustworthy orders of magnitude.

In order to suppress this second order contribution to $\Delta\rho_{(2)}$, ETC models in which m_t is generated by a strong "top condensate" self-interaction rather than the usual "sideways" interaction have been proposed [28]. In those models the strong interaction that generates the top self-interaction must be tuned to keep m_t at the weak scale. However the weak-isospin-violating interactions, acting only on the t quark, not the technifermions, then suppress the above $\Delta\rho$ contributions by at least $(m_t/\Lambda_{TC})^2$.

III CONCLUSIONS

To conclude, I have noted that realistic ETC models are chiral. Models with separate chiral groups, which I have argued are most natural, contain massive ETC bosons in the technicolor adjoint representation. In models that generate m_t perturbatively, these gauge bosons give rise to a contribution to $\Delta\rho$ of about 5%, an order of magnitude above the experimental limits. That contribution may be suppressed, by raising the lowest ETC scale and tuning the ETC couplings to generate the large top mass.

REFERENCES

1. T. Appelquist and C. Bernard, Phys. Rev. D22 (1980) 200.
2. A. Longhitano, Phys. Rev. D22 (1980) 1166; Nucl. Phys. B188 (1981) 118.
3. T. Appelquist and G-H Wu, Phys. Rev. D48 (1993) 3235.
4. Some recent references are:
    D. Espriu and M.J. Herrero, Nucl. Phys. B373 (1992) 117;
    A. De Rújula, M.B. Gavela, P. Hernández and E. Massó, Nucl. Phys. B384 (1992) 3;

P. Hernandez and F. Vegas, "one loop Effects of Non-Standard Triple Gauge Boson Vertices", CERN-TH 6670, LPTHE-Orsay 92/56, FTUAM-92/34, Phys.Lett.B307:116-127,1993;

C. Grosse-Knetter and R. Kögler, "Unitary Gauge, Stückelberg Formalism and Gauge Invariant Models for Effective Lagrangians", BI-TP 92/56, Phys.Rev.D48:2865-2876,1993.

5. M.E. Peskin and T. Takeuchi, Phys. Rev. Lett. 65 (1990) 964; Phys. Rev. D46 (1992) 381.
6. K. Hagiwara, R.D. Peccei, D. Zeppenfeld and K. Hikasa, Nucl. Phys. B282 (1987) 253.
7. T. Appelquist, Proceedings of the Fourth Mexican School of Particles and Fields, Mexico City, December 3-14, 1990, and the VI Jorge Andre Swieca Summer School, Sao Paulo, Brazil, January14-26, 1991, World Scientific Publishing, 1991.
8. R.S. Chivukula, A. Cohen, and K. Lane, Nucl. Phys. B343, 554 (1990)
9. T. Appelquist, J. Terning and L.C.R. Wijewardhana, Phys. Rev. D44, 871 (1991).
10. T. Appelquist and J. Terning,, Physics Letters B315 (1993) 139.
11. F. Feruglio, lectures given at the second National Seminar of Theoretical Physics, Parma, Italy, 1-12 September, 1992.
12. L. Susskind, Phys. Rev. **D20** (1979) 2619;
S. Weinberg, Phys. Rev. **D19** (1979) 1277;
E. Farhi and L. Susskind, Phys. Rept. **74** (1981) 277.
13. S. Dimopoulos and L. Susskind, Nucl. Phys. **B155** (1979) 237;
E. Eichten and K. Lane, Phys. Lett. **B90** (1980) 125.
14. R.S. Chivukula, S.B. Selipsky and E.H. Simmons, hep-ph/9204214, Phys. Rev. Lett. **69** (1992) 575; R.S. Chivukula, E. Gates, E.H. Simmons and J. Terning, hep-ph/9305232, Phys. Lett. **B311** (1993) 157.
15. B. Holdom, hep-ph/9407311, Phys. Lett. **B339** (1994) 114;
G. Wu, hep-ph/9412206, Phys. Rev. Lett. **74** (1995) 4137;
K. Hagiwara and N. Kitazawa, hep-ph/9504332, Phys. Rev. **D52** (1995) 5374;
T. Yoshikawa, hep-ph/9506411, Hiroshima preprint HUPD-9514;
R.S. Chivukula, B. Dobrescu and J. Terning, hep-ph/9506450.
16. P. Antilogus *et al.* (LEP Electroweak Working Group), LEP-EWWG/95-02.
17. T. Appelquist, M.J. Bowick, E. Cohler and A.I. Hauser, Phys. Rev. **D31** (1985) 1676.
18. R.S. Chivukula and H. Georgi, Phys. Lett. **B188** (1987) 99;
S.F. King, Phys. Lett. **B229** (1989) 253;
K. Lane and M.V. Ramana, Phys. Rev. **D44** (1991) 2678;
M.B. Einhorn and D. Nash, Nucl. Phys. **B371** (1992) 32;
H. Georgi, hep-ph/9209244 , Nucl. Phys. **B416** (1994) 699;
L. Randall, hep-ph/9210231, Nucl. Phys. **B403** (1993) 122.
19. T. Appelquist, N. Evans and S. Selipsky, Phys. Lett. **B 374** (1996) 145.
20. J. Terning, hep-ph/9410233, Phys. Lett. **B344** (1995) 279.

21. R.S. Chivukula, E.H. Simmons and J. Terning, hep-ph/9404209, Phys. Lett. **B331** (1994) 383; hep-ph/9506427; hep-ph/9511439.
22. M.E. Peskin and T. Takeuchi, Phys. Rev. **D46** (1992) 381.
23. T. Appelquist and O. Shapira, Phys. Lett. **B249** (1990) 83;
    N. Evans, hep-ph/9403318, Phys. Lett. **B331** (1994) 378.
24. H. Pagels and S. Stokar, Phys. Rev. **D20** (1979) 2947.
25. T. Appelquist and N. Evans, hep-ph/9509270, Phys. Rev. **D53** (1996) 2789.
26. J. Erler and P. Langacker, hep-ph/9411203, Phys. Rev. **D52** (1995) 441;
    and Phys. Rev. **D50** (1994) 1173 (Review of Particle Properties), p. 1312.
27. When extended to include the (ν_τ, τ) doublet this becomes a generalized Pati-Salam model: J.C. Pati and A. Salam, Phys. Rev. Lett. **31** (1973) 661; Phys. Rev. **D8** (1978) 1240.
28. Y. Nambu, "New Theories In Physics", Proc. XI Warsaw Symposium on Elementary Particle Physics (ed. Z. Adjuk *et al.*, World Scientific, Singapore, 1989);
    V.A. Miranskii, M. Tanabashi and K. Yamawaki, Phys. Lett. **B221** (1989) 177;
    R.R. Mendel and V.A. Miranskii, Phys. Lett. **B268** (1991) 384, erratum **B275** (1992) 512;
    W.A. Bardeen, C.T. Hill and M. Lindner, Phys. Rev. **D41** (1990) 1647;
    C.T. Hill, hep-ph/9411426, Phys. Lett. **B345** (1995) 483;
    R.S. Chivukula, B.A. Dobrescu and J. Terning, hep-ph/9503203, Phys. Lett. **B353** (1995) 289;
    K. Lane and E. Eichten, hep-ph/9503433, Phys. Lett. **B352** (1995) 382.

LEP Status Report

Luigi Rolandi

CERN- CH 1211 Geneva 23
Rolandi@CERN.CH

Abstract. Recent results on W physics and on searches for new particles from analyses of the full 1996 data set collected at LEP are reported.

INTRODUCTION

In 1995, after 6 years of very successful data-taking at $\sqrt{s} = M_Z$, superconducting cavities were installed in the LEP ring and the LEP2 program started with a pilot run of 5 pb^{-1} at a centre-of-mass energy of 135 GeV.

In spring 1996 the total RF voltage installed on LEP was in excess of 1600 MV and the beams collided for the first time at a centre-of-mass energy above the W$^+$W$^-$ production threshold. The total amount of integrated luminosity delivered to each experiment was about 12 pb^{-1}. In August and September 1996 more superconducting RF cavities were installed and LEP centre-of-mass energy increased to 172 GeV. About 11 pb^{-1} were delivered by LEP to each experiment before the Winter shut-down. More cavities will be installed in the forthcoming years and the centre-of-mass energy of LEP should eventually exceed 192 GeV in June 1998.

This report summarizes the most recent results of the analyses of the 161 and 172 GeV data that were presented by the four LEP Collaborations in a jamboree held at CERN on February 25 1997. More details on the analyses can be found in the copies of the transparencies of these talks [1–4] that are available on WWW and in the references therein.

These results supersede those presented in my talk given in Santa Barbara in October 1996 since they include also the results of the run at 172 GeV that ended in November 96. The large majority of these results are unpublished and should be considered as preliminary.

The measured cross sections for the dominant s-channel process at LEP2 ($e^+e^- \to f\bar{f}(\gamma)$) are shown in Fig. 1, for center of mass energies from 90 GeV to 170 GeV. They agree very well with the prediction from ZFITTER. It is interesting to remark that the cross section for q$\bar{q}$ production at 161 GeV is 200 times smaller than at the Z peak. Moreover, about 80% of this cross section is due to radiative return to the Z, i.e. to the emission of a high energy photon from the initial state resulting in an effective centre-of-mass energy of the e^+e^- collision equal to the Z mass.

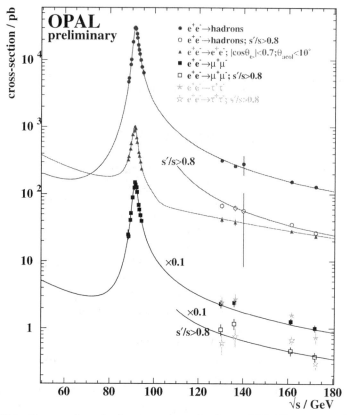

FIGURE 1. Cross sections for fermion pair production at LEP. The cross sections above 120 GeV are measured inclusively (full dots) and for $s'/s > 0.8$ (open dots) to exclude the strong radiative return to the Z peak. The cross section for Bhabha scattering is measured for polar angles θ in the interval $-0.7 < \cos\theta < 0.7$

The LEP2 physics program is focused on the study of the W pair production and on searches for new particles. In the following I summarize the results of these analyses.

W PAIR PRODUCTION

The visible cross section for four-fermion production above 161 GeV is dominated by real WW production that goes via the so-called CC03 diagrams shown in Fig. 2. Many other diagrams contribute to the same final state but they have a small visible cross section and, after the selection cuts, they contribute only a few % to the selected sample.

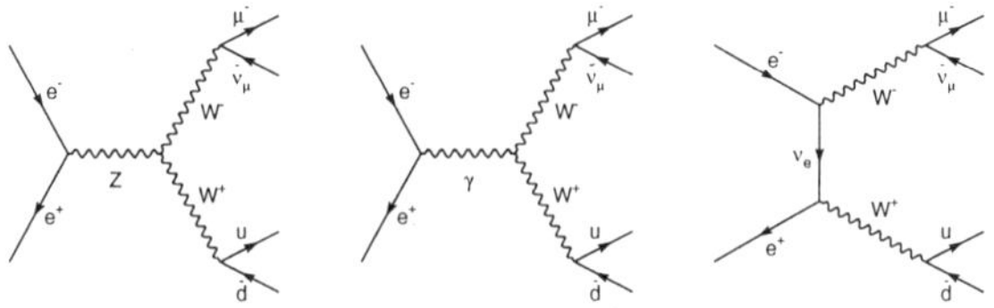

FIGURE 2. Diagrams for four-fermion production with two resonant W's (CC03).

The measurement of the cross section at threshold ($\sim 2m_W + 0.5$ GeV) provides a sensitive measurement of the W mass within the framework of the Standard Model, with very small dependence on other parameters. At 172 GeV the W mass is measured from direct reconstruction of the final state.

The event selection is simple for the final state where at least one W decays into lepton plus neutrino. The final state $\ell\ell\nu\nu$ (11%) has two acoplanar leptons and missing energy, while the final state $q'\bar{q}\ell\nu$ (43%) has an isolated lepton, two hadronic jets and isolated missing energy. These channels can be selected with high efficiency and very low background. The totally hadronic channel (46%) has a large QCD background: the events are selected using many topological properties that are combined in a single distribution (neural net, likelihood function, weights) which is compared with the predicted distributions for the signal and the background obtained with Monte Carlo simulation. Each experiment has selected typically 5 events in the channel $\ell\ell\nu\nu$ and 15 events in the $q'\bar{q}\ell\nu$ channel in the 161 GeV data sample. The cross sections measured in each channel are combined using the SM branching ratios. The combined results of each experiment [5–8] are shown in Fig. 3 together with the expected dependence of the cross section on the W mass.

The average cross section of 3.69 ± 0.45 pb corresponds to $m_W = 80.40 \pm 0.22$ GeV. The main systematic error on the W mass comes from the uncertainty on the beam energy and is smaller than 100 MeV.

At 172 GeV the W mass is measured by direct reconstruction using the $q'\bar{q}\ell\nu$ and the fully hadronic channels. A value of the W mass is extracted for each event using the energies and the directions of the reconstructed jets and leptons, applying the constraints of energy and momentum conservation and imposing, in some cases, the equality of the two W masses. The measured distribution is compared with the Monte Carlo expectations for many W masses to fit m_W.

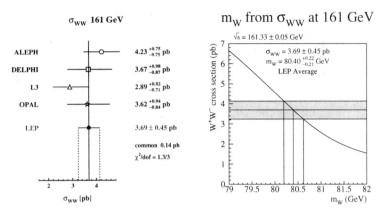

FIGURE 3. Measurements of the WW cross section at 161 GeV (left) and its dependence on m_W (right).

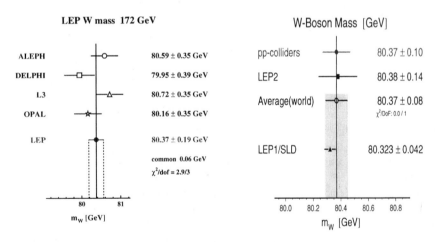

FIGURE 4. Direct measurement of W mass at LEP (left) and comparison of the LEP measurement with the $p\bar{p}$ colliders measurement and with the result of Standard Model fit of LEP1/SLD data (right).

The simulation predicts a bias of about 200 MeV, mainly due to the combined effects of the constraints and initial state radiation. Each Collaboration selected about 80 WW pairs for this measurement that is statistically limited. The main systematic errors come for the simulation of the jets and, in the fully hadronic final state, from the interaction of the four quarks from the decays of the two W's (color reconnection). The results are shown in Fig. 4a. The LEP average mass with direct reconstruction is $m_W = 80.37 \pm 0.19$ GeV.

The two values of the W mass measured at LEP with two different techniques are comparable in precision. Their average is $m_W = 80.38 \pm 0.14$ GeV. This average is in good agreement with the average of the measurements done at $p\bar{p}$ colliders [9-11] (see Fig. 4b).

Using the full LEP2 statistics (~ 500 pb^{-1}) the W mass will be measured at LEP with an error between 25 and 35 MeV depending on the understanding of the systematic errors.

HIGGS SEARCH

The combination of the analyses of the LEP collaboration on the whole data sample collected at the Z peak excluded a SM Higgs with a mass smaller than 65.6 GeV at 95% confidence level.

At LEP2, above the Z peak, the SM Higgs is produced associated with a real Z in the reaction $e^+e^- \to ZH$. Many final state topologies are possible, depending on the specific decays of the Z and of the H. The most sensitive ones are those where the Higgs decay into $b\bar{b}$ and the Z decays into $q\bar{q}$ or $\nu\bar{\nu}$. While the latter can be kinematically separated from the background exploiting the large missing energy and missing mass from the Z decay, the four-jet final state has a prohibitive background from QCD and four-fermion processes. This background can be reduced by exploiting the b-tagging capabilities of the LEP experiments.

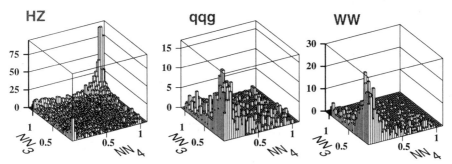

FIGURE 5. Combined b-tag variable for the two selected jets (see text) in a four jet final state. The signal (left) is peaked at (1,1) while the backgrounds peaks at (0,0).

The main ingredient for b-tagging is the precise measurement of the impact parameter performed with the vertex detector. The average b-hadron decay length from H decay is a few millimeters, resulting in impact parameters of charged tracks (wrt the primary vertex) that are typically larger than the detector resolution ($\sim 20\mu$m). Other properties of b-hadron decays, as the presence of a high p_T lepton or the fatness of b-jets, are also used. This information is combined in an estimator that is applied to each jet of a four-jet event. Tagging efficiency of 85% on a b-jet can be obtained rejecting 85% of the non-b jets. Fig. 5 shows the global performance of the b-tagging in four-jet events. In this analysis, the four jets are paired in all possible combinations. The pair with the mass closest to the Z mass is excluded and the tagging is applied to the remaining two jets allowing a good separation of the signal with respect to the background.

The signal for SM Higgs production was searched for by the four Collaborations in the 161 GeV and 172 GeV data samples and no significant excess above the predicted background was found. Limits on the Higgs mass have been set, assuming the Standard Model cross section and decays. Fig. 6 shows the result obtained by the L3 collaboration and Table 1 summarizes the results obtained by each analysis. The analyses have not yet been combined; the combination will be sensitive to masses of $\sim$ 75 GeV.

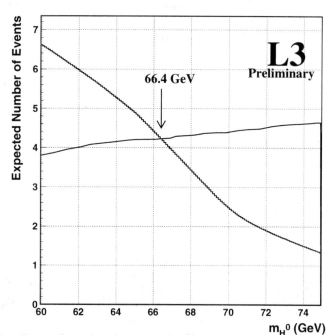

FIGURE 6. Expected number of events and 95% confidence line taking into account the masses of the observed candidates.

TABLE 1. Summary of the Standard Model Higgs Boson Search.

	Overall Efficiency	Expected Background	Number of Candidates	95% C.L. (GeV)
ALEPH	29%	0.98	0	70.7
DELPHI	28%	4.26	2	65.0
L3	28%	5.79	4	66.4
Opal	26%	4.00	3	68.8

In 1998 LEP will run at a centre-of-mass energy of 192 GeV. With an integrated luminosity of 200-300 pb^{-1}, these analyses combined can discover a SM Higgs lighter than 95 GeV or can exclude a SM Higgs lighter than 100 GeV.

Search for Higgs Bosons of the MSSM

In the Miminal Supsersymmetric extention of the Standard Model (MSSM) there are two Higgs doublets and five Higgs physical states of which one (h) is lighter than the Z at tree level. Radiative corrections, which mainly depend on the top mass, increase the h mass to a maximum of about 130 GeV. Two processes could be observed at LEP2: $e^+e^- \to hZ$ and $e^+e^- \to hA$ (where A is the CP-odd state), depending on m_h, on the mixing α between the two CP-even states and on the ratio $\tan\beta$ between the vacuum expectation values of the two doublets. These two processes are complementary since the cross section of the first is proportional to $\sin^2(\beta - \alpha)$ and that of the second to $\cos^2(\beta - \alpha)$.

The first process is studied with the same analysis as the SM Higgs, simply reducing the expected cross section by $\sin^2(\beta - \alpha)$. The second is studied searching for a $b\bar{b}b\bar{b}$ final state, exploiting again the b-tagging. The results of these analyses are summarized in Table 2.

TABLE 2. Summary of the h A searches. The limits are on both m_h and m_A assuming $\tan\beta > 1$.

	Overall Efficiency	Expected Background	Number of Candidates	95% C.L. (GeV)
ALEPH	54%	0.84	0	62.5
DELPHI	40%	1.99	0	58.0
L3	31%	1.32	1	45.0

Since no excess of events is seen, new limits can be set, improving substantially on those already obtained using Z peak data. The exclusion plot produced by the Aleph collaboration in the $m_h \tan\beta$ plane is shown in Fig. 7.

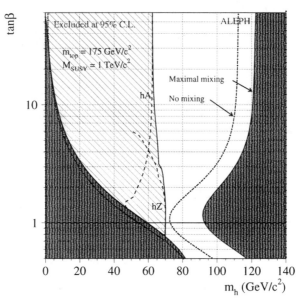

FIGURE 7. Exclusion region in the plane $m_h \tan\beta$. The dark regions are not allowed by the model.

SEARCH FOR OTHER SUPERSYMMETRIC PARTICLES

In the hypotheses of conserved R-parity and that the neutralino (χ_1^0) is the lightest SUSY particle, all supersymmetric particles decay eventually to χ_1^0 that is invisible. Charginos and neutralinos decay to χ_1^0 with the emission of a virtual W or Z that materializes in a pair of fermions. Sleptons decay into lepton and χ_1^0. The visible energy depends on the difference in mass ΔM between the produced particle and the neutralino. For very small ΔM (small visible energy) the topology of the final state is similar to those of the so-called $\gamma\gamma$ events, where two photons emitted by the initial state particles interact at small momentum transfer while the electron and the positron escape undetected in the beam pipe.

The searches for pair produced SUSY particles select the events requiring missing energy and missing mass produced by the two undetected neutralinos and typically acoplanar leptons and/or acoplanar jets. The searches are optimized for each final state and for small, moderate and large ΔM. The analyses performed by the LEP Collaborations have not shown a significant excess above the expected background. The results of the OR of these analyses are shown in Table 3.

TABLE 3. Summary of SUSY searches

	Expected Background	Number of Candidates
ALEPH	13.4	14
L3	9.3	8
OPAL	8.2	7

For each channel the result of these analyses is presented in a model independent way, plotting the excluded cross sections as a function of the masses of the supersymmetric particles (see Fig. 8a). It can be presented also as a mass exclusion plot, assuming the couplings of the MSSM (see Fig. 8b) for fixed values of the model parameters.

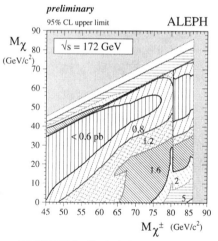
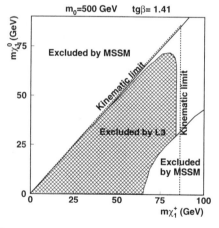

FIGURE 8. Excluded cross sections at 95% C.L. (left) for chargino pair production as a function of the chargino and neutralino masses. Excluded masses (right) in the MSSM for $\tan\beta$=1.4 and m_0= 500 GeV

The various searches can be combined to exclude regions in the space of the parameters of the model as shown in Fig. 9.

The mass of χ_1^0, the lightest supersymmetric particle, can be computed for

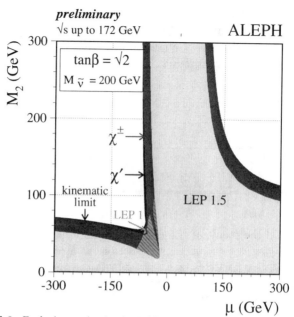

FIGURE 9. Exclusion region in the MSSM parameter plane M_2 vs μ for $\tan\beta=1.4$

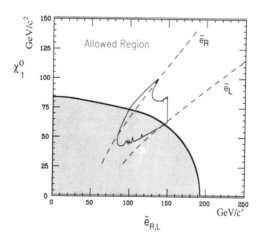

FIGURE 10. Region determined from the kinematics of the CDF event assuming the reaction $q\bar{q} \to ee \to ee\chi_1^0\chi_1^0 \to ee\tilde{G}\tilde{G}\gamma\gamma$. The hatched area is excluded by the ALEPH search for acoplanar photons.

each allowed set of values of the parameters of the model, assuming in addition the M_1, M_2 unification [12] (M_1 and M_2 are the common U1 and SU2 gaugino masses at the unification scale). An absolute lower limit on the mass of χ_1^0 around 23 GeV is found for sneutrino masses larger than 100 GeV.

Assuming an almost massless gravitino ($\tilde{G}$) as the lightest SUSY particle, the neutralino χ_1^0 decays into a gravitino emitting a photon. The production of a neutralino pair results then into a final state with two acoplanar photons. This channel has been searched at LEP and no significant excess of events has been found above the predicted background. An upper limit around 70 GeV have been set on the neutralino mass under this hypothesis. This search excludes (see Fig. 10) part of the region of the plane selectron-mass vs neutralino-mass compatible with the SUSY interpretation of a single anomalous event with large missing transverse energy, reported by CDF in 1995, which contains only two electrons and two photons [13].

ALEPH 4-JET EVENTS

In November 1995 ALEPH reported [14] an excess of events with peculiar properties in the four-jet events selected on the data sample collected at 130 and 136 GeV. The four reconstructed jets are paired in the three possible combinations and for each combination the two invariant masses of the two di-jet pairs are computed. The combination with the smallest mass difference is chosen and the sum of the two di-jet masses is plotted. The experimental resolution of this quantity is very good (~ 1.6 GeV) as a consequence of the energy and momentum conservation constraints imposed on the measured kinematics. The plot of the di-jet mass sum (see Fig. 11a.) shows a significant excess of events (9) compared to the expected background (0.7) at about 105 GeV. The events in the peak differ from standard QCD events also in the jet-charge and the QCD matrix element distributions.

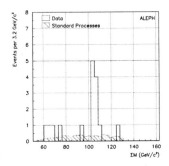
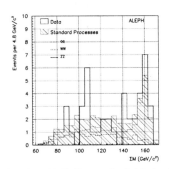

FIGURE 11. Di-jet mass sum (see text) in the 131-136 GeV sample (left), in the 161-172 GeV sample (right).

The data collected by ALEPH at 160 and 172 GeV and selected with the same criteria also show an excess, although less significant, at the same di-jet mass (see Fig. 11b): 9 events are observed where 3.2 are expected. Fig. 12 shows the two data sets combined after a WW rejection is applied to the high energy data : 18 events are observed where 3.1 are expected and the region outside the peak is well reproduced by the simulation.

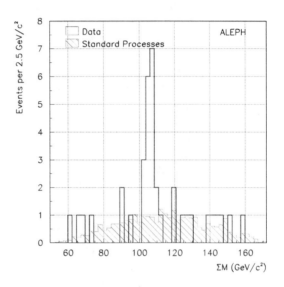

FIGURE 12. Di-jet mass sum (see text) in the combined 130-136-161-172 sample after WW rejection cuts.

The distribution of the mass difference for the events in the peak is more compatible with a finite difference (~ 10 GeV) than with zero mass difference. It is also compatible, besides the overall number of events, with the SM expectation. Similar analyses have been performed by Delphi, L3 and Opal and have not shown any excess compared with the SM predictions: the combinations of the three analyses see 9 events where 9.2 are predicted.

The reconstructed four vectors of all particles of the events selected by ALEPH have been used as an input to the simulation programs of the other three detectors. The result of this simulation has confirmed the capability of the other three detectors of producing a peak of adequate resolution with that input. The conclusion of this analysis is that the other three detectors would have seen the peak if the events were produced at their interaction point.

Accepting this conclusion, the ALEPH excess seen at 133 GeV and confirmed at higher energy is incompatible with the observations of the other three experiments and has to be interpreted as a statistical fluctuation with probability smaller than 10^{-4}. More data are needed to clarify the experimental discrepancy.

CONCLUSIONS

The production of pairs of W bosons has been observed at LEP and the W mass has been measured with an error of about 150 MeV. At the end of the LEP2 program this error will decrease below 35MeV.

The searches for supersymmetric particles and for the Higgs bosons have been negative. The lower limit on the mass of the Standard Model Higgs is now at 71 GeV. New regions of masses and parameters space will be explored when the LEP centre-of-mass energy will be increased to 183 and 192 GeV in 1997 and 1998.

More data are needed to clarify the experimental discrepancy seen in the 4-jet events.

REFERENCES

1. Glen Cowan, Aleph results at 172 GeV, talk given at CERN on February 25 1997
   http://alephwww.cern.ch/ALPUB/seminar/Cowan-172-jam/cowan.html
2. Francois Richard, Delphi results at 172 GeV, talk given at CERN on February 25 1997
   http://delphiwww.cern.ch/delfigs/richard970225
3. Marco Pieri, L3 physics results at 172 GeV Lep run, talk given at CERN on February 25 1997
   http://hpl3sn02.cern.ch/conferences/talks97.html
4. Sachio Komamiya, Opal at 170, 172 GeV, talk given at CERN on February 25 1997
   http://www.cern.ch/Opal/plots/komamiya/koma.html
5. ALEPH Collaboration, CERN-PPE 97-025 (1997).
6. DELPHI Collaboration, CERN-PPE 97-009 (1997).
7. L3 Collaboration, CERN-PPE 97-014 (1997).
8. OPAL Collaboration, Phys. Lett. B389 416 (1996).
9. UA2 Collaboration, Phys. Lett. B276 354 (1992).
10. CDF Collaboration, Fermilab-Pub-95/033-E (1995).
11. D0 Collaboration, talk given at ICHEP'96, Warsaw by M. Rijssenbeek.
12. ALEPH Collaboration, CERN-PPE 96-083 (1996)
    J.Ellis, T. Falk, K. Olive and M. Schmitt, Phys. Lett B388 97 (1996).
13. CDF Collaboration, Fermilab-Conf-96/240-E (1996).
14. ALEPH Collaboration, Z. Phys. C71 179 (1996).

The Tev 2000 Report: Electroweak Physics with High Luminosity at the Fermilab Tevatron

Dan Amidei
Randall Laboratory of Physics
University of Michigan,
Ann Arbor, MI, 48109-1120 USA

Abstract. The addition of the Main Injector and the Antiproton Recycler to the Fermilab accelerator complex create the prospect of integrating samples in excess of 20 fb^{-1} at the Tevatron collider. The Tev2000 report is a study of the "High-P_T" physics that may be realized with such data sets. We find a rich catalog of measurements, including characterization of the top quark, sensitivity to a light Higgs boson and low energy SUSY, and a precision electroweak program with sensitivity comparable and complementary to that of LEP-I.

I INTRODUCTION

For the next decade, the Fermilab Tevatron Collider remains the high energy frontier of particle physics. Luminosity enhancements planned or possible with the Main Injector will dramatically increase the discovery reach, and, in conjunction with upgrades to the collider detectors, will move the experimental program into a regime of *precision* hadron collider physics.

The Tev2000 Working group has completed a preliminary study of the physics potential for an extended Tevatron program, and this paper is a summary of that work [1]. I concentrate here on top physics, a completely new arena for particle physics at an existing U.S. facility. I also include brief summaries of the prospects for detecting a light Higgs boson or low energy SUSY, as well as the outline of a global precision electroweak program.

In the Tev2000 report, we avoid the vagaries of operating point or schedule, and simply specify the physics reach as a function of integrated luminosity, assuming detector performance maintained at the level of the CDF and DØ upgrades for Tevatron Run 2, beginning in 1999. This summary presents expectations for 2 luminosity goals: the present Run 2 plan of 1-2 fb^{-1}, and a Run 3 of greater than 25 fb^{-1}.

CP397, *Future High Energy Colliders*, edited by Z. Parsa
© 1997 American Institute of Physics

II TOP PHYSICS

The top quark is strongly coupled to the electroweak symmetry breaking sector, and decays into a b quark and a real W before hadronizing. The top quark is an obvious new avenue for study of the fermion sector, and offers a number of experimental techniques not previously available for quarks.

A Event Samples in $t\bar{t}$

The standard $t\bar{t}$ selection is based on the expected decay chain $t\bar{t} \rightarrow (W^+b)(W^-\bar{b})$ and the subsequent decays of the W's into fermion pairs [2]. At least one W is tagged in the mode $W \rightarrow l\nu$ by requiring an isolated high lepton (e or μ) with large "transverse momentum" (P_T) to the beamline and large "missing transverse energy", $\not{E}_T$. In the "dilepton" analysis the leptonic decay of the other W is identified with a loose lepton selection; this mode has small backgrounds but small branching fraction of just 4/81.

In the "lepton+jets" mode, the second W decays to quark pairs, giving larger branching fraction of $24/81 \approx 30\%$ (lepton = e or μ). The final state of (lνb)(jjb) is separated from the primary background, W+jets, by requiring a large multiplicity of high E_T jets and also evidence of a B decay, using either secondary vertex identification (SVX) or a tag of the "soft lepton" from $b \rightarrow l\nu_l X$ (SLT). The largest sample results from requiring at least 3 jets with $E_T \geq 15$ GeV and $| \eta | \leq 2.4$. The constrained fit technique used in the top mass measurement requires a "completely reconstructable" event having at least 4 final state jets.

Future $t\bar{t}$ event yields at the Tevatron can be inferred with reliable precision by extrapolating from the presently understood situation. The Upgrades for Run 2 will have improved lepton acceptance and much improved b-tagging. The efficiency for tagging at least 1 b will be 85% per event, and the efficiency for tagging *both* b's will be 42% per event. We will assume that selection efficiencies can be maintained at high luminosity by staged evolution in detector technologies. The total efficiency including branching fraction $\epsilon \cdot B$ for some pertinent final states is shown in Table 1.

We calculate absolute yields at the Run II operating point of $\sqrt{s} = 2.0$ TeV, using $\sigma_{t\bar{t}} = 6.8$ pb for $m_t = 175$ GeV/c^2 [3]. The future yields are shown for the benchmark data sets in Table 1. At $m_t = 175$ GeV/c^2 each inverse femtobarn at the Tevatron will yield approximately 600 identified, b-tagged events and approximately 250 double b-tagged, completely reconstructable events.

TABLE 1. Top yields

Mode	$\epsilon \cdot B$	$1 fb^{-1}$	$10 fb^{-1}$
produced		6.8K	68K
dilepton	1.1%	82	820
W + 3j	10%	680	6.8K
W + 3j * b	8.6%	584	5.8K
W + 4j	8.9%	605	6.0K
W + 4j * b	7.6%	517	5.2K
W + 4j * bb	3.8%	258	2.6K

B Measurement of the Top Mass

The mass of the top quark is a fundamental Standard Model parameter and should be measured as accurately as possible. The best understood mass measurement technique at present is complete reconstruction in $t\bar{t} \to W + 4$ jets [4]. Events with at least four jets are selected as above, and the lepton and the four highest E_T jets in the event are fit to the hypothesis $t\bar{t} \to (Wb)(Wb) \to (l\nu b)(jjb)$. Each jet is extrapolated back to a parton energy by correcting on average for instrumental and QCD effects. The 2-C fit has multiple solutions in each event, and the configuration with lowest χ^2 and consistency with b-tagging is chosen. The shape of the mass spectra for various values of m_t, as well as for the expected backgrounds, are derived from Monte Carlo samples, and a maximum likelihood fit to the data spectrum yields the best estimate of the top mass.

Almost all of the systematic uncertainties in the top mass measurement are coupled to the reliability of the Monte Carlo models for the spectrum of fit masses in signal and background. Assuming the theory model is accurate, most of the uncertainty concerns resolution effects. Instrumental contributions include calorimeter nonlinearity, losses in cracks and dead zones, and absolute energy scale. A larger and more intractable part of the energy resolution concerns the reliability of the extrapolation to parton energies, and it is the understanding of QCD, not the detector, which limits the mass resolution. All of these issues can be addressed by *in situ* calibration procedures such as E_T balancing in $Z/\gamma + 1$ jet events and isolation of $W \to jj$ in top events.

The statistical error for the three event classes W+4j, W+4j+b, and W+4j+bb are shown in Table 2, as understood from the present measurement, and extrapolated to the future scenarios assuming the event yields above. Note that in large samples the double tagged events have good statistical precision.

We have also made a careful accounting of the probable evolution of the systematic uncertainty for the single tagged analysis. The *in situ* energy calibration techniques have been benchmarked with simulation studies for 1 fb^{-1},

TABLE 2. Expected m_t precision, all entries in GeV/c^2.

Mode	δm_t	$1 fb^{-1}$	$10 fb^{-1}$
stat. W+4j	$38/\sqrt{N}$	1.6	0.5
stat. W+4j+b	$35/\sqrt{N}$	1.5	0.5
stat. W+4j+bb	$27/\sqrt{N}$	1.7	0.5
sys. W+4j+b	see text	3.7	1.2
total W+4j+b	$\delta^2_{\text{stat}} + \delta^2_{\text{sys}}$	4.0	1.3

and the resolution improvement from these is slightly worse than $\sqrt{N}$. The b-tag bias can be addressed in control sample studies, and should scale as $1/\sqrt{N}$. Control of the small but significant uncertainty due to background modelling can be done in a large Z+jets sample, and this has also been benchmarked for 1 fb^{-1}. The net effect of this more careful consideration of the systematic error is listed in the fourth row of Table 2, and, compared to the row above, is seen to be slightly degraded from simple $1\sqrt{N}$ scaling.

Adding these systematic errors in quadrature with the statistical error leads to the top mass precision listed at the bottom of Table 2. With 10 fb^{-1} at the Tevatron, the experimental contributions to the top mass uncertainty will be limited to the order of 1.3 GeV/c^2 per experiment.

Other techniques under study include likelihood fits to the kinematics in the dilepton sample, complete reconstruction in the 6 jet mode, and accurate measurement of the mean decay length in b's from top decay, and all of these techniques project to errors of $\sim 1-2$ GeV for 10 fb^{-1} [5-7]. Since these techniques are statistically independent, we believe that 10 fb^{-1} will allow a combined measurement of the top mass with the control of the experimental uncertainties at the level of 1 GeV/c^2.

Ultimately, the theoretical inputs must also be verified. For instance, the distribution of fit masses used as input templates to the likelihood fit are derived from a theoretical calculation, and several studies raise questions concerning the modelling of hard gluon radiation and other subtleties in the final state [8]. Resolution of these concerns requires more statistical precision than presently available, but we expect that the theoretical uncertainties will ultimately be controlled at the level of the experimental precision, of order 1 GeV/c^2.

Taking all of the above into account, we believe that 10 fb^{-1} at the Tevatron will allow a **measurement of the top mass with a precision of 2 GeV/c^2 per experiment**.

If the W mass is known to 20 MeV/c^2, a precision of 2 GeV on the top mass will fix the Higgs mass to within 50% of itself [1]. Improvement beyond 2 GeV in top mass precision does not affect the Higgs precision unless the W mass precision is likewise improved.

C Top Pair Production

An accurate measurement of the $t\bar{t}$ production cross section is a precision test of QCD. A cross section significantly higher than the theoretical expectation would be a sign of non-Standard Model production mechanisms, for example the decay of a heavy resonant state into $t\bar{t}$ pairs [9], or anomalous couplings in QCD [10].

Future samples will have large statistics in the l+jets mode and systematic uncertainties will be the limiting factor.

For the acceptance, the reliability of jet counting and b-tagging are at issue. Initial state radiation can be examined using a sample of Z+jets, while the jet energy threshold uncertainty can be addressed as in the top mass discussion. With 1 fb^{-1} of data it will be possible to measure the b-tagging efficiency *in top events*, using dilepton events selected without a b-tag and the ratio of single to double tags in lepton plus jets events. We assume that these studies will give uncertainties that scale with $\sqrt{N}$.

With large samples, one can measure the bottom and charm content as a function of jet multiplicity in W + jet events using the $c\tau$ distribution of the tagged jets and use this to tune the Monte Carlo models for W + 3 or more jet backgrounds. Finally, in Run 2 and beyond, the luminosity will be measured either through the $W \to l\nu$ rate, or the mean number of interactions per crossing, and we will assume 5% for the future precision of the luminosity normalization.

Accounting for all effects we find that the total $t\bar{t}$ cross section can be measured with a precision of 11% for 1fb^{-1} and 5.9% for 10 fb^{-1}. This will challenge QCD, and provide a sensitive test for non-Standard production mechanisms and decay width to Wb.

D "Single Top" Production

At a hadron collider, the top quark decay width $\Gamma(t \to X)$ cannot be directly measured in the $t\bar{t}$ sample, but its main component can be accessed through single top processes [11]. The single top cross section is directly proportional to the partial width $\Gamma(t \to b + W)$ and, assuming there are no anomalous couplings, this is a direct measure of $|V_{tb}|^2$.

The principal processes leading to single top production are $q'\bar{q} \to t\bar{b}$, via an s-channel "W^*" and $qb \to q't$, via t-channel "W-gluon fusion". For $m_t = 175$ GeV/c^2, W-gluon fusion, at 1.6 pb, is twice as large as W^* at 0.8 pb, and the combined rate for single top production by these two processes, ~ 2.4 pb, is only a factor of 3 down from the $t\bar{t}$ rate at this energy [12].

We have simulated the measurement of electroweak single top production using the ONETOP Monte Carlo with $m_t = 170 GeV$ [13]. We model detector performance for b tagging and jet energy resolutions based on the top mass and cross section analyses described above. B tagging backgrounds are modelled on the current CDF analysis. Our data selection criteria were similar to the $t\bar{t}$ selection, except to ask for just 2 and only 2 jets, with at least one b-tag.

The signal for single top production is a peak in the Wb invariant mass plot. We find that the shape of the signal-plus-background curve is easily distinguished from the background shape alone. We calculate the fractional statistical uncertainty in the cross section as $\sqrt{S+B}/S$, where the size of signal (S) and background (B) are the numbers of each kind of event in the a mass peak window of 50 GeV around the generated top quark mass. We find a signal yield of approximately 100 events per fb^{-1}, above a background about twice as large, mainly from $W + b\bar{b}$. With 10 fb^{-1} it will be possible to measure the single top cross section with a statistical precision of 5.5%.

Many of the sources of systematic uncertainty in the single top cross section are common to the $t\bar{t}$ cross section discussed above. We assume that systematic uncertainties related to selection efficiencies and backgrounds will shrink as $\sqrt{N}$ and find that for large samples the dominant uncertainty is that of the luminosity normalization. For the case of 10 fb^{-1} we find that the measurement of the single top cross section will have a total uncertainty of approximately 10%.

We have made a detailed study of the extraction of $\Gamma(t \to b + W)$, and $|V_{tb}|^2$ from the combined single top cross section [14]. The constant of proportionality between the cross section and the width has theoretical uncertainties originating in α_s, the parton distribution functions, and the choice of scale Q^2. These are estimated to total roughly 10% at present [15], and we assume that better measurements of parton distributions and $\sigma_{t\bar{t}}$ will improve this to 7%. Combining all uncertainties, we find that a measurement of the inclusive single top cross section with 10 fb^{-1} will yield the partial width $\Gamma(t \to Wb)$ with precision of 12%, and therefore V_{tb} with a precision of 6%.

E Search for $t\bar{t}$ Resonances

Several models have been proposed for extensions of the Standard Model which would produce enhancements or resonances in the $t\bar{t}$ invariant mass ($M_{t\bar{t}}$) spectrum [16,17].

We have studied sensitivity to this kind of object via direct search for resonant structure in the $M_{t\bar{t}}$ distribution. We reconstruct $M_{t\bar{t}}$ on an event-by-event basis using the same event sample and constrained fitting techniques used in the top mass measurement, with an additional constraint that the t and $\bar{t}$ decay products have a mass equal to the measured M_{top}.

For definiteness, we use the example of a topcolor Z' decaying to a $t\bar{t}$ pair, assuming that the resonance width is less than the detector resolution on $M_{t\bar{t}}$ ($\approx$ 6% at $M_{t\bar{t}}$ = 800 GeV/c^2). The cross section, $\sigma \cdot B(X \to t\bar{t})$, is determined by theory, and the Pythia Monte Carlo provides the decay $X \to t\bar{t}$. The acceptance including all BR's, detector, and selection effects is 6.5% and approximately flat vs. $M_{t\bar{t}}$.

We add the distribution for a Z' ($M_{Z'}$ = 800 GeV/c^2, $\Gamma_{Z'}$ = 1.2%) to the $M_{t\bar{t}}$ distribution from standard model $t\bar{t}$ production, quantify the excess by fitting the $M_{t\bar{t}}$ distribution below the resonance to estimate the background in the region 700-900 GeV/c^2. We define the discovery limit as the minimum $\sigma \cdot B(X \to t\bar{t})$ for the production of $X \to t\bar{t}$ in order to observe a $\geq$5 sigma excess. With 10 fb^{-1}, we will be able to observe a narrow Z' resonance decaying to $t\bar{t}$ out to approximately 800 GeV.

F Top Decay

In the Standard Model with 3 generations, existing experimental constraints and the unitarity of the CKM matrix require $V_{tb} \simeq 1$, predicting that the weak decay of the top will proceed almost exclusively through W + b. The $t \to Wb$ decay vertex is completely fixed by the universal V-A coupling to the SU(2) bosons. For m$_t$ = 175 GeV/c^2, the partial (but almost total) decay width is $\sim$ 1.8 GeV [18], cutting off the long distance part of the strong interaction. There is no hadronization: all strong interaction issues for the top quark should be well described by perturbative QCD, and charge and helicity information should flow directly to the final state. The top is the first opportunity to study the decays of a naked quark, with experimental techniques and advantages familiar from muon decay.

The detailed experimental issues pertinent to the study of top decays are similar to the efficiency and background issues discussed above in relation to the cross section measurement, and the energy scale issues discussed above in relation to the mass measurement. The comments below are brief, and further detail can be found in the Tev2000 Report [1].

1 The W-t-b Vertex

It is possible that the physics of an underlying theory at a high mass scale may couple to the large top mass, appearing as new non-universal top interactions [19]. One manifestation of anomalous couplings would be a departure from the predicted mixture of W helicity states in the decay. Since the top decay precedes hadronization, the W polarization information persists in the final state, and is experimentally accessible through the charged lepton helicity angle, $\cos \theta_e^*$, which is measured in the lab frame [20] as

$$\cos\theta_e^* \approx \frac{2M_{eb}^2}{m_{eb\nu}^2 - M_W^2} - 1 \qquad (1)$$

A general analysis of the W helicity states through the $\cos\theta_e^*$ distribution can then be performed.

The Standard Model predicts only left-handed or longitudinal W's in top decay, with the branching fraction to the longitudinal component depending only on the top mass. For $m_t = 170$ GeV/c^2, $BF(t \to b + W_{\text{long}}) = 69.2\%$. Non-universal top couplings will, in many cases, appear as a departure of $BF(t \to b + W_{\text{long}})$ from the value expected for the measured m_t.

We have studied $Br(t \to b + W_{\text{long}})$ as a sensitivity benchmark for anomalous top couplings. We use a four vector level Monte Carlo [13] with selection bias and resolution smearing modeled on the CDF experience. We assume that the constrained mass fit will allow us to know perfectly which b jet belongs to the semi-leptonic top decay. We correct the $\cos\theta_e^*$ distribution for the bias imposed by the selection cuts. We fit the distribution the Standard Model hypothesis for the admixture of W_{long} and W_{left} and get a good fit with $BF(t \to b + W_{\text{long}}) = 0.708 \pm 0.030$, as shown.

After accounting for the effect of combinatorics, energy scale uncertainties, and backgrounds, we find that for sample sizes expected to be available at a high luminosity Tevatron, of order 10K and above, the top quark decay branching fraction to longitudinal W bosons may be measured with a statistical precision approaching 1%, and is systematically limited. We have also studied the effect of a V+A term, and find that it would be discernible with similar sensitivity.

2 Measurement of a t→ b Branching Fraction

In the Standard Model, a 176 GeV top quark decays almost exclusively to b's. This is easy to test. One looks at top events containing W's, and measures the branching fraction into b's:

$$BF(t \to (W)b) = \frac{|V_{tb}|^2}{|V_{td}|^2 + |V_{ts}|^2 + |V_{tb}|^2} \qquad (2)$$

The notation is meant to remind that this is the fraction of top decays to W's that *also* contain b's. Since the standard analysis identifies $t\bar{t}$ events by requiring at least 1 W and 1 b, $BF(t \to (W)b)$ is measured from the number and distribution of tagged b-jets in top events. There are three methods which can be employed:

• The ratio of double b-tagged to single b-tagged events in the b-tagged lepton plus jets sample.

• The number of b tagged jets in the dilepton sample.

- The number of times that events tagged by both the secondary vertex and soft lepton algorithms have both tags in the same jet vs. the number of times the tags are in different jets.

These techniques are not exclusive, and can be combined. This analysis has been carried out in the present CDF data set, and the precision can be scaled to the expected size of the Run 2 samples. The systematic uncertainty is dominated by the uncertainty on the tagging efficiency, which is measured from the data using b semileptonic decays, and will fall as $1/\sqrt{N}$, provided that the experiments can continue to record these events. The small non-$t\bar{t}$ backgrounds will be measured to high accuracy by Run 2. A sample of 10 fb^{-1} at the Tevatron allows the measurement of the branching fraction $BF(t \to (W)b)$ with a precision of 1%.

3 Measurement of a $t \to W$ Branching Fraction

If all top decays proceed through W emission, the ratio of dilepton to single lepton events is $R_l = 1/6$. If t decays include a non-W state with no leptonic decays, the branching fraction to W's is given in terms of the ratio R_l as

$$BF(t \to W(b)) = \frac{9R_l}{1+3R_l} \qquad (3)$$

The expected precision on this ratio can be scaled from the present understanding of cross section measurements in the l+jets and dilepton modes. We find that with 10 fb^{-1} the ratio of dilepton to single lepton rates in top events will allow determination of the top branching fraction to W's in association with b to a precision of 3.5%.

This analysis is obviously model dependent, but consistent with the popular non-standard model that $t \to Wb$ may be augmented with $t \to H^+b$ where in this case $BF(H^+ \to c\bar{s}) = 100\%$. The W branching fraction limit can be turned around to give a limit on $t \to H^+b$; we find that 10 fb^{-1} at the Tevatron will allow observation of a charged Higgs in top decay down to branching fractions of 6%. The limits attainable by 10-20 fb^{-1} of $p\bar{p}$ data, in conjunction with the CLEO $B(b \to s\gamma)$ measurement will be enough to exclude $m(H^+) \lesssim m(t)$ for *any* value of $\tan\beta$.

4 Rare Top Decays to W, Z, and γ

Standard Model predictions for the branching fractions of FCNC decays are around 10^{-10}, [21], so any observation of such decays will signal new physics. As illustration, we consider the signal for a flavor changing neutral current decay $t \to c\gamma$ in a $t\bar{t}$ event. If the other top in the event decays in the leptonic channel, the acceptance is almost the same as the standard model l+jets

TABLE 3. A Top Physics Program: Summary of expected precision vs integrated luminosity at the Tevatron

Measurement	1 fb^{-1}	10 fb^{-1}
Yields		
$N_{3jet+b-tag}$	580	5.8K
$N_{4jet+2b-tag}$	260	2.6K
δm_t	3.5	2.0
Production		
$\delta\sigma_{t\bar{t}}$	11%	6%
$\delta\sigma_{ll}/\sigma_{l+j}$	14%	4.8%
$\delta\sigma_{t\bar{b}}$	26%	10%
$\delta\sigma \cdot B(Z' \to t\bar{t})$	100 fb	25 fb
Decay		
$\delta BF(t \to (W)b)$	3%	1.0%
$\delta BF(t \to W(b))$	10%	3.5%
$\delta BF(W_{V+A})$	2%	0.6%
$\delta BF(W_{long})$	4%	1.3%
$\delta\Gamma(t \to Wb)$	28%	12%
δV_{tb}	14%	6%
Rare Decays		
Br(c+γ)	3.0×10^{-3}	4.0×10^{-4}
Br(c+Z)	1.5×10^{-2}	3.8×10^{-3}
Br(Hb)	15%	6%

mode, and it then becomes a simple matter to scale from present results. The background from $W + \gamma +$ two jets is about 1 fb. Although it is unlikely that this background will be kinematically consistent with $t\bar{t}$ (for example, that $m(\gamma + j) = m(t)$), we take the very conservative assumption that this background is irreducible. We find that 10 fb^{-1} will probe branching fractions for this decay down to 4.0×10^{-4}

Sensitivity to other rare decays can be scaled from this estimate. For the case $t \to Z + c$, where the Z decays to leptons, after adjusting for branching ratios and different backgrounds, we find sensitivity down to 3.8×10^{-3}.

G Conclusions for Top Physics

We have reviewed the prospects for top physics at the Fermilab Tevatron in the Main Injector Era. The conclusions are preliminary, but have the strength of being extrapolations from real measurements in the well understood environments of the present day Collider experiments. We believe that this is only the beginning of the catalog of top physics measurements at the Tevatron, and that this report is best interpreted as a survey of *sensitivities* in each of the categories of mass reconstruction, cross sections, branching ratios, decay dynamics, and rare decays. In the not unlikely event that this 180 GeV/c^2 fermion harbors surprises, this study benchmarks the capability to explore the new physics at the Tevatron facility.

III OTHER HIGH P_T PHYSICS OPPORTUNITIES

In addition to the prospects for top physics, large integrated data sets at the Tevatron will allow incisive measurements in the electroweak sector, discovery of a Light Higgs boson, and a probe of a significant part of the parameter space of the Minimal Supersymmetric Model. These prospects are summarized briefly here, detailed discussion can be found in the Tev 2000 report.

A Light Higgs Physics

A light intermediate-mass scalar in the mass region $80 GeV/c^2 < m_H < 130 GeV/c^2$ is predicted by minimal supersymmetric models, and current precision electroweak data also show a slight preference for a low mass Higgs. This study confirms recent theoretical speculation that there is a **luminosity threshold for the detection of a light Higgs boson at the Tevatron**, and suggests that this threshold varies from 5 to 25 fb^{-1} as m_H varies from 60 to 120 GeV/c^2.

- The process $q\bar{q} \to WH$, with $H \to b\bar{b}$, is the best single mode for the detection of light Higgs boson at the Tevatron, and leads to the luminosity thresholds stated above. The analysis relies heavily on the understanding of b-tagging, the "W + flavor" backgrounds, and mass fitting with jets, and is therefore a natural complement and extension of the top physics program.

- The process $p^+p^- \to (W,Z)H$, with $H \to \tau^+\tau^-$ and $(W,Z) \to jj$, is difficult at the Tevatron due to the large $(Z \to \tau^+\tau^-)jj$ background, but may add to the overall significance of the observation. Other channels, such as ZH with $Z \to \nu\bar{\nu}$ and $H \to b\bar{b}$, have not been investigated, and should be. Because a set of combined channels is likely to have better significance than our single studied channel of WH with $H \to b\bar{b}$, the luminosity thresholds above are probably conservative, and the mass reach may be slightly higher.

- We have studied the potential of the $W + H \to b\bar{b}$ measurement at the LHC, assuming equivalent detection efficiencies, etc and find that it is difficult there because of large top backgrounds. It may be that the intermediate mass region is accessible at the LHC only via the rare decay mode $H \to \gamma\gamma$. Since the branching fraction to $\gamma\gamma$ varies with the choice of SUSY parameters, the LHC cannot prove that the light Higgs boson of SUSY does not exist if it is not found there.

- The process $q\bar{q} \to WH$ is complementary to the LEP II/NLC process $q\bar{q} \to ZH$, since they involve the coupling of the Higgs boson to different weak bosons. The ratio of these couplings can vary in multi-Higgs models with multiplets other than doublets (e. g., Higgs triplets).

Although further study is needed, the opportunity to detect a light Higgs boson at the Fermilab Tevatron appears to be real.

B Supersymmetric Physics

Supersymmetry is an elegant and comprehensive extension to the Standard Model. It solves the gauge hierarchy problem, provides a candidate for cold dark matter, is consistent with gauge unification, and is naturally decoupled from Standard Model particles. It is a feature of most string theories.

The experimental consequence of supersymmetry at the weak scale is the presence of 32 new particles in the mass range of about 100 to 1000 GeV/c². It is not surprising that none of these particle have been discovered yet, since current facilities are only sensitive below this mass range. However, with the increased luminosity available at TeV33, the Tevatron will be able to probe a significant fraction of the expected SUSY mass range for the first time.

Using a constrained minimal supersymmetric model (MSSM) with four free parameters and an unknown sign, a detector similar to the upgraded DØ/CDF detectors, and an integrated luminosity of 25 fb^{-1} at TeV33, our preliminary conclusions are:

- We will be able to search for charginos with masses up to 250 GeV/c². The mass reach depends on the exact value of the unknown SUSY parameters, but a significant fraction of the possible parameter space has light charginos accessible at the Tevatron.

- The Tevatron can find gluinos with masses below about 275 GeV/c² for any choice of the SUSY parameters. For some parameter values the Tevatron will be sensitive to gluinos with masses up to 400 GeV/c².

- The Tevatron can search for light supersymmetric top quarks in various decay modes up to about 180 GeV/c² mass.

The Tevatron enjoys a window of opportunity to discover the first evidence for a highly motivated theory beyond the Standard Model, or to significantly constrain that theory. The increased luminosity available at TeV33 is necessary to exploit this opportunity during the next decade.

C   Intermediate Vector Boson Physics

With very large integrated luminosities at the Tevatron, the electroweak sector sector of the SM can be probed in great detail. Our preliminary studies arrive at the following conclusions:

- With 10 fb^{-1} it should be possible to measure the mass of the W boson with a precision of at least 30 MeV/c^2, and 20 MeV/c^2 may well be within reach. This is about a factor of 2 better than what one expects for LEP II. With a precision of 20 MeV/c^2 (30 MeV/c^2) for the W mass, and 2 GeV/c^2 for the top quark mass, the Higgs boson mass can be predicted with an uncertainty of about 40% (50%) of itself. This prediction may be very useful for direct Higgs searches at the Tevatron, LHC, or NLC, and comparison with the results of a direct search may constitute an ultimate test of the SM.

- The W width can be measured with an uncertainty of about 15 MeV. This is an improvement of almost one order of magnitude of the current uncertainty. At LEP II Γ_W can only be measured with a precision of a few hundred MeV.

- The W charge asymmetry will be a very powerful tool in constraining the parton distribution functions. In many processes the error in the parton distribution functions currently constitutes a major source of uncertainty. The forward backward asymmetry, A_{FB} in Z boson decays provides a useful cross check on the Higgs boson mass extracted from the W mass measurement.

- With 10 fb^{-1}, the WWV and $Z\gamma V$, $V = \gamma$, Z, vertices can be determined with a precision of $\mathcal{O}(10\%)$ and $\mathcal{O}(10^{-2} - 10^{-3})$, respectively, at the Tevatron. The expected accuracy for the WWV couplings is comparable or better than that of LEP II. However, since the methods used to extract limits on anomalous couplings at the two colliders are different, data from the Tevatron and LEP II yield complementary information. Tevatron experiments will be able to place limits on the $Z\gamma V$ couplings which are up to a factor 100 better than those which can be achieved at LEP II. At the LHC, with 100 fb^{-1}, it will be possible to place limits on anomalous WWV and $Z\gamma V$ couplings which are a factor 3 to 100 better than those one can expect for the Tevatron with 10 fb^{-1}.

- The Tevatron offers a unique chance to search for the SM "radiation zero" in $W\gamma$ production, which provides an additional test of the gauge theory nature of the SM. At the LHC, due to the large qg luminosity, QCD corrections obscure the dip in the photon lepton rapidity difference distribution which is caused by the radiation zero. This is not the case at Tevatron energies. Currently, the experimental results are statistically limited. With integrated luminosities of 2 fb^{-1} or more, it should be possible to conclusively establish the existence of the radiation zero.

- With an integrated luminosity of 10 fb^{-1}, limits on the branching ratios of rare W decays of $\mathcal{O}(10^{-5})$ to $\mathcal{O}(10^{-7})$ can be obtained. W decays into two pseudoscalar mesons offer an opportunity to probe meson decay form factors at a very high momentum transfer where these form factors have not been tested so far.

- The Tevatron offers a unique opportunity to search for CP violation in W boson production and decay since it collides protons and antiprotons, ie. the initial state is a CP eigenstate. The extremely large number of W boson events expected at a superluminous Tevatron will make it possible to search for small CP-violating contributions to W boson production, at the level of $\mathcal{O}(10^{-3} - 10^{-4})$.

- An integrated luminosity of 10 fb^{-1}, will produce a sufficient number of $W\gamma\gamma$, $Z\gamma\gamma$ and $WW\gamma$ events to extract direct information on the quartic gauge boson couplings.

IV CONCLUSIONS

As previously noted, we believe that this work is not complete. We have shown that a future path of increasing luminosity at the Tevatron will lead to a full program of measurements in Top, IVB, Higgs, SUSY, and Exotic physics. Our general conclusions are as follows:

1. Fermilab will be the top factory for many years. As with other heavy quarks, the top may be entering the first of many decades of serious scrutiny. U.S. HEP program planning should recognise this major scientific opportunity.

2. Our study confirms recent theoretical speculation that there is a luminosity threshold for the detection of a light Higgs boson at the Tevatron. The most promising single detection technique relies on detailed understanding of b-tagging and the "W + flavor" backgrounds, and is therefore a natural complement and extension of the top physics program.

   (a) the presence of the detection threshold and its value be confirmed in more detailed simulation, including b-tagging in the presence of multiple interactions.

   (b) a Tevatron strategy for crossing the luminosity threshold be developed and implemented.

3. The Tevatron program can either discover SUSY or significantly constrain a large fraction of current theoretical prejudice. The actual sensitivity

and discovery potential for supersymmetric states at the Tevatron deserves significantly more study.

4. High instantaneous luminosity conditions need to be understood better with perhaps both simulation and actual detector research and development. This is especially true for the top and Higgs studies. Will it blind the current detectors? If so, how? We urge the Laboratory to initiate an active program to investigate these questions and to engage the high energy physics community in the effort. We believe the effort will benefit from computing, R&D, and possibly even test beam resources.

5. There may be significant luminosity capability beyond the "classic" Main Injector scenario during Run II. If such incremental increases in peak luminosity cannot be handled by the detectors, could this capability be channeled into a significant increase in the useful longevity of $> 10^{32}$ stores? After all, integrated luminosity is the key.

6. The physics overall is tantalizing, and we believe that waiting for the LHC is neither cost effective nor prudent. The Laboratory and the experimental collaborations should make every effort to maximize the physics return of the Tevatron. This implies the need for **an overall plan for the long term Tevatron Program** including the accelerator, the detectors, and physics simulation.

V ACKNOWLEDGEMENTS

I acknowledge the dedication and work of the Tev2000 authors, and I thank the organizers of this conference for a stimulating survey of our future.

This work is supported in part by the U.S. Department of Energy under Grant DE-FG02-95ER40899.

REFERENCES

1. "Future Electroweak Physics at the Tevatron", Report of the Tev2000 Working Group, Fermilab-Pub-96/046, 1996.
2. F. Abe et al., Phys. Rev. Lett. **74**, 2626 (1995).
3. E. Laenen, private communication.
4. F. Abe et al., Phys. Rev. D **50**, 2966, 1994.
5. U. Heintz, "Top Mass Analysis of Dilepton events," DØ Note 2658.
6. P. Azzi et al., "Evidence for $t\bar{t}$ production in the All-hadronic Channel", CDF Internal Note 3187, June, 1995.
7. J. Incandela, "Using the B Decay Length Distribution in Top Events to Measure m_t, CDF Internal Note 1921, December 1992.

8. L. Orr, T. Stelzer, and W.J. Stirling, University of Durham Preprint DTP/94/112, University of Rochester Preprint UR-1397; L. Orr and W.J. Stirling, University of Durham Preprint DTP/94/60, University of Rochester Preprint UR-1365; S. Mrenna and C.-P. Yuan, Phys. Rev. D. **46**, 1007-1021, 1992; Yu. L. Dokshitzer, V.A. Khoze and S.I. Troyan, Proceedings of the 6th International Conference on Physics in Collision, Singapore 1987; G. Marchesini and B.R. Webber Nucl.Phys. **B330**, 261-283.
9. C.T. Hill, "Topcolor Assisted Technicolor", Fermilab-Pub-94/395-T.
10. P. Cho and E.H. Simmons, Phys. Rev.**D51**, 2360, (1995). D. Atwood, A. Kagan and T. Rizzo, "Constraining Anomalous Top Quark Couplings at the Tevatron", SLAC-PUB-6580, July, 1994.
11. S. Dawson, Nucl. Phys. **B249**, 42 (1985); S.S.D. Willenbrock and D.A. Dicus, Phys. Rev. D **34**, 155 (1986); S. Dawson and S.S.D. Willenbrock, Nucl. Phys. **B284**, 449 (1987); C.-P. Yuan, Phys. Rev. D **41**, 42 (1990); S. Cortese and R. Petronzio, Phys. Lett. **B253**, 494 (1991); G.V. Jikia and S.R. Slabospitsky, Phys. Lett **B295**, 136 (1992), R.K. Ellis and S. Parke, Phys. Rev. D **46**, 3785 (1992); G. Bordes and B. van Eijk, Z. Phys. **C57**, 81 (1993); G. Bordes and B. van Eijk, Nucl. Phys. **B435**, 23 (1995): T. Stelzer and S. Willenbrock, Phys. Lett. **B357**, 125-130 (1995).
12. A.P. Heinson, A.S. Belyaev and E.E. Boos, "Electroweak top quark production at the Fermilab Tevatron", in the proceedings of the Workshop on Physics of the Top Quark, Iowa State University, Ames, Iowa, May 1995, hep-ph/9509274.
13. E. Malkawi, C.-P. Yuan, "A Global Analysis of the Top Quark Couplings to Gauge Bosons," Phys. Rev. D. **50**, 4462, 1994.
14. A.P. Heinson and P. Baringer, DØ Note 2777 (1995).
15. H. Weerts, private communication.
16. C. Hill, *Topcolor Assisted Technicolor*, Fermilab-Pub-94/395-T
17. K. Lane, *Top Quarks and Flavor Physics*, BUHEP–95–2.
18. I. Bigi *et al.*, Phys Lett. 181B, 157, (1986).
19. For example, see G. Kane, C.-P. Yuan, and D. Ladinsky, Phys. Rev. D. **45**, 124, 1992; D. Atwood, A. Aeppli, and A. Soni, Phys. Rev. Lett **69**, 2754, 1992; R.S. Chivukula, S.B. Selipsky, E.H. Simmons, Phys. Rev. Lett **69**, 575, 1992; M.Peskin and P. Zerwas, talks presented at the *First International Workshop on Physics and Experiments at a Linear e+e- Collider*, Saariselka, Finland, September 1991; D. Carlson, E. Malkawi, C.-P. Yuan, Phys. Lett. **B337**, 145, 1994; K. Fujikawa and A. Yamada, Phys. Rev. Lett **49**, 5890, 1994.
20. G. Kane, C.-P. Yuan, and D. Ladinsky, Phys. Rev. D. **45**, 124, 1992.
21. S. Parke, *Summary of Top Quark Physics*, FERMILAB -Conf-94/322-T. Presented at DPF'94, University of New Mexico, Albuquerque, NM, August 2-6, 1994.
22. K. Fuji, "Top Physics at Future e^+e^- Colliders, Experimental Aspects", *First International Workshop on Physics and Experiments at a Linear e+e- Collider*, Saariselka, Finland, September 1991;
23. P. Igo-Kemenes, M. Martinez, R. Miquel, and S. Orteau in *Second International Workshop on Physics and Experiments at a Linear e+e- Collider*, Waikoloa,

HI, April 1993.
24. P. Igo-Kemenes in *Second International Workshop on Physics and Experiments at a Linear e+e- Collider*, Waikoloa, HI, April 1993.
25. G. Bagliesi *at.*, Top Quark Physics, Experimental Aspects, in *First International Workshop on Physics and Experiments at a Linear e+e- Collider*, Saariselka, Finland, September 1991.
26. M. Peskin *at.*, Physics Issues for the Next Linear Collider in *First International Workshop on Physics and Experiments at a Linear e+e- Collider*, Saariselka, Finland, September 1991.
27. G. A. Ladinsky and C.-P. Yuan, in *Second International Workshop on Physics and Experiments at a Linear e+e- Collider*, Waikoloa, HI, April 1993.
28. See for example D. Froidevaux, Top Quark Physics at LHC/SSC CERN/PPE/93-148

Status of LEP2 and LHC

Eberhard Keil

CERN, Geneva, Switzerland

Abstract. This paper discusses the status of the two colliders at CERN near Geneva, Switzerland. The electron-positron collider LEP2 now operates above the threshold for $W^\pm$ pair production. The Large Hadron Collider Project LHC, a proton-proton collider with a planned luminosity of 10 $\text{nb}^{-1}\text{s}^{-1}$ at a centre-of-mass energy of 14 TeV, is under construction and will be installed in the LEP tunnel.

INTRODUCTION

This paper discusses the status of the two colliders at CERN near Geneva, Switzerland. The status of the electron-positron collider LEP2, operating above the threshold for $W^\pm$ pair production, is presented in the next Chapter. The status of the Large Hadron Collider Project LHC which is under construction and will be installed in the LEP tunnel, is presented in the Chapter thereafter. The conclusions are in the last Chapter.

STATUS OF LEP2

The electron-positron collider LEP has been operated at a beam energy of about 45.6 GeV, half the Z^0 mass, from 1989 to 1995. In June 1996, enough super-conducting RF cavities had been installed to reach a beam energy of 80.5 GeV, exceeding the threshold of $W^\pm$ pair production. This phase of operation is called LEP2 [1].

Achieved 1996 LEP2 Parameters

Tab. 1 shows the list of achieved LEP2 parameters up to the end of the run in August 1996. The LEP2 circumference C is largest circumference of any storage ring in operation. The maximum beam energy is above the $W^\pm$ threshold, and determined by the amount of super-conducting RF system which was installed during the run from June to August. We operate LEP2 with four trains of two bunches each. The spacing between the bunches in a train is $118\lambda_{RF}$. We have limits to the bunch and beam currents. The bunch current is limited at injection by the transverse mode coupling instability [2]. A second limit on the bunch current may be multi-bunch beam breakup in bunch trains [3], caused by wakefields, excited by earlier bunches in a train acting on later ones. Depending on the relative magnitude of these limits, we have different scenarios, including operating with individual bunches instead of bunch trains. The luminosity L follows from the currents and beam radii.

TABLE 1. Achieved 1996 LEP2 Parameters

Circumference C	26659	m
Maximum operating energy E	80.5	GeV
Number of bunch trains	4	
Number of bunches in a train	2	
Bunch spacing s	$118\lambda_{RF}$	
Bunch radii $\sigma_x : \sigma_y$	300 : 3	μm
Maximum bunch current I	0.35	mA
Beam-beam tune shifts $\xi_x = \xi_y$	0.03	
Maximum luminosity L	30	$\mu b^{-1} s^{-1}$

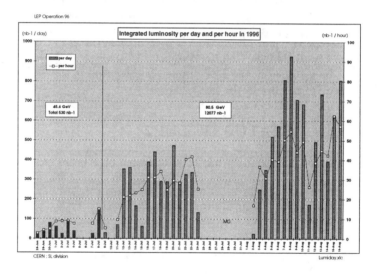

FIGURE 1. Daily LEP2 Luminosity 1996

Fig. 1 shows the daily integrated luminosity with the left scale in nb^{-1}/d and the right scale in nb^{-1}/h. In the period from 28 June to 9 July we were running at the Z^0 for detector calibration. On 5 July we changed the lattice back to the 1995 lattice. On 9 July we made $W^\pm$ pairs. Up to 24 July we ran with four equidistant bunches, and till 16 August with four trains of two bunches each. The gap in luminosity is caused by a faulty septum magnet in the PS and by a machine development period of about a week.

Bunch Trains

Since LEP came into operation in 1989, the bunch current has been limited by transverse mode coupling instability. This predicted limit made us look for ways of increasing the number of bunches beyond four in each beam. In 1988, Rubbia [4] proposed to install a pretzel scheme in LEP, similar to that in CESR at Cornell University [5]. The number of bunches was severely limited by the LEP experiments, unless their electronics was rebuilt to take much shorter bunch spacings. LEP was operated with pretzels and eight bunches from 1992 to 1994. In 1990, Meller [6] proposed to replace the equidistant individual bunches in CESR by trains of bunches. In 1992, I made a proposal for LEP [7]. The bunches were to collide with a horizontal crossing angle and a concomitant horizontal offset in the super-conducting quadrupoles next to the interaction points. This scheme was dead by the end of 1993 because of the synchrotron radiation background from these quadrupoles [8]. Fortunately enough, Herr [9] proposed another scheme without a crossing angle and without an offset in the super-conducting quadrupoles in early 1994. LEP was running with one train in each beam, colliding in ALEPH and DELPHI by the end of 1994, and with four trains in each beam since 1995 [10].

Fig. 2 shows a typical bump in an even-numbered pit, where the two beams should collide head-on, and there should be no offset between them in the first quadrupole doublet from the interaction points IP. The logical place for the first electrostatic separator which launches the vertical separation bump between the two beams is behind the first quadrupole doublet where such separators already exist. The vertical orbit will then oscillate freely, crossing the vertical axis every half vertical betatron wavelength. A logical place to close the bump is in the neighbourhood of one of these zero crossings. We decided to install new separators near the seventh quadrupole QS7 from the IP, in order to avoid vertical orbit offsets in the Cu cavities of the RF system. These offsets are one driving mechanism for synchro-betatron resonances. We use the existing separators near the fourth quadrupoles from the interaction points for closing the bumps. During injection and energy ramping, the beams are separated at the interaction points by not exciting the the separators behind the first quadrupole doublet.

LEP2 Lattice

The limit on the bunch current implies that the beam radii should become smaller as the beam energy increases, contrary to the natural behaviour, beam

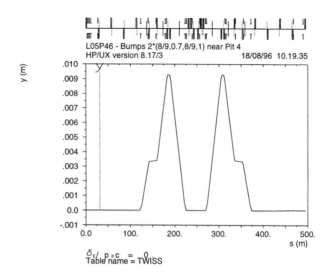

FIGURE 2. Bunch Train Bump near IP4. The interaction point is close to the centre of the abscissa. The diagram above the graph schematically shows the horizontally focusing (defocusing) quadrupoles as rectangles above (below) the axis, and the separating dipoles as centred rectangles. The curve is the vertical orbit y of the positron bunches at 45.6 GeV in collision.

radii increasing like the energy, in a machine with a given lattice. In order to achieve the desired variation, we need lattices with smaller natural emittances than we have now, i.e. lattice with a higher horizontal phase advance and smaller dispersion than the lattices used in the past. Since 1992, we have been using a lattice with phase advances $\mu_x = 90°$ and $\mu_y = 60°$ in the arc cells. This year, we started with a lattice having $\mu_x = 108°$ and $\mu_y = 60°$. This lattice did not have enough dynamic aperture and we had to abandon it, returning to the lattice with $\mu_x = 90°$ and $\mu_y = 60°$. In parallel, another lattice with $\mu_x = 108°$ and $\mu_y = 90°$ has been developed [11] which in simulations has a better dynamic aperture than the abandoned lattice. Putting it into LEP is more difficult than it sounds at first sight, because it involves three day shifts of re-cabling the vertically focusing sextupoles. For the remainder of 1996, we foresee three weeks of running at $\mu_x = 90°$ and $\mu_y = 60°$, followed by the re-cabling, and another three weeks of running at $\mu_x = 108°$ and $\mu_y = 90°$.

Super-Conducting RF System in LEP2

By June 1996, enough super-conducting RF cavities had been installed to exceed the threshold of $W^{\pm}$ pair production. Tab. 2 shows the numbers of RF cavities, installed now and in the future. A total of 128 Cu cavities has been in LEP since the beginning. Eight of them were removed to make space for the pretzel separators. The first batch of super-conducting RF cavities was made of Nb sheet, the later batches of Cu sheet coated with a thin layer of Nb. The super-conducting RF cavities come in modules which are about 10 m long, contain 4 cavities, and deliver about 40 MV and 0.5 MW. The first line describes the situation between June 1996 and the beginning of the shutdown in August 1996. The second line shows the RF system that was installed in August and September 1996. In the shutdown from November 1996 to May 1997, we will start removing Cu cavities, and install more super-conducting ones, raising the energy to 94 GeV. In the shutdown from November 1997 to May 1998, we shall continue removing Cu cavities and complete the installation of super-conducting ones, raising the energy to 96 GeV. The cavity producting will be over in Spring 1997, and the module assembly in Summer 1997. Removing Cu cavities reduces the impedance of LEP, and should allow increasing the beam current.

TABLE 2. LEP2 Cavity Installation Schedule

Date	Cu	Nb	NbCu	GeV	Status
Jun 96	120	4	140	84	Installed
Oct 96	120	16	160	86	Installed
May 97	86	16	224	94	Approved
May 98	52	16	256	96	Approved

The super-conducting RF cavities reached their design field of 6 MV/m for the NbCu cavities and 5 MV/m for the Nb cavities [12]. So far, no deterioration of the cavities with time has been observed, which might have been caused by cryo-pumping on the cavity surfaces, by dust from opening and closing valves, and by the Nb layer peeling off. However, there were many trips of modules and of the RF power plant. Avoiding them requires improved diagnostics and controls which will be installed. We observe mechanical oscillations of the RF cavities at frequencies of about 100 Hz. They change the resonant frequency of the accelerating RF mode. If the cavity is tuned to the flank of the resonance curve, wild voltage oscillations occur. There are two causes of these mechanical vibrations: the cryogenic systems and the electromagnetic forces, i.e. the vector product of surface current and magnetic field in the accelerating RF mode. The wild voltage oscillations are avoided by tuning the cavities on resonance, and accepting the increase in reflected RF power and in the RF power flowing through the couplers, and the reduction in efficiency.

TABLE 3. LEP2 Performance Goals

Year	1996	1997	1998
Maximum energy E/GeV	86	94	96
Bunch trains	4	4	4
Bunches in a train	2	2	2
Maximum bunch current I/mA	0.35	0.5	0.65
Peak luminosity $L/\mu\text{b}^{-1}\text{s}^{-1}$	50	70	110

LEP2 Performance Goals

The goal of LEP2 is accumulating an integrated luminosity of 500 pb^{-1} in three years. Tab. 3 shows the performance predictions for the last part of 1996, and for the years 1997 and 1998. The maximum beam energy increases as discussed before. We expect to run with four trains of two bunches each. More bunches are not useful because we reach the RF power limit with the eight bunches, if we can achieve the bunch currents listed. We expect higher bunch currents in later years, because of the removal of Cu cavities. The last line shows the estimated peak luminosities.

STATUS OF LHC

The LHC project was approved by Council in December 1994. Big contracts start being placed. In September 1996, the Finance Committee approved contracts for the whole 50000 t of steel supply, the supervision of the civil engineering work, and eight magnet measuring benches. I assume for the remainder of the milestones that enough funds are available to construct the LHC in a single stage. By the end of 1999 most of the big contracts will have been placed, the prices will be known, and a final decision on the configuration can be taken. We assume that LEP will stop operation at the end of 1999. Continuing LEP operation into 2000 must be justified on scientific grounds and money must be found. Dismantling LEP will start in October 2000 when the civil engineering work for LHC is advanced such that it becomes necessary to break into the tunnel, in particular for the ATLAS and CMS caverns. Injection tests are foreseen from October 2003. Commissioning with beam will start in the second half of 2005. The latest conceptual design report [13] for the LHC was issued in October 1995.

LHC Parameters

Tab. 4 shows the LHC parameters. The circumference C is that of LEP, and known to even more digits than shown. The maximum energy E is a round figure, achieved at the dipole field B listed. The bunch spacing s corresponds to 10 RF wavelengths. Together with the distance from the interaction point IP to the separating dipoles it determines the number of parasitic collisions, about 15 on either side of the IP. The bunch population N and the bunch radii σ_x and σ_y are shown at the interaction point IP. They are adjusted such that the beam-beam tune shift parameter ξ falls into range believed to be achievable from experience with other hadron colliders which were or are in operation [14], and that the luminosity L is in a range which the experiments believe they can handle. The total beam-beam tune spread from the nearly head-on collisions and from all parasitic collision should be small enough to fit between nonlinear resonances of order up to twelve. Not all the space ℓ_Q between the IP and the front face of the nearest quadrupole is available to the experiments. At the assumed inelastic non-diffractive cross section $\sigma_{pp} = 60$ mb, the number of events in a single collision is $n_c = 19$.

TABLE 4. LHC Parameters

Circumference C	26659	m
Energy E	7	TeV
Dipole field B	8.4	T
Bunch spacing s	25	ns
Bunch population N	10^{11}	
Bunch radius $\sigma_x = \sigma_y$	16	μm
Bunch length σ_s	75	mm
Beam-beam parameter ξ	0.0034	
Luminosity L	10	nb^{-1}s^{-1}
Distance to nearest quadrupole ℓ_Q	± 23	m
Events/collision n_c	19	

Layout and Experiments

Fig. 3 shows the layout of the LHC. The pits are at the centres of the octants. The two LHC rings cross in Pits 1, 2, 5, and 8. The circumferences of the two rings are the same, since both rings have four inner and four outer arcs. The two large experiments, ATLAS and CMS, are diametrically opposite in Pits 1 and 5, respectively. Both are approved experiments with a cost ceiling of 475 MCHF each. Technical proposals were published in 1994 [15,16]. The LHCC expects technical proposals for the subsystems. The heavy-ion experiment ALICE will be in Pit 2. The technical proposal for the core experiment [17]

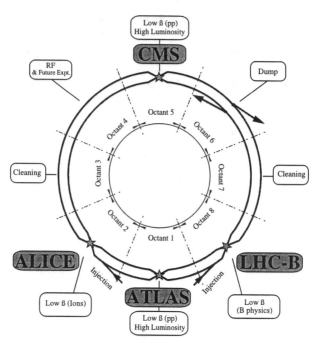

FIGURE 3. LHC Layout and Experiments

was published in 1995. The LHCC is waiting for the technical proposal for the muon arm. The LHC-B experiment is dedicated to the study of CP violation and other rare phenomena in the decay of Beauty particles. It uses colliding beams and a forward detector, contrary to the HERA-B experiment which uses a single beam and a gas jet target. A letter of intent [18] has been submitted to the LHCC. The LHCC wants an R&D programme for the detector. The two beams are injected into LHC into outer arcs upstream of Pits 2 and 8. The beam cleaning insertions to steer the beam halo into staggered sets of collimators rather than the super-conducting magnets are in Pits 3 and 7. Pit 4 houses the RF system. Pit 6 is reserved for the beam dumping system.

Layout and Optical Functions near Pit 5

The mimic diagram in Fig. 4 shows the LHC layout schematically over about 500 m in the neighbourhood of Pit 5. The low-β interaction point IP5 and CMS are close to the centre. On either side of IP5 is a quadrupole triplet, actually consisting of four quadrupoles. Because of the antisymmetry designed into LHC, the first quadrupole of the triplet focuses horizontally on the left, and defocuses on the right, and similarly for all other quadrupoles. The boxes

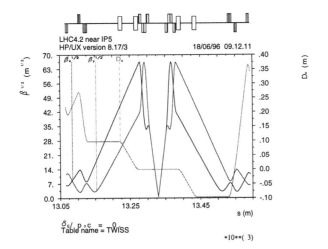

FIGURE 4. Layout and Optical Functions near Pit 5. The abscissa is distance along the beam in km. The interaction point is close to the centre of the abscissa. The diagram above the graph schematically shows the horizontally focusing (defocusing) quadrupoles as rectangles above (below) the axis, and the separating dipoles as centred rectangles. The curves are the square roots of the horizontal and vertical amplitude functions in black and red and the horizontal dispersion in green.

behind the first triplet are the dipole magnets which first separate the two beams, and then make them parallel again at the correct distance of 194 mm at 1.9 K. Just before the dispersion suppressors there is a second triplet. The graph shows the optical functions $\sqrt{\beta_x}$ in black, and $\sqrt{\beta_y}$ in red. They are proportional to the horizontal and vertical beam radii. The antisymmetry of LHC makes them swap values when passing through IP5. Their values at IP5, $\beta_x = \beta_x = 0.5$ m, are too small to show clearly. Their maximum values in the first triplet are quite large. The green curve shows the horizontal dispersion D_x. It is matched to $D_x = D'_x = 0$ at IP5, and has asymmetric nonzero values behind the first separating dipoles.

Injection

The proton bunches for LHC are injected first into the PS booster which operates with just one bunch in each of the four rings. For this, the PS booster will be equipped with two new RF systems, one operating at $h = 1$ and the other at $h = 2$, superimposed such that the bunches are made longer and the direct Laslett space charge detuning is reduced. The four bunches from the four booster rings are injected into the PS. Then the PS booster cycles again, and another batch of four bunches is injected into the PS, and accelerated to 26 GeV. At that energy, the PS beam is adiabatically debunched, and captured by a new RF system, operating at $h = 84$ and 40 MHz. From this moment onwards, the bunches have the LHC bunch spacing of 25 ns. Three of these PS beams are successively injected into the SPS, filling only 3/11 of the SPS circumference, and accelerated by a new RF system at 80 MHz with a circumferential voltage of 0.7 MV to 450 GeV. Just before ejection towards the LHC, another new RF system at 400 MHz with a circumferential voltage of 6 MV is adiabatically turned on to match the SPS bunches to the LHC buckets. One such cycle takes about 16.8 s. Repeating this sequence twelve times fills one of the LHC rings, and takes about three minutes. The acceleration in the LHC from 450 GeV to 7 TeV takes about 20 minutes.

LHC Dipoles

The LHC dipoles [19] occupy about 2/3 of the circumference. The two apertures are in the same iron yoke and cryostat because the space in the tunnel does not allow two independent magnets and because a 2:1 design is cheaper. The space between the two apertures was increased from 180 to 194 mm in order to make their fields more independent and to reduce the collaring forces during their manufacture. Contrary to earlier designs, the cryogenic distribution line is no longer in the magnet cryostat. Only the cooling pipes with gaseous He at 50 K and 4.2 K for the intermediate heat shields, and the heat exchanger pipe at 1.9 K remain in the magnet cryostat. Its outer diameter was reduced to 914 mm. Cooling to this low temperature of 1.9 K is necessary to achieve the dipole field $B = 8.4$ T with super-conducting NbTi cable. The non-magnetic collars are made of aluminium. The field quality and its consequences for the LHC performance, in particular the dynamic aperture, are hotly debated between the magnet designers and my colleagues in accelerator physics [20].

So far, industry built seven and we tested six long dipole prototypes with 50 mm coil aperture and a length of 10 m, the nominal length when the orders were placed. The three best magnets, which had their first quench above 8.4 T and trained rapidly up to 9.6 T, are in the test string. The last one is being tested. Two prototypes with 56 mm coil aperture and the nominal magnetic length of 14.2 m are being constructed. The first magnet will be assembled in industry, the second at CERN, using collared coils produced in industry. The cold mass of the first of these magnets is expected in March 1997. Tests are foreseen in June 1997. A further four 10 m dipoles with 56 mm coil aperture are also in the pipeline, because the tooling for 10 m dipoles exists at several companies. In parallel to the long magnets, a programme of 1 m models is under way at CERN [21]. Its aim is studying the influence of individual coil parameters on the magnet behaviour with a fast turn around rate and qualifying possible design solutions. So far, eight single-aperture models were produced, at a rate of about one per month. The models tested so far show that at 2 K a field of 8.9 T can be reached for the first natural quench. Common to all models is a gradual training above this field level. A twin-aperture model with the same cross section as the long dipoles, now being fabricated in industry, will be completed and tested by the end of 1996. The cable insulation with polyimide tape was decided. Research and development programmes are under way on the super-conducting cable to increase the uniformity of the inter-strand contact resistance [22] and on current leads made of high T_c super-conductors [23,24].

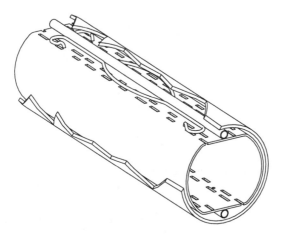

FIGURE 5. Cut-away drawing of the LHC vacuum chamber, perforated beam screen, cooling pipes and support springs.

Vacuum Chamber

Fig. 5 shows one of the beam apertures. The coil aperture is now 56 mm. Just inside, still at 1.9 K, is the vacuum chamber with outer and inner diameters of 52 and 49 mm, respectively, made of stainless steel. Just inside the vacuum chamber is the beam screen which now has a racetrack shape with an inner horizontal diameter of 44 mm and a height of 36 mm. The purpose of the beam screen is absorbing the synchrotron radiation power, about 0.2 W/m from one nominal beam, at 5 to 20 K, using gaseous helium flow through the cooling pipes shown in Fig. 5, since the heat load at 1.9 K would be excessive. The beam screen is supported every 1.7 m. The synchrotron radiation photons will desorb gas molecules from the beam screen which will then be deposited on the screen by cryopumping. To avoid building up a gas layer on the beam screen, and deteriorating the vacuum, the beam screen has pumping slots in the straight top and bottom parts through which the desorbed gases can diffuse to the vacuum chamber walls where they are cryopumped again, but this time at 1.9 K where vapour pressure effects are negligible and the risk of being desorbed by the synchrotron radiation is absent. The electro-magnetic impedance of these slots has been the subject of intense studies [20,25]. The beam screen is made of a high Mn content stainless steel to give a low magnetic permeability, and coated on the inside with 50 μm of copper in order to reduce its resistive impedance and associated heat load.

Cryogenics

The super-conducting magnets are immersed in a bath of super-fluid He at 1.9 K, pressurised at about 1 bar. The heat is transferred by heat exchangers, consisting of a tube containing flowing saturated He II, and running through a half cell of the LHC arcs. It was checked in the string tests, that the liquid and gaseous He may flow in the *same* or in the *opposite* direction in this tube. The LHC tunnel is inclined with respect to the average vertical by at most 14.2 mr. Allowing He flow in either direction avoids the complication of changing the orientation of the cooling loop whenever the tunnel slope changes sign. Contrary to earlier designs, the cryogenic fluids are supplied by a separate cryogenics line which runs along the tunnel walls and is connected to the cryostats every half cell. The cryogenics supply line is fed from the four even pits where much equipment is available from LEP2 [26].

In collaboration with CEA in France, key technologies for high-capacity refrigeration at 1.8 K are being developed [27]. This includes very low pressure heat exchangers, cold volumetric and hydrodynamic compressors to be used as components of practical and efficient thermodynamic cycles. Prototypes of such machines from European industry were tested in the laboratory.

Test String

The LHC test string [28] consists of a quadrupole of 3 m length and three dipoles of 10 m length each, the cryogenic equipment needed to supply the magnets with cryogenic fluids and gases, the cryogenic valves and short circuits for the electrical bus-bars, the power converters for the magnets, and control and diagnostic equipment. The test string has been in operation for two years, with more than 7500 hours at 1.8 K. It was cycled 2150 times, simulating several years of routine LHC operation. Its purpose is the experimental validation of the cryogenic cooling scheme and the development of the quench detection and magnet protection systems. The 1.8 K cooling with super-fluid helium was tested in steady state conditions and during transients. Much was learned on quench detection and magnet protection from the 20 natural and 64 provoked quenches so far; 35 of them occurred at or above the nominal field. In addition, there were about 15 quenches in the magnets *before* they were installed in the string. The temperature increases during ramping upwards at 10 A/s and downwards at 130 A/s were 6 mK and 50 mK, respectively, small enough not to quench the magnets. Simulating a heat load due to particle losses at 1 W/m caused temperature increases less than 30 mK.

The pressure rise during quenches was measured for various configurations of pressure relief valves, and for various delays in opening them. The observed pressure remains less than 14 bar. Therefore the number of pressure relief valves was reduced from four to two in a half cell. The delays in opening the pressure relief valves are long enough that commercially available valves can be used. The string is installed on a slope simulating the slope of the tunnel. By swapping the string feed and string return boxes, flow of liquid and gaseous He in the same and opposite direction was tested, and both were found to be possible. This makes the layout of the cooling loops independent of the local slope of the tunnel.

New Civil Engineering

The new civil engineering, shown in Fig. 6, is concentrated around the interaction points. New caverns and access shafts are needed for the ATLAS and CMS experiments in Pits 1 and 5. The ATLAS cavern is so large that the CERN Main Building would fit into it. Less work is needed in Pits 2 and 8 where caverns and shafts already exist. New transfer tunnels are needed from the SPS to the LHC; TI2 for the clockwise and TI8 for the anti-clockwise beam, respectively. The transfer lines will be equipped with room temperature magnets. The tunnels for the beam dumps near Pit 6 are also new.

CONCLUSIONS

In LEP2, the installation of super-conducting RF cavities has continued as planned (cf. Tab. 2 since the Symposium, and a beam energy of 86 GeV has been reached. With the 90°/60° lattice, an integrated luminosity of 8.3 pb^{-1} was achieved in 17 scheduled days. Then the lattice was changed, to phase advances of 108° and 90°, and an integrated luminosity of 3.0 pb^{-1} was achieved in the remaining 10 scheduled days of the 1996 running period.

Since the publication of the Yellow Book [13] progress was made in many areas of the LHC design. There still are many ongoing studies of which I only mention a few. My colleagues in accelerator physics study the effects of the errors of the magnetic fields in the super-conducting magnets on the dynamic aperture, mostly by computer simulation, i.e. tracking [20]. These errors are caused by the arrangement of the coils, by fabrication tolerances,

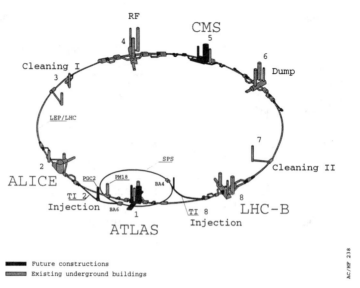

FIGURE 6. New Civil Engineering. The existing LEP tunnel, experimental halls and access shafts are shown in bright grey. The new experimental halls for ATLAS and CMS with their access shafts, the new transfer tunnels and the new beam dump tunnels are shown in red (or dark grey in the printed version).

amplified by the 2:1 design, and by persistent currents at the injection field of only about 0.5 T. etc. We are also concerned about designing an LHC lattice which is robust enough to be operated with ease. Apart from the dipoles, other magnets need to be built. The insertion quadrupoles are particularly challenging, because the large beam size and the high sensitivity of the beam to their errors [29–31]. Studies of the super-conducting cable continue to find a cable which is mechanically stable and has a high and uniform inter-strand contact resistance [22].

ACKNOWLEDGEMENT

This status report is based on the work of the very many people who are now working on the LEP2 and LHC projects.

REFERENCES

1. The LEP2 Team, LEP Design Report – Vol. III LEP2, CERN-AC/96-01 (LEP2) (1996).
2. B. Zotter, Proc. Eur. Part. Accel. Conf. (Berlin 1992) 273.

3. R.L. Gluckstern, H. Okamoto and B. Zotter, Phys. Rev. **E52** (1995) 1026.
4. C. Rubbia, Proc. Eur. Part. Accel. Conf. (Rome 1988) 290.
5. R. Littauer, IEEE Trans. Nucl. Sci. **NS-32** (1985) 1610.
6. R.E. Meller, "Proposal for CESR Mini-B", Cornell University Report CON 90-17 (1990)
7. E. Keil, Lecture Notes in Physics **425** (Springer, 1994) 106.
8. R. Bailey et al., Proc. 4th Eur. Part. Accel. Conf. (London 1994) 445.
9. W. Herr, Bunch Trains without a Crossing Angle, CERN SL/94-06 (DI) (1994) 323.
10. P. Collier, B. Goddard and M. Lamont, Proc. 5th Eur. Part. Accel. Conf. (Sitges 1996) 876.
11. Y. Alexahin et al., Proc. Part. Accel. Conf. (Dallas 1995) 560.
12. D. Boussard, Proc. 5th Eur. Part. Accel. Conf. (Sitges 1996) 187.
13. The LHC Study Group, LHC – The Large Hadron Collider – Conceptual Design, CERN/AC/95-05 (LHC) (1995).
14. E. Keil, Particle Accelerators **27** (1990) 165.
15. ATLAS: Technical Proposal for a General-Purpose pp Experiment at the Large Hadron Collider at CERN, CERN/LHCC/94–43 (1994).
16. CMS: The Compact Muon Solenoid – Technical Proposal, CERN/LHCC/94–38 (1994).
17. ALICE – Technical Proposal for A Large Ion Collider Experiment at the CERN LHC, CERN/LHCC/95-71 (1995).
18. LHC-B: A Dedicated LHC Collider Beauty Experiment for Precision Measurements of CP-Violation, CERN/LHCC/95-5 (1995).
19. R. Perin, 10th General Conference of the EPS (Sevilla 1996).
20. J. Gareyte, Proc. 5th Eur. Part. Accel. Conf. (Sitges 1996) 168.
21. N. Andreev et al., Proc. 5th Eur. Part. Accel. Conf. (Sitges 1996) 2258.
22. D. Richter et al., DC Measurements of Electrical Contacts between Strands in Super-Conducting Cables for the LHC Main Magnets, ASC Pittsburgh '96 (LHC Project Report 67, October 1996).
23. A. Ballarino et al., Design of 12.5 kA Current Leads for the Large Hadron Collider Using High Temperature Superconductor Material, ICEC'96 (Kitakyshu, Japan, 1996) (LHC Project Report 62, 23 October 1996).
24. A. Ballarino et al., Design and Tests on the 30 to 600 A HTS Current Leads for the Large Hadron Collider, ICEC'96 (Kitakyshu, Japan, 1996) (LHC Project Report 78, 5 November 1996).
25. F. Ruggiero, Particle Accelerators **50** (1995) 83.
26. V. Benda et al., Proc. 5th Eur. Part. Accel. Conf. (Sitges 1996) 361.
27. Ph. Lebrun, L. Tavian, G. Claudet, Proc. KRYOGENIKA'96, Praha (1996) 55.
28. A. Bézaguet et al., Proc. 5th Eur. Part. Accel. Conf. (Sitges 1996) 358.
29. G.A. Kirby et al., Progress in the development of the 1 m model of the 70 mm aperture quadrupole for the LHC low-β insertions, ASC Pittsburgh '96 (LHC Project Report 68, October 1996).
30. G.A. Kirby et al., Design study of a super-conducting insertion quadrupole magnet for the large hadron collider, ASC Pittsburgh '96 (LHC Project Report 75, November 1996).
31. A. Zlobin et al., Development of High Gradient Quadrupoles for the LHC Interaction Regions, ASC Pittsburgh '96.

Precision Physics at LHC

Ian Hinchliffe

*Ernest Orlando Lawrence Berkeley National Lab
Berkeley CA 94720*[1]

Abstract. In this talk I give a brief survey of some physics topics that will be addressed by the Large Hadron Collider currently under construction at CERN. Instead of discussing the reach of this machine for new physics, I give examples of the types of precision measurements that might be made if new physics is discovered.

INTRODUCTION

The LHC machine is a proton-proton collider that will be installed in the 26.6 km circumference tunnel currently used by the LEP electron-positron collider at CERN [1]. The 8.4 tesla dipole magnets each 14.2 meters long (magnetic length) are of the "2 in 1" type; the apertures for both beams have common mechanical structure and cryostat. These superconducting magnets operate at 1.9K and have an aperture of 56 mm. They will be placed on the floor in the LEP ring after removal and storage of LEP. The 1104 dipoles and 736 quadruples support beams of 7 TeV energy and a circulating current of 0.54 A.

Bunches of protons separated by 25 ns and with an RMS length of 75 mm intersect at four points where experiments are placed. Two of these are high luminosity regions and house the ATLAS [2] and CMS [3] detectors. Two other regions house the ALICE detector [4], to be used for the study of heavy ion collisions, and LHC-B [7], a detector optimized for the study of B-mesons and B-Baryons. The beams cross at an angle of 200μrad resulting in peak luminosity of 10^{34} cm^{-2} sec^{-1} which has a lifetime of 10 hours. At the peak luminosity there are an average of $\sim 20 pp$ interactions per bunch crossing. Ultimately, the peak luminosity may increase to 2×10^{34} cm^{-2} sec^{-1}. The machine will also be able to accelerate heavy ions resulting in the possibility of Pb-Pb collisions at 1150 TeV in the center of mass and luminosity up to 10^{27} cm^{-2} sec^{-1}.

The rest of this talk will concentrate on the physics of the ATLAS and CMS detectors which are designed to exploit the high luminosity pp mode of the LHC and to perform measurements that will lead to an understanding of the mechanisms behind electroweak symmetry breaking. The great success of recent experiments in hadron and electron colliders has confirmed the validity of the standard model to a very high degree of precision and have brought into sharper focus the need to perform experiments that can provide insight into the sector of the model that is responsible for the generation of the W and Z masses. There are many possible options for this mechanism and while some, such as super-symmetry, might be favored by a majority of theorists, there is little experimental guidance at present.

---

[1] This work was supported in part by the Director, Office of Energy Research, Office of High Energy and Nuclear Physics, Division of High Energy Physics of the U.S. Department of Energy under Contracts DE-AC03-76SF00098

Most discussions of this physics in the context of LHC experiments focus on the huge energy range opened up by this facility and the consequent opportunity to discover particles over a very large mass range. In this talk, I will have a rather different emphasis and will give examples of the types of detailed measurements that can be done once new physics is discovered. I will first discuss the measurements that might be made to determine the properties of Higgs bosons and will then discuss and example of a scenario in which supersymmetric particles are produced. In his talk at this meeting, Frank Paige, will give another, more detailed, example of the latter.

HIGGS PHYSICS

The standard model Higgs boson could be observed in several channels at LHC depending upon its mass. For a very detailed discussion of Higgs physics at LHC see ref [6]. The following decay channels have been discussed

- $pp \to H \to \gamma\gamma$
- $pp \to H \to \ell^+\ell^-\ell^+\ell^-$
- $pp \to H(\to ZZ(\to \ell^+\ell^-\nu\bar{\nu}))X$ at large mass
- $pp \to H(\to W(\to jet+jet)W(\to \ell^+\nu))$ at large mass
- $H \to b\bar{b}$ for a Higgs boson produced in association with a W or $t\bar{t}$ pair.

The first of these is useful for masses immediately above those that can be probed at LEP. The signal to background ratio is poor. Excellent photon energy resolution is required to observe this signal, and this process is one that drives the very high quality electromagnetic calorimetry of both experiments. A simulation from the atlas collaboration is shown in Figure 1 [8] This mode can discover the Higgs if its mass is too high to be detected at LEP and below about 140 GeV. At larger masses the branching ratio becomes too small for a signal to be extracted. If the Higgs is produced in association with a W or $t\bar{t}$, the cross section is substantially reduced, but the presence of additional particles proportionally larger reduction in the background. Observation in this channel will provide important information regarding the Higgs boson couplings.

The search for the Standard Model Higgs relies on the four-lepton channel over a broad mass range from $m_H \sim 130$ GeV to $m_H \sim 800$ GeV. Below $2m_Z$, the event rate is small and the background reduction more difficult, as one or both of the Z-bosons are off-shell. In this mass region the Higgs width is small ($\lesssim 1$ GeV) and so lepton energy or momentum resolution is of great importance in determining the significance of a signal [9].

For $m_H < 2m_Z$, the main backgrounds arise from $t\bar{t}$, $Zb\bar{b}$ and continuum $Z(Z/\gamma)^*$ production. Of these, the $t\bar{t}$ background can be reduced by lepton isolation and by lepton pair invariant mass cuts. The $Zb\bar{b}$ background cannot be reduced by a lepton pair invariant mass cut but can be suppressed by isolation requirements. The ZZ^* process is an irreducible background. Both CMS and ATLAS studied the process for $m_H = 130$, 150 and 170 GeV.

At larger values of the Higgs boson mass where it decays to two on-shell Z's, the signal to background ratio is excellent and the observability is limited only by the available statistics. One can turn to decay channels that have a larger branching ratio. The first of these is $H \to ZZ \to \ell\ell\nu\bar{\nu}$. Here the signal involves looking for a Z decaying to lepton pairs and a large amount of missing energy. The signal appears as a Jacobian peak in the missing E_T spectrum. There are more potentially important sources of background in this channel than in the 4ℓ final state. In addition to the irreducible background from ZZ final states, one has to worry about $Z + jets$ events where the missing E_T arises from neutrinos in the jets or from cracks and other detector effects that cause jet energies to be mismeasured. At

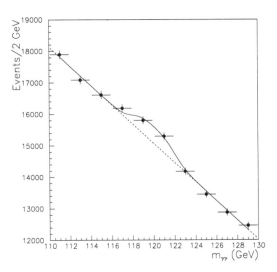

FIGURE 1. A simulation of the possible observation of a Higgs boson via its decay to the $\gamma\gamma$ final state. Shown is the reconstructed $\gamma\gamma$ invariant mass distribution, with a signal showing evidence for a Higgs boson decay. Figure from the ATLAS collaboration.

high luminosity the background from the pile up of minimum bias events produces a E_T^{miss} spectrum that falls very rapidly and is completely negligible for $E_T^{miss} > 100$ GeV, provided the calorimeter extends to $|\eta|< 5$ [10].

The CMS analysis of this process [3,11] uses a central jet veto requiring that there be no jets with $E_T > 150$ GeV within $|\eta|< 2.4$. By requiring a jet in the far forward region (see below), most of the remaining ZZ background can be rejected. A study by CMS requiring a jet with $E > 1 TeV$ and $2.4 <|\eta|< 4.7$, produces an improvement of approximately a factor of three in the signal to background ratio at the cost of some signal.

Substantially larger event samples are available if the decay modes $H \to WW \to \ell\nu + jets$ and $H \to ZZ \to \ell\ell + jets$ can be exploited efficiently. Extraction of a signal is more difficult due to the larger background that arises from $t\bar{t}$, $W + jet$ and $Z + jet$ events. Nevertheless one can expect that these channels could be exploited to confirm a discovery in the purely leptonic final state [3,12,2]

Depending upon its mass, the Higgs boson might be observed in several channels simultaneously. For example, a mass 110 GeV could result in the following measurements, M_h with a precision of order 100 MeV, and the following combinations of cross-sections and branching ratios

- $\sigma(pp \to H + X) BR(H \to \gamma\gamma)$
- $\sigma(pp \to H + W) BR(H \to \gamma\gamma)$
- $\sigma(pp \to H + W) BR(H \to b\bar{b})$

At a mass of order 135 GeV, the following should be measurable

- $\sigma(pp \to H + X) BR(H \to \gamma\gamma)$

- $\sigma(pp \to H + W)BR(H \to ZZ^*)$

If the Discovered particle is a Higgs boson, then the production rates and branching ratios are predicted once the mass is known. These measurements would therefore enable consistency checks to be performed.

There could be Higgs bosons other than the one predicted by the standard model and the LHC will be able to search for these also. Most of the decay modes already discussed can be used in this case, but other modes might become available. The simplest modification to the Higgs sectors is that in the minimal super-symmetric model. Here there are three neutral and one charged Higgs bosons; h, H, A and $H^\pm$. If one assumes that the masses of all the other super-symmetric particles are too heavy to influence the properties of these bosons, the masses and decay properties are given by two independent variables which can be taken to be m_A and $\tan\beta$. Possible new observations of these particles include

- H and $A \to \tau^+\tau^-$
- H and $A \to \mu^+\mu^-$
- $A \to Z(\to \ell^+\ell^-)h(\to b\bar{b})$
- $H \to h(\to b\bar{b})h(\to \gamma\gamma)$
- $t \to H^+ b$

The first of these is particularly important as it enables the mass of the particle to be measured well and is applicable over a large region of parameter space. A simulation is shown in Figure 2 from the CMS collaboration [13]. Events are selected by requiring that an isolated electron or muon be observed. A further selection requiring that the events contain a jet with a single charged track with $25 < p_t < 40$ GeV and no other track with $p_t > 2.5$ GeV. The dominant background is then from the Drell-Yan production of τ pairs. Using the measurement of $ETmiss$, the momenta of the τ candidates can be reconstructed and the invariant mass of the $\tau\tau$ system formed. This is shown in Figure 2. A signal is clearly visible. A similar conclusion is reached by ATLAS [14].

These many channels can be combined to probe the whole of the parameter space in the model. For a very detailed and exhaustive discussion see [6]. The conclusions can be summarized as follows. The modes are sufficient for either experiment to *exclude* the entire $tan\beta - M_A$ plane at 95% confidence with 10^5 pb^{-1}. Over a significant fraction of the parameter space at least two distinct modes will be visible. For example, if h is observed at LEP II and M_A is small the LHC will see the H^+ in top quark decay, $H \to ZZ^*$, and possibly $H/A \to \tau\tau$.

SUPER SYMMETRY

If super symmetry proves to be accessible at LHC, then it is most likely that the first new particles to be exploited will be the squarks and gluions. Since these have strong couplings, their production rates are much larger than those relevant to direct Higgs boson production. Indeed it is possible that Higgs bosons could be produced in the decays of these super-symmetric particles and that this source of Higgs bosons would be the largest one and the one that leads to their discovery.

The mass spectrum and detailed decay properties of the super symmetric particles are very model dependent making a general study rather difficult. The situation is complicated by the real possibility that the LHC may be a factory for super-symmetric particles; many different ones are produced at the same time. Early studies of super-symmetric signals concentrated on a specific particle and a particular decay mode demonstrating that cuts could be made that ensure that the signal from this decay stands out above the standard

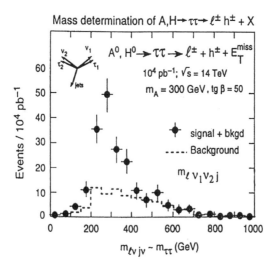

FIGURE 2. A simulation of the possible observation of a Higgs boson via its decay to the $\tau\tau$ final state. Shown is the reconstructed $\tau\tau$ invariant mass distribution, with a signal showing evidence for a Higgs boson decay. Plot from the CMS collaboration

model background. These studies provide a convincing case that super-symmetry could be discovered at the LHC. The next level of work addresses the question of how the masses and couplings of the particles could be determined and the underlying theory constrained. Here one faces the problem that the dominant background for super-symmetry is super-symmetry itself. I will make a few general remarks about super-symmetry phenomenology at the LAC and will then discuss one case study. Frank Paige will discuss others in his talk [15]

The following features are characteristic of most super-symmetric models

- Squarks are heavier than slept ons

- The stop and possibly bottom squarks are the lightest squarks.

- The gluino is heavier than the charged and neutral "ino"'s that are partners of the electroweak gauge bosons.

- The lightest super symmetric particle (LSP) is stable or sufficiently long lived that it leaves the detector. This particle is almost always electrically neutral.

The following generic signals arising from the production and decay of sparticles are

- E_T^{miss} from the loss of the LSP.

- High Multiplicity of large p_t jets from the decay of heavy objects.

- Many leptons from decays of charged and neutral inos.

- Copious b production from the decays of sbottom and stop squark and Higgs bosons.

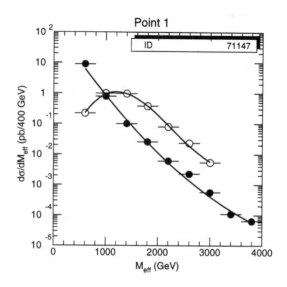

FIGURE 3. Distribution in the variable M_{eff} (see text). The closed circles represent the distribution in the standard model. The open circles show the possible contribution from super-symmetry.

The relative importance of these signals will depend upon the model. The first could be absent in models where the LSP is unstable. Additional signals could also be present. for example in the dynamically broken models [16], the LSP may be unstable and may decay to $\gamma + \tilde{G}$, reducing the missing E_T rate ($\tilde{G}$ exits unobserved) but providing **every** super-symmetry event with an additional pair of isolated photons.

These general features will be used to determine that something new physics is seen at LHC. In order to be more concrete, I will proceed with a specific example in the context of the SUGRA model [17] This model has the advantage that rather few parameters specify it completely. The model is assumed to unify at some high scale where a common gaugino mass $m_{1/2}$ is defined. All scalar particles are assumed to have a common mass m_0 at this scale. Three other parameters then fully specify the model: $\tan\beta$, a variable A with dimension of mass that affects mainly the splitting between the partners of the left and right handed top quark, and the sign of μ. For example, events are selected which have at least 4 jets one of which has $E_t > 100$ GeV and the others have $E_T > 50$. An additional requirement of $E_T^{miss} > 100$ GeV and sphericity $S > 2$ is made, and the event rate is plotted against M_{eff} defined as the scalar sum of the E_T of the four jets and E_T^{miss}. The distribution in this variable is shown in figure 3 From this figure one can see that at low values of M_{eff} the standard model contributions will dominate. At larger values the contribution from super-symmetric particles of large mass will begin to dominate. There is a strong correlation between the mass scale of super-symmetry and the position of the peak in this distribution. This can be used to determine the scale, defined as the lesser of the squark and gluino masses, to about 10/

Once super-symmetry has been discovered and its mass scale established approximately. More detailed studies will be carried out in order to constrain the underlying super-symmetric model. There is a large rate of production for the second lightest neutralino from the decays of heavier super-symmetric particles. This is occurs because this particle

(χ_2) has a component that is the partner of the W gauge boson and is therefore produced with substantial rates in the decays of super-symmetric particles that are too light to decay by strong interactions. This particle will then decay to the LSP. If kinematic-ally allowed, the decay $\chi_2 \to LSP + h$ is dominant. This results in a production of Higgs bosons that is much larger than that expected in the standard model. In this case the h will likely be discovered from its dominant decay to $b\bar{b}$ in super-symmetry events! If this channel is closed then the dominant decay $\chi_2 \to \ell^+\ell^- LSP$ will be substantial. Events will have two isolated leptons of opposite charge. This characteristic signature can be used the measure the mass difference of χ_2 and LSP and can be used as a hook to work up the decay chain to discover other super symmetric particles.

I will illustrate this last comment by a specific example. For this purpose a particular point in the parameter space was selected for simulation [23]. The mass spectrum is as follows: Gluino $m_{\tilde{g}} = 298$ GeV $m_{\tilde{q}_r} = 312$ GeV, $m_{\tilde{q}_l} = 317$ GeV $m_{\tilde{t}_1} = 263$ GeV, $m_{\tilde{t}_2} = 329$ GeV
$m_{\tilde{b}_1} = 278$ GeV, $m_{\tilde{b}_2} = 314$ GeV Sleptons $m_{\tilde{e}_l} = 215$ GeV, $m_{\tilde{e}_r} = 206$ GeV, Neutralinos $m_{\chi_1} = 44$ GeV, $m_{\chi_2} = 98$ GeV, $m_{\chi_3} = 257$ GeV, $m_{\chi_4} = 273$ GeV Charginos $m_{\chi_1^\pm} = 96$ GeV, $m_{\chi_2^\pm} = 272$ GeV Higgs $m_h = 68$ GeV, $m_H = 378$ GeV, $m_A = 371$ GeV, $m_{H^+} = 378$ GeV.

At this point the total production rate for gluino pairs is very large, and many other supersymmetric particles are produced in the decay of gluinos. Of particular significance is χ_2 which decays to $\chi_1 e^+ e^-$ and $\chi_1 \mu^+ \mu^-$ with a combined branching ratio of 32%. The position of the end point of this spectrum determines the mass difference $m_{\chi_2} - m_{\chi_1}$ [18]. Backgrounds are negligible if the events are required to have two such dilepton pairs, which can arise from the pair production of gluinos with each decaying to $b\bar{b}(\to \chi_2(\to \chi_1 \ell^+ \ell^-))$ which has a combined branching ratio of 24%. The event rate is so large that the statistical error in the determination of the mass difference is very small and the total error will be dominated by systematic effects. The enormous number of $Z \to \ell^+ \ell^-$ decays can be used to calibrate, and an error of better than 50 MeV on $m_{\chi_2} - m_{\chi_1}$ is achievable[2]. In the context of the model, this measurement constrains $M_{1/2}$ with an error of order 0.1%.

The small mass difference between the gluino and the sbottom can also be exploited to reconstruct a the masses of these particles [19]. Here a partial reconstruction technique is used. Events are selected where the dilepton invariant mass is close to its maximum value. In the rest frame of χ_2, χ_1 is then forced to be at rest. The momentum of χ_2 in the laboratory frame is then related to the momentum of the $\ell^+ \ell^-$ pair by $p_{\chi_2} = (1 + m_{\chi_1}/m_{\ell^+\ell^-})p_{\ell^+\ell^-}$. χ_2 can then be combined with an additional $b - jet$ to reconstruct the *tildeb* mass. An additional $b_j et$ can then be added to reconstruct the $\tilde{g}$ mass. Figure 4 shows the scatterplot on these two invariant masses together with a projection onto $m_{\tilde{b}}$ and $\delta m = m_{tildeg} - m_{\tilde{b}}$. Peaks can clearly be seen above the combinatoric background. This method can be used to determine $m_{\tilde{g}}$ and $m_{\tilde{b}}$. The values depend on the assumed value of m_{χ_1}: $m_{\tilde{b}} = m_{\tilde{b}}^{true} + 1.5(m_{\chi_1}^{assumed} - m_{\chi_1}^{true}) \pm 3$GeV and $m_{\tilde{g}} - m_{\tilde{b}} = m_{\tilde{g}}^{true} - m_{\tilde{b}}^{true} \pm 0.5$ GeV. Note that the event rates in this example are enormous; there are approximately 2.3 Million events in 10 fb^{-1} that have two isolated leptons four b-jets and missing energy!

Once several quantities have been measured, one will attempt to constrain the parameters of the SUSY model by performing a global fit much as the standard model is tested at LEP [21]. To get and indication of how well this might work, many choices of parameters within the SUGRA model were made and those that resulted in masses within the expected error were retained [22]. Measurements of m_h, $m_{\chi_2} - m_{\chi_1}$ and $m_{\tilde{g}} - m_{\tilde{b}}$ with errors of ± 5 GeV, ± 0.50GeV(10σ) and ± 3 GeV (1.5σ) respectively result in the constraints $\delta m_{1/2} = 1.5$ GeV, $\delta m_0 = 15$ GeV and $\delta \tan \beta = 0.1$. It is clear from this example that precise measurements of

---

[2] Recall that the current error on M_W from CDF/D0 [20] comes from an analysis involving E_T^{miss} has far fewer events and has an error of order 150 MeV

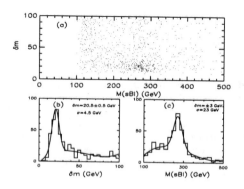

FIGURE 4. The reconstruction of gluino and sbottom decays from the decay chain $\tilde{g} \to \chi_2(\to \chi_1 \ell^+ \ell^-)\tilde{b}$. Events are selected near the end point of the $\ell^- \ell^+$ mass distribution and the momentum of χ_2 reconstructed. Two b-jets are then required and the mass of $b + \chi_2$ ($= m_{\tilde{b}}$) and the mass difference $\delta m = m_{bb\chi_2} - m_{b\chi_2}$ are computed. The scatterplot in these two variables and the projections are shown.

SUSY parameters will be made at LHC if supersymmetric particles exist. For more details and other examples of the types of measurements that might be done at LHC, see ref [23].

CONCLUSIONS

In this talk, I have not tried to give an overview of the new physics that the LHC experiments might see. The reader can refer to other documents for more details [2,3,24]Instead I have tried to give a sense of some of the measurements of the new physics that might be performed once the new physics is seen. Emphasis is often given to searches for the Higgs boson. There are good reasons for this; it is the one missing particle in the standard model, but, more importantly, its properties are fully predicted once its mass is given. Very detailed simulations of signals and backgrounds can be performed therefore. However this emphaisis on the Higgs boson may be misleading. Production rates are very small and consquently experiments need the full luminosity of the LHC.

Most theorists would be very dissapointed if the LHC discovered a standard model Higgs boson of mass 400 GeV. Varients of the standard model offer are larger opportunity to understand the origins of mass generation and reason for the size of the elextroweak scale. At present, supersymmetry is the most popular option. In this case the Higgs sector is richer and SSC experiments will probe it. Again event rates for the **direct** production of Higgs bosons are small and the experiments are challenging. If superysmmetry is relavent, the first observations at LHC will come, not from the Higgs sector, but rather from the decays of squarks and gluinos which have vastly larger production rates. I have given an example of some of the detailed measurements that will await LHC experimenters if supersymetry is seen. Indeed a theorist /citeellis jested that the LHC might be more approporaitly called the "bevatrino".

ACKNOWLDEMENT

The work was supported in part by the Director, Office of Energy Research, Office of High Energy Physics, Division of High Energy Physics of the U.S. Department of Energy under Contracts DE-AC03-76SF00098. Accordingly, the U.S. Government retains a nonexclusive, royalty-free license to publish or reproduce the published form of this contribution, or allow others to do so, for U.S. Government purposes.

REFERENCES

1. P. Lefevre, et al., CERN/AC/95-05.
2. Atlas Technical proposal, CERN/LHCC/94-43.
3. CMS Technical proposal, CERN/LHCC/94-38.
4. ALICE Technical Proposal, CERN/LHCC/95-71.
5. LHC-B Technical Proposal, CERN/LHCC/95-XX.
6. E. Richer-Was, et al., ATLAS Note PHYS-NO-074 (1996).
7. LHC-B Technical Proposal, CERN/LHCC/95-XX.
8. D. Froidevaux. F. Gianotti and E. Richter Was, ATLAS Note PHYS-NO-064 (1995),
9. J.-C. Chollet, et al., ATLAS internal note PHYS-NO-17 (1992), L. Poggioli, ATLAS Note PHYS-NO-066 (1995).
10. M. Bosman and M. Nessi, ATLAS Note PHYS-NO-050 (1994).
11. N.Stepanov, CMS-TN/93-87 (1993); S. Abdullin and N. Stepanov, CMS-TN/94-179 (1994).
12. S. Abdullin and N. Stepanov, CMS-TN/94-178 (1994).
13. R. Kinnunen, J. Tuominiemi, and D. Denegri, CMS-TN/93-98 (1993) and CMS-TN/93-103; C. Seez, CMS-TN/93-84.
14. D. Cavalli, et al., ATLAS Note PHYS-NO-025 (1993).
15. F. Paige, these proceedings.
16. M. Dine, A. Nelson and Y. Shirman, *Phys. Rev.* **D51**, 1362 (1995).
17. L. Alvarez-Gaume, J. Polchinski and M.B. Wise, Nucl. Phys. **B221**, 495 (1983);
    L. Ibañez, Phys. Lett. **118B**, 73 (1982);
    J.Ellis, D.V. Nanopolous and K. Tamvakis, Phys. Lett. **121B**, 123 (1983);
    K. Inoue et al. Prog. Theor. Phys. **68**, 927 (1982);
    A.H. Chamseddine, R. Arnowitt, and P. Nath, Phys. Rev. Lett., **49**, 970 (1982).
18. J. Soderqvist, To appear in Proceedings of 1996 DPF Summer study Snowmass CO
19. W-M Yao, To appear in Proceedings of 1996 DPF Summer study.
20. F. Abe, et al., *Phys. Rev. Lett.* **75**, 11 (1995), S. Abachi, et al.FERMILAB-PUB-96-177-E.
21. LEP Electroweak working group.
22. A. Bartl, et al., To appear in Proceedings of 1996 DPF Summer study;
23. I. Hinchliffe et al., hep-ph/9610544, Phys Rev D (to appear).
24. I Hinchliffe and J Womersley (eds) hep-ex/9612006,Presented at 1996 DPF / DPB Summer Study on New Directions for High-Energy Physics (Snowmass 96), Snowmass, CO, 25 Jun - 12 Jul 1996.
25. J. Ellis, private communication.

Very Large Hadron Colliders*

Mike Harrison**

**Brookhaven National Laboratory, RHIC Project*
P.O. Box 5000, Upton, NY 11973-5000

Issues pertaining to the next generation (post LHC) Hadron Colliders have been addressed over the past several years at workshops at Indiana [1], Indianapolis[2], and more recently at Snowmass[3]. Although no attempt has been made to produce a detailed parameter set, most work has addressed energies in the range of 80-100 TeV centre-of-mass with a peak luminosity of 10^{34} cm^{-2} s^{-1}. This is sufficient to illuminate the potential problems associated with this class of machines. There have been two distinct design concepts were examined; the low-field and high-field options. It is significant to note that while an SSC-like approach to access this energy range is technically feasible the cost of such a device is deemed prohibitive. There is general agreement that new technologies are necessary to achieve a cost breakthrough and as such that the dominant technical challenges for the future are driven by cost considerations, unlike linear or muon colliders, where cost is merely important. The organizing committee at Snowmass challenged the sub-group with a cost goal of $50M per Tev, and although neither option is sufficiently mature at this point to attempt any meaningful cost estimate, this figure gives an indication of the potential challenge facing the machine designers. The basic accelerator concept for an VLHC is perceived to be similar to today's machines: long repetitive arcs in a 2-in-1 magnet scheme with a few interaction regions encompassing the experimental regions and accelerator utilities. While a wide variety of design have been examined it has not been deemed terribly important to produce any integrated VLHC design at this point in time.

The low-field option [4] is essentially a direct attack on the problem on lowering the unit costs of the repetitive structures. Based around a novel 1.5 -> 2 T superferric 'double-C transmission line magnet', this approach offers the possibility of dramatically reducing the magnet costs per Tesla with a highly efficient use of superconductor and a simple magnet structure with most of the magnet warm. Since the field strength in a superferric magnet is limited, a long machine circumference is unavoidable (~600 km) low cost tunneling and magnet installation become very important. Other issues associated with a very large machine size involve the stored energy in the circulating beam arising from the large number of bunches, beam stability problems driven by the high impedance, and unit costs associated with the non-magnetic arc elements (instrumentation, vacuum, corrections etc.). Applying the 'Snowmass cost criteria' of $50M per

---

* Work performed under the auspices of the U.S. Department of Energy.

TeV together with a 3-4 fold increase in length from an SSC like object, then relative cost reductions of a factor of ~20 are needed from todays techniques. It is undemonstrated at this time that simplifications of this scale can be achieved, and indeed it can be observed that simple concepts tend to become more complex (and thus costly) as reality sets in.

The high-field option [5] follows a different line of attack from the low-field one. Based on an as yet unproven magnet operating at 12.5 T, this design relies on exploiting the emittance damping resulting from synchrotron radiation. With damping times in the 2-4 hr range the integrated luminosity during a 10-20 hr store is essentially independent of the initial beam parameters. Figure 1 shows an emittance evolution during a store. Very bright beams, which can not be achieved by conventional techniques arise quite naturally in this dynamical framework. Since the low field performance requires little more than getting the beam to 'stay in the pipe', cost savings are therefore hoped to arise from tolerance reduction in the emittance preservation requirements, field quality requirements, the use of a lower injection energy, and the shorter tunnel. Issues associated with this approach involve handling the radiated power into the bore tube (~1-2 W/m), ensuring acceptable vacuum in the presence of desorbed gases, and, of course, the dominant fact that there is no accelerator magnet capable of achieving the necessary 12.5 T field. The desired field strength and synchrotron heating lead naturally to a magnet approach based on high temperature superconductor (HTS) technology, but lower temperature approaches based on existing superconductors are possible though probably much too expensive.

A major topic of discussion during the past several years involved the potential use of HTS technology for future hadron machines. During the last several years great progress has been made in this area, and the possibilities of R&D efforts to develop these materials for magnet uses, received much attention. The low-field magnet design could potentially use a BSSCO material in a relatively straightforward way. This material is starting to become available in engineering quantities and is well matched to the lower fields and current densities in the superferric environment. A different material, YBCO, has promise for the high field magnets but is less advanced than BSSCO at this time and to date has only been produced in very thin films. The use of HTS technology is compelling. In addition to the increased cryogenic efficiency arising from a higher operating temperature, these materials have less sensitivity to operating temperature variations which can be exploited to greatly simplify the cryogenic system. The increased Carnot efficiency could also be exploited to simplify cryostat designs by tolerating a higher heat leak. The area of HTS was certainly identified as a key technology within the charge to this sub group for future efforts.

Figure 1. Emittance evolution at 50 TeV

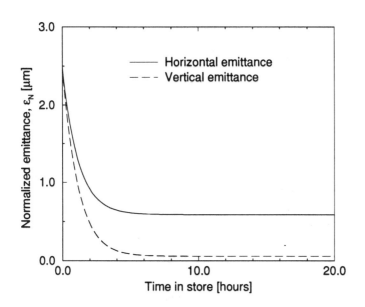

In addition to work focused on the two machine options, other more generic topics were worked on which would be suited to either approach. An analysis of interaction regions indicated that beta-star values lower than the normally assumed 20 cm appear feasible with a free space of ± 25 m and a quadrupole gradient of 300 T/m. With the inclusion of crab crossing schemes beta-star's as low as 5 cm appear possible with second order chromatic correction. Machine lattices with long arc cells have potential advantages including cost savings. During the workshop it was concluded that adequate beam dynamics performance could be achieved with cell lengths of the order of 500 m with the optimum length determined primarily by systematic rather than random field harmonics in the magnets. Beam stability appears to be a problem for the low-field option where the small bore tube and large circumference result in rapid resistive wall growth times. Single bunch stability may also prove to be an issue. One of the more interesting aspects of the machine parameters involves bunch spacing and intensity. For a given luminosity the machine design would prefer fewer bunches with more intensity to minimize stored energy and synchrotron radiation power. For reasonable parameter sets, the number of interactions per crossing varies in the range of 20 - 40, with the accelerator preferring the latter and the experiments presumably preferring the former. At this point in time there has been no consensus reached on the optimum values.

Various options for the vacuum system have been investigated [6]. Distributed pumping is needed in all cases but acceptable vacuum performance can be achieved. The warm bore tube of the low-field option simplifies this system. In the high field case cryo-pumping is needed for the synchrotron desorbed gas which limits the operating temperature to 20K or below. In the absence of HTS technology large cryogenic systems are unavoidable. While there does not appear to be any technical problems with very large cryogenic facilities the fundamental issue is, once again, cost. Other aspects of the machine design such as power supplies, quench protection, beam handling, and installation are covered in the sub group report.

In conclusion, one can say that while no technical problems have been found which would preclude the construction of a 50 x 50 Tev Hadron Collider, present day techniques are simply too expensive to be deemed viable in today's cost cutting climate. Fortunately there appears to be no shortage of ideas on potentially cheaper approaches which would benefit from R&D support. The single most important R&D topic at this time is the use of HTS technology in the area of accelerator magnets.

References

[1] Proceedings of the Workshop on Future Hadron Facilities in the U.S., Bloomington, Indiana, Fermilab TM-1907, 1994.

[2} New Low Cost approaches to High Energy Hadron Colliders at Fermilab, APS Spring meeting, Indianapolis, 1996.

[3] A wide variety of contributed papers on these topics can be found in the Snowmass 96 proceedings.

[4] The 'Pipetron', A Low Field Approach to a Very Large Hadron Collider, E. Malamud editor, Snowmass 96 proceedings.

[5] Beyond the LHC, A Conceptual Approach to a Future High Energy Hadron Collider, M. Syphers, M. Harrison, S. Peggs, International Conference on High Energy Accelerators, Dallas 1995.

[6] Beam Tube Vacuum in Very Large Hadron Colliders, W. Turner, Snowmass 96 Proceedings.

e^+e^- Physics

Hitoshi Murayama[1]

Department of Physics, University of California, Berkeley, CA 94720
and
Ernest Orlando Lawrence Berkeley National Laboratory, Berkeley, CA 94720

Abstract. I discuss the physics capability of a future e^+e^- linear collider and comment on how such a facility would complement the physics program at the LHC.

I INTRODUCTION

As repeatedly emphasized already in this workshop, the main issue in particle physics is how the electroweak symmetry is broken. All the other questions in particle physics which are not answered within the Standard Model cannot be addressed without knowing how the electroweak symmetry is broken. Such questions include: why are there three generations, why is CP symmetry broken, why is $m_e \ll m_t$, why is $m_W \ll M_{\text{Planck}}$ etc., and all are very important questions. In Table 1 I gave examples of possible answers to fundamental questions in the standard model to demonstrate how future directions of particle physics will differ depending on the outcome of our future study on the electroweak symmetry breaking.

There will be, fortunately, LHC, which is an already approved facility to study the mechanism of the electroweak symmetry breaking. The capability of the LHC experiments is already reviewed beautifully in the previous talk by Ian Hinchliffe. In view of the above motivation, however, I believe that it is necessary to understand the mechanism of electroweak symmetry breaking with an absolute clarity for particle physics to move further forward, and an e^+e^- linear collider with $\sqrt{s} =$ 300–1500 TeV and $\int dtL =$ 50–200 fb^{-1} will be an ideal facility as a great and necessary compliment to LHC for this purpose. I will briefly review the physics capability of such a machine in this talk. Many of the discussions can be found in [1] in more detail; see also references therein.

Possible mechanisms of electroweak symmetry breaking can be broadly classified as either "weakly coupled" (with a light Higgs boson) or "strongly coupled" (with no Higgs boson); I refer more details of theoretical discussions to the talks by Scott Willenbrock, Jack Gunion, and Tom Appelquist. I will discuss the physics capability and the manner how a linear collider would complement the LHC program for both cases separately in the following sections.

---

[1] This work was supported in part by the U.S. Department of Energy under Contracts DE-AC03-76SF00098 and DE-FG-0290ER40542 and in part by the National Science Foundation under grant PHY-95-14797, and also by the Alfred P. Sloan Foundation.

TABLE 1. "Explanations" given to basic questions on physics of electroweak symmetry breaking in various theory scenarios. The contents in this table are meant to be examples, rather than representative ones. Especially those on fermion masses are controversial.

	Standard Model	Supersymmetry	Technicolor
Existence of Higgs boson	Only scalar boson introduced just to break EW symmetry	Just one of many scalar bosons, nothing special	No Higgs boson. There are only fermions and gauge bosons.
Why electroweak symmetry is broken	by an ad hoc choice $m^2 < 0$	m^2 driven negative dynamically by top quark Yukawa coupling	new strong force binds fermions to let them condense
pattern of quark, lepton masses	choose size of Yukawa couplings to reproduce them	sequential breaking of flavor symmetry just below the Planck scale	further new forces at 1 to 1000 TeV scales
new phenomena	only a Higgs boson	superpartners of all known particles below TeV scale	resonances at 1–10 TeV, PNGBs and new fermions at 0.1–1 TeV

II LIGHT HIGGS BOSON

Let me take the Minimal Supersymmetric Standard Model (MSSM) as an example below. There are five Higgs bosons in this model,

$$h^0, H^0, A^0, H^+, H^-,$$

and the mass of the lightest neutral scalar h^0 has to be smaller than $m_{h^0} \lesssim 130$ GeV including radiative corrections. It decays primarily into $b\bar{b}$, and into $\gamma\gamma$ with a branching fraction of $\sim 10^{-3}$ or less.

The LHC will see the signal of a light neutral scalar decaying into $\gamma\gamma$ with an impressive capability even in the high luminosity environment. The ATLAS and CMS experiments will discover the Standard Model Higgs boson over the entire mass range above LEP2 reach up to 600 GeV or so. The $\gamma\gamma$ rate is in general lower in the Minimal Supersymmetric Standard Model than in the Standard Model, but still they will cover the entire MSSM parameter space. This is a highly significant capability of these experiments.

However I have a worry if there were only LHC and no electron positron collider. It is not the fact that there may remain a hole in the Higgs sector parameter space, as sometimes emphasized (such a hole in the Next-to-Minimal Supersymmetric Standard Model was also discussed by Gunion in this workshop). Such a hole may be filled by running experiments longer at high luminosity and combine two experiments. My worry is that it is not clear what we will learn either by seeing this signal or by not seeing it. For instance, just seeing a peak in $\gamma\gamma$ invariant mass would leave many possible interpretations: technipion, existence of vector-like quarks with a singlet scalar [2], etc. Such a signal many not necessarily tell us the answer to the most basic question: is the electroweak sector strongly coupled of weakly coupled.

The discovery of a Higgs boson is relatively easy at an e^+e^- collider, even if it decays invisibly. However this is not the end of the story. The most interesting questions is how we know that it is *the* Higgs boson once it is discovered, namely, how we can establish that the newly discovered particle is responsible for generating W, Z, top, and all other quark, lepton masses. What are its crucial characteristics? There are three of them. (1) It has to be a scalar particle. (2) It has a condensate in the vacuum. (3) It generates m_W and m_Z. Can we test these characteristics experimentally?

It is not difficult to test the crucial characteristics of *the* Higgs boson at an electron positron linear collider once it is found (Fig. 1). The most promising production process for

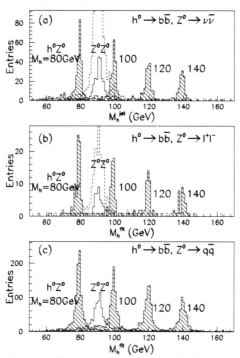

FIGURE 1. The invariant mass distributions in Zh events for the standard model Higgs boson [3]. All decay modes of Z, (a) $Z \to \nu\bar{\nu}$, (b) l^+l^-, (c) $q\bar{q}$ can be used.

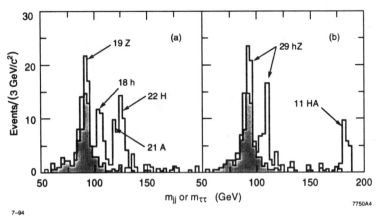

FIGURE 2. The invariant mass distributions in dijet or tau-pair for MSSM Higgs bosons and Standard Model backgrounds [4].

a light Higgs boson which we discuss here is $e^+e^- \to Zh$. For a Higgs boson of mass 130 GeV, a collider with $\sqrt{s} = 300$ GeV and 50 fb^{-1} would produce about ten thousand events, which is enough statistics to do a completely model-independent study to determine the nature of the discovered new particle. First of all, the angular distribution of the Higgs boson shows that it is produced by an s-channel process rather than t-channel. The polarization asymmetry of Higgs boson production can be determined as -0.14 ± 0.06 using leptonic Z only. The smallness of the asymmetry tells us that there is no significant γ-Z interference, whose relative sign is roughly opposite for different electron polarization. Therefore we learn that either γ or Z dominates in the process. A small but finite asymmetry then confirms it is Z-dominated, and hence the production is due to ZZh coupling. One can check that the final Z-boson is mainly longitudially polarized by reconstructing Z decay distribution, and hence the Z-boson can be regarded effective as a scalar boson. The angular distribution of the Higgs boson is $\sin^2\theta$ in the high energy limit, which tells us the discovered particle is a scalar, CP-even particle. Combining these observations, it establishes that the production occurs via $Z_\mu Z^\mu h$ operator. Since usual scalar fields without a condensate have only $Z_\mu Z^\mu \phi^*\phi$ coupling but no $Z_\mu Z^\mu \phi$ coupling, the existence of $Z_\mu Z^\mu h$ coupling implies that h has a vacuum expectation value. Finally the total production rate independent from the decay modes can be measured using leptonic decay of Z, which gives us 4% level measurement of the ZZh coupling with 50 fb^{-1} integrated luminosity [4]. If the observed particle is *the* Higgs boson, the ZZh coupling has to be $g_Z m_Z = em_Z/\sin\theta_W \cos\theta_W$. Having ZZh coupling with the right strength establishes that it is responsible for generating m_Z. In this way, one can unambiguously establish that the observed new particle is *the* Higgs boson. If the coupling is less, it contributes to a part of the Z mass, and there should be more Higgs boson(s) to generate the entire Z mass, which would show up in dijet or tau-pair mass distributions as shown in Fig. 2.

Furthermore, one can test whether the discovered particle is responsible for generating quark and lepton masses. The measurement of the relative branching fraction to $b\bar{b}$ and $\tau^+\tau^-$ tests whether the coupling of the Higgs boson to various fermion species is indeed proportional to their masses as required. One can also measure the cross section of $e^+e^- \to t\bar{t}h$, which tests the coupling of the top quark to the Higgs boson. These measurements can be done at 10–20 % levels.

A truly interesting strategy is to use information from all possible experiments, pp, e^+e^- and $\gamma\gamma$ colliders. The LHC measures the product $\Gamma(h \to gg)\Gamma(h \to \gamma\gamma)/\Gamma_h$, while a $\gamma\gamma$

collider measures $\Gamma(h \to \gamma\gamma)^2/\Gamma_h$. An e^+e^- linear collider will give us Γ_h indirectly, knowing the ZZh vertex from the total production rate and the relative branching fraction into WW^*. Combination of all three experiments will give us a model-independent determination of gg and $\gamma\gamma$ partial widths. Such a measurement is of a great interest since *any* charged or colored particles which obtain their masses from electroweak symmetry breaking contribute to these partial widths and do not decouple even when they are heavy. For instance, one can convincingly exclude a fourth generation in this method. This is a wonderful example how different colliders cooperate to give us useful information on physics of electroweak symmetry breaking and beyond.

III SUPERSYMMETRY

The search for and study of superparticles offer us the best example where the LHC and an e^+e^- linear collider play different roles, which combine to give us a coherent picture of physics at yet deeper level.

Discovery of supersymmetry at the LHC is regarded to be relatively easy. In the ordinary framework of supersymmetry,[2] the dominant signature of supersymmetry is large missing E_T with many jets, possibly also with leptons and/or b-jets. For instance, the gluon fusion produces a pair of gluinos, gluinos decays into a squark and a quark, the squark decays into a chargino and a quark, the chargino decays into the lightest neutralino and W, and W into jets or a lepton and a neutrino. Since the lightest neutralino and the neutrino escape detectors, one sees large missing E_T with many jets (and leptons). Ian Hinchliffe and Frank Paige demonstrated that many precision measurements can be done at LHC as well once supersymmetry is discovered.

Similar to the case of the Higgs boson, again the interpretation of the signal is not foolproof. If one sees, in addition to the missing E_T signal, like-sign dileptons, it is consistent with the "Majorana-ness" of gluino and the interpretation becomes more solid. However, missing E_T, multi-leptons, multi-b-quarks might well occur in technicolor-type scenarios as well. Numerous consistency checks can be done at LHC. However, it is still favorable if one can directly see that (1) there are new particles with the same quantum numbers as those of known particles, (2) their spin differ by 1/2, and (3) their interactions respect relations required by supersymmetry. All three are possible at an e^+e^- linear collider.

First important point to recall is that the colored superparticles tend to be about a factor of 3 or 4 heavier than the colorless superparticles in many theoretical scenarios. Therefore, the discovery reach of LHC on squark, gluino up to 2 TeV or above roughly corresponds to an e^+e^- collider at 1–1.5 TeV. Once there are superparticles within the kinematic reach, discovery is typically not difficult. Again the main issue here is how well we can establish that the newly discovered particles are superpartners of known particles. The crucial characteristics of superparticles are: (1) they have the same quantum numbers as their partners, (2) their spins differ by 1/2 from their partners, and (3) their interactions have the same strengths as the interactions of their partners.

Let us suppose we see sleptons at a future e^+e^- linear collider. It is easy to determine that the sleptons have the same quantum numbers as the leptons, just by counting the number of events. For instance the production of $\tilde{\mu}$ is due to s-channel γ, Z exchange. The total production cross section and the left-right asymmetry completely determines the coupling of $\tilde{\mu}$ to γ and Z. This establishes the point (1). Even though $\tilde{\mu}$ decays into μ and the lightest neutralino which escapes detection, the angular distribution of the $\tilde{\mu}$ can be also reconstructed up to a two-fold ambiguity. Fortunately, the "wrong" solution has a flat distribution which can be subtracted. Then one sees $\sin^2\theta$ distribution (Fig. 3) which shows that $\tilde{\mu}$ is a scalar particle. This establishes the characteristics (2). The point (3) is

---

[2] It is assumed that the R-parity is exact and the lightest superparticle is a stable neutralino.

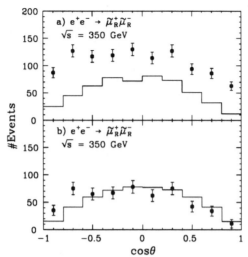

FIGURE 3. The angular distribution of $\tilde{\mu}_R^-$ in $e^+e^- \to \tilde{\mu}_R^+\tilde{\mu}_R^-$ reconstructed from final-state $\mu^\pm$ four-momenta, knowing the $\tilde{\mu}_R$ and $\tilde{\chi}_1^0$ masses: (a) with the two solutions unresolved and (b) with the "background" due to the wrong solutions subtracted. The histogram in (a) is the distribution of the right solutions, while that in (b) is the distribution of the original sample before selection cuts [5].

more difficult to establish but is possible [6,7]. Fig. 4 shows a result of a case study how well one can establish the equality between two different couplings, the usual U(1) gauge coupling e-e-B_μ and its supersymmetric cousin, e-$\tilde{e}$-$\tilde{B}$. We label the former by g' and the latter by $g_{\tilde{B}\tilde{e}e}$. Two couplings are related by supersymmetry, $g' = g_{\tilde{B}\tilde{e}e}/\sqrt{2}$. The figure shows how well one can determine the ratio of the coupling constants experimentally from a pair production of $\tilde{e}_R$ [7]. In this way, one can unambiguously establish that the new phenomenon observed is indeed due to supersymmetry.

There is even more excitement after the discovery. Measurement of superparticle masses will tell us physics at very high scales, like GUT- or Planck-scales. The best example is the following test of grand unified theories (GUT) using the masses of gauginos. It is now well-known that the measured value of $\sin^2\theta_W$ is remarkably consistent with the prediction of supersymmetric GUTs. The reason why we can test a theory at a very high scale in this case is because GUTs predict that the three gauge coupling constants are the same, $\alpha_1 = \alpha_2 = \alpha_3$ at the GUT-scale. We can extrapolate the measured gauge coupling constants to higher energies, and test whether they meet at a single point. Some people take this seriously, others think it is just a numerical accident. Now supersymmetric GUTs predict further relations. The masses of three gauginos, M_1, M_2, M_3 of bino, wino, and gluino, respectively, also have to be the same at the GUT-scale. This prediction of GUTs is independent of the symmetry breaking pattern [8] and does not receive logarithmically enhanced threshold corrections [9]. Once superparticles are discovered, we can measure their masses, and extrapolate the measured values to higher energies. Then we can see whether they meet at a point. This gives us an independent test of GUTs from that using gauge coupling constants, and if verified, it can hardly be an accident.

For such a measurement of gaugino masses, an e^+e^- linear collider with polarized electron beam is crucial. Since gauginos $\tilde{B}$ and $\tilde{W}$ mix with higgsinos to form two chargino and four neutralino mass eigenstates, one needs to disentangle the mixing to measure the masses of gauginos. In our paper [5], four experimental observables, $m(\tilde{\chi}_1^0)$, $m(\tilde{\chi}_1^\pm)$, $\sigma(e_R^-e^+ \to \tilde{e}_R^-\tilde{e}_R^+)$

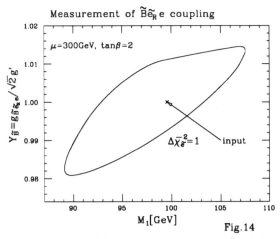

FIGURE 4. The test whether the U(1) gauge coupling constant g' and the coupling e-$\tilde{e}$-$\tilde{B}$ are related as dictated by supersymmetry. The test here is done by a fit to the data of selectron-pair production.

and $\sigma(e_R^- e^+ \to \tilde{\chi}_1^- \tilde{\chi}_1^+)$ were used to extract four parameters M_1, M_2, μ and $\tan\beta$. The Fig. 5 shows the accuracy how well one can extract M_1 and M_2 consistent with the inputs, which satisfy a simple relation from GUTs. Therefore, one can make an important test of grand unified theories by measuring cross sections and masses of superparticles. As we heard from Ian Hinchliffe, a similar test can be done at LHC between M_3 and $M_2 - M_1$. Combining two colliders would test whether of M_1, M_2, and M_3 all meet at the same scale where we have seen the gauge coupling constants to meet.

The combination of LHC and an e^+e^- linear collider could determine the full superparticle spectrum. We would love to see a "Precision Supersymmetry Working Group" formed between LHC and linear collider experimentalists to pass the data to each other to determine the full superparticle spectrum. It is important that such a determination of the spectrum does not depend on particular theoretical assumptions. With an e^+e^- collider, we can be certain that the observed new particles are indeed superpartners of known particles. The mass of the lightest neutralino can be easily measured at e^+e^- which fixes the ambiguity in measurements at LHC discussed by Ian Hinchliffe. Once we are sure that the new phenomena are due to supersymmetry and have determined its spectrum, we can move forward to the next questions in particle physics with a solid confidence.

IV STRONG ELECTROWEAK SYMMETRY BREAKING SECTOR

Now we come to the discussion of the other type of scenario, where the electroweak symmetry is broken by a new strong force. A representative model is technicolor, where this new gauge interaction attracts pairs of technifermions very strongly with each other and let them condense, $\langle \bar{f}f \rangle \neq 0$. The generic signatures of this scenario are: (1) no light Higgs boson (below, say, 600 GeV), (2) the scattering between two longitudinal W bosons become strong at higher energies (say, $E \gtrsim 1.8$ TeV), and (3) there possibly are resonances due to new strong interactions (techni-ρ decaying into W^+W^- or $W^\pm Z^0$, techni-ω into $Z^0\gamma$, etc).

First general statement on this scenario is that all experimental signatures are rather rare

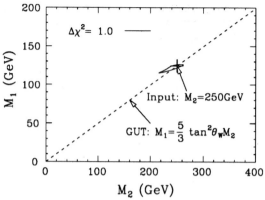

FIGURE 5. The $\Delta\chi^2 = 1$ contour in the M_1-M_2 plane obtained from a global fit to simulated data [5]. The dotted line indicates the GUT prediction: $M_1 = (5/3) \tan^2 \theta_W \, M_2$.

and weak, and it will be difficult for experiments to see the effects of new strong interaction, even at LHC. Even though it is likely that one can see certain excess in like-sign dilepton with large missing E_T, it may not be easy to directly interpret it as a signal of strong WW interaction.

If one also has an electron positron linear collider in addition to the LHC, this difficult signal becomes more convincing. A linear collider can unambiguously establish the absence of any kinds of light Higgs boson below its kinematic reach, $m_h \lesssim 0.9(\sqrt{s} - m_Z)$. The *absence* of a Higgs boson, combined with an excess in like-sign dilepton, implies a strongly interacting electroweak symmetry breaking sector. Recall that it is not easy to establish the absence of a light Higgs boson at the LHC alone. Also, the Higgs boson may decay mainly invisibly, which reduces the $\gamma\gamma$ signature substantially. The invisible decay is not specific to the supersymmetric models, in which Higgs bosons may decay into a pair of neutralinos, but also possible in other models as well. For instance if the fourth generation exists with little mixing to lower generations, and if $2m_{\nu_4} < m_H < 2m_{l_4} < 2m_{q_4}$, the Higgs boson decays mainly into $\bar{\nu}_4 \nu_4$ and is hard to be detected. One can also look for associate production processes like $t\bar{t}H$, WH even in this case [10]; but it seems to be not easy to convince ourselves that there is *no* Higgs boson. On the other hand, an invisibly decaying Higgs boson can be easily seen at a linear collider using Zh production with Z decaying leptonically [4].

So far the role a linear collider plays may seem secondary, just to give a supportive evidence by proving there is no light Higgs boson. But there are more active roles an electron positron linear collider can play as well.

Table 2 shows the significance of strong WW scattering studied at an $e^\pm e^-$ collider with $\sqrt{s} = 1.5$ TeV and an integrated luminosity of 200 fb^{-1}. The statistical significance is comparable to that at the LHC. Similarly, a strong $WW \to t\bar{t}$ scattering can be studied as well.

If there is a techni-ρ resonance, an electron positron collider will have an ideal signal. The production of W-pairs has one of the biggest cross sections at a future e^+e^- linear collider. If the W-bosons in the final state are longitudinally polarized, they can rescatter due to a tail of the techni-ρ resonance. The rescattering modifies various final state distributions of the W decay products. Studies show that one can see the effects of a techni-ρ up to 2 TeV at 95 % confidence level at a linear collider with $\sqrt{s} = 500$ GeV and 50 fb^{-1} [12], which is already comparable to the reach at the LHC. Fig. 6 has confidence level contours for the real and imaginary parts of the rescattering amplitude F_T at $\sqrt{s} = 1.5$ TeV with 190 fb^{-1} [13].

TABLE 2. Total numbers of $W^+W^-, ZZ \to$ 4-jet signal S and background B events calculated for a 1.5 TeV $e^{\pm}e^-$ linear collider with integrated luminosity 200 fb^{-1} after cuts. The statistical significance $S/\sqrt{B}$ is also given. The hadronic branching fractions of WW decays and the $W^{\pm}/Z$ identification/misidentification are included. S/N is Improved by using 100% polarized e_L^- beams in a 1.5 TeV e^+e^-/e^-e^- collider [11].

channels	SM $m_H = 1$ TeV	Scalar $M_S = 1$ TeV	Vector $M_V = 1$ TeV	LET
$S(e^+e^- \to \bar{\nu}\nu W^+W^-)$	330	320	92	62
B(backgrounds)	280	280	7.1	280
$S/\sqrt{B}$	20	20	35	3.7
$S(e^+e^- \to \bar{\nu}\nu ZZ)$	240	260	72	90
B(backgrounds)	110	110	110	110
$S/\sqrt{B}$	23	25	6.8	8.5
$S(e^-e_L^- \to \nu\nu W^-W^-)$	54	70	72	84
B(background)	400	400	400	400
$S/\sqrt{B}$	2.7	3.5	3.6	4.2
$S(e_L^-e_L^- \to \nu\nu W^-W^-)$	110	140	140	170
B(background)	710	710	710	710
$S/\sqrt{B}$	4.0	5.2	5.4	6.3

Shown are the 95% confidence level contour about the light Higgs boson value of $F_T = 1$, as well as the 68% confidence level contour about the value of F_T for a 4 TeV techni-ρ. Even the non-resonant LET point is well outside the light Higgs boson 95% confidence level region.

The signatures of strong electroweak symmetry breaking sector discussed so far are WW scattering and are relatively model-independent. There are other signatures relevant at lower energies, though more model-dependent. Since our aim is to sort out the correct model which describes the electroweak symmetry breaking, such model-dependence is of great interest. Now we turn our discussions to the model-dependent signatures.

First of all, one needs to recall that the scenario of strongly interacting electroweak symmetry breaking sector has many problems. Just to name a few, Peskin–Takeuchi S-parameter, flavor-changing neutral currents, typically too small m_t, large isospin splitting $m_b \ll m_t$, R_b, etc. Since it may not make much sense to discuss experimental signatures of models which are already excluded, I would like to discuss several attempts to cure some of the above problems. Interestingly enough, such attempts tend to give us signatures at lower energies than a model-independent discussion gives.

The first example is the $Zt\bar{t}$ vertex. Suppose technicolor theory is correct, in the sense that the source of W, Z and all fermion masses originate from a single technifermion condensate $\langle \bar{T}T \rangle$. Since m_t is large, $\simeq 175$ GeV, there needs to be a fairly strong four-Fermi interaction, $t\bar{t}\bar{T}T$. Such an operator can be generated by an exchange of Extended Technicolor (ETC) gauge boson which converts a standard model fermion (top quark in this case) to a technifermion. Exchange of such an ETC gauge boson gives an interesting contribution to the $Zb\bar{b}$ vertex. The naive ETC model reduces R_b by about 5 % which is not acceptable . There are two modified ETC models, a diagonal ETC boson and a non-commuting ETC gauge boson, which have two independent and cancelling contributions to R_b. In each case, one can choose a parameter such that the additional contribution is consistent with the current value of R_b. Interesting point is that these models tend to give a rather large correction to $Zt\bar{t}$ vertex. An analysis [14] shows one can measure vector and axial form factors of the top quark at 10 % level with 50 fb^{-1} for each electron beam polarization at $\sqrt{s} = 400$ GeV. The

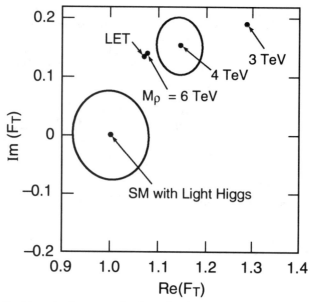

FIGURE 6. Confidence level contours for the real and imaginary parts of F_T at $\sqrt{s} = 1.5$ TeV with 190 fb^{-1}. The initial state electron polarization is 90%. The contour about the light Higgs boson value of $F_T = (1,0)$ is 95% confidence level and the contour about the $M_\rho = 4$ TeV point is 68% confidence level.

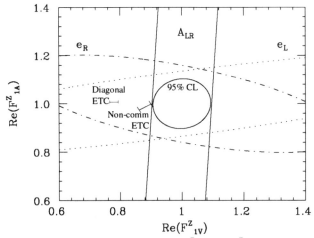

FIGURE 7. 95% confidence level contours for F_{1V}^Z and F_{1A}^Z, obtained from the maximum-likelihood analysis using a sample of 50 fb^{-1} each of right- and left-polarized electrons [14]. Predictions from diagonal ETC and non-commuting ETC put in [15].

predicted values of the vector form factor falls typically outside the 95 % confidence level contour.

Another interesting model is an attempt to reduce the S-parameter which tends to be too large. Since the minimal model of technicolor, one-doublet model with $N_{TC} = 2$, is now excluded at more than 99 % confidence level, one needs to find a mechanism to reduce the S-parameter. An attempt by Appelquist and Terning [16] is to introduce a large isospin splitting, thereby sacrificing the T-parameter a little, to reduce the S-parameter even in a one-family model. Their point is that one can have techni-leptons to be rather light; then the contribution to the T-parameter can be small enough even when there is a large isospin splitting between techni-electron and techni-neutrino. Their sample spectrum of techni-fermions is

N	50	GeV,
E	150	GeV,
Q	600	GeV.

As apparent from the spectrum, this model predicts light techni-ρ, $\overline{N}N$ at 100–300 GeV and a light charged pseudo-Nambu-Goldston boson $N\overline{E}$ at 50–150 GeV. This techni-ρ does not contribute much to the WW rescattering because the techni-neutrino contributes little to the W and Z masses. However it can appear as a narrow resonance in e^+e^- collision. $N\overline{E}$ can be produced similar to a charged Higgs boson whose main decay mode is $P^+ \to \nu_\tau \tau^+$. A search for it is straight forward, looking for acoplanar τ-pairs using right-handed electron polarization to suppress the WW background. They appear to be the same as $\tilde{\tau}$ in supersymmetry, but one can distinguish them after kinematic fit to the data and determine the mass of the neutrino (neutralino) for P^+ ($\tilde{\tau}$). On the other hand there are many colored pseudo-Nambu-Goldstone bosons at 250–500 GeV, which are targets of experiments at the LHC.

There are also attempts to solve the problem of flavor-changing neutral currents which have typically too large rates in extended technicolor models. The mechanism called techni-GIM [17] is one of such attempts. It requires a very complicated gauge structure and needs many new fermion fields below 1 TeV to cancel anomalies. There typically are many pseudo-Nambu-Goldstone bosons as well, whose masses arise due to gauge interactions. Since colorless pseudo-Nambu-Goldstone bosons are typically lighter than colored one, the situation is quite similar to the supersymmetry. The LHC will look for colored ones, a linear collider for colorless ones.

Here is the summary of this section. A combination of two observations: (1) the absolute absence of a Higgs boson at a linear collider and (2) a slight excess in WW-scattering at the LHC can be a convincing signature of strong electroweak sector. Moreover, the excess in WW-scattering observed at the LHC can be cross-checked with a linear collider at $\sqrt{s} = 1.5$ TeV; if the excess is due to a techni-ρ, $\sqrt{s} = 500$ GeV may be already enough. There are other model-dependent signatures like $Zt\bar{t}$ coupling, pseudo-Nambu-Goldstone bosons, light techni-resonances, etc, which help sorting out the correct model. Here again it is clear that the combination of both types of colliders is important to understand physics of electroweak symmetry breaking.

V CONCLUSION

An e^+e^- linear collider with $\sqrt{s} = 300$–1500 TeV would establish the mechanism of electroweak symmetry breaking with little model assumptions. It will sort out parameters and models convincingly. Such a facility would be a wonderful complement to the LHC to clear the dark clouds of electroweak scale physics which blocks our sight towards the most fundamental questions in particle physics. Once the sky clears up, we may even peek at the Planck-scale world as emphasized by David Gross at the beginning of the workshop.

ACKNOWLEDGMENTS

I express sincere thanks to the organizers of this workshop. This work was supported in part by the Director, Office of Energy Research, Office of High Energy and Nuclear Physics, Division of High Energy Physics of the U.S. Department of Energy under Contract DE-AC03-76SF00098 and in part by the National Science Foundation under grant PHY-90-21139, and also by Alfred P. Sloan Foundation.

REFERENCES

1. H. Murayama and M. Peskin, *Ann. Rev. of Nucl. and Part. Sci.* **46**, 533 (1996).
2. H. Murayama, talk given at "Third International Workshop on Physics and Experiments with e^+e^- Linear Colliders," Iwate, Japan, Sep 8–12, 1995. Proceedings, eds. by A. Miyamoto et al., World Scientific, 1996.
3. JLC-I, JLC Group, KEK-92-16, Dec 1992.
4. Patrick Janot, Invited talk at 2nd International Workshop on Physics and Experiments with Linear e^+e^- Colliders, Waikoloa, HI, 26-30 Apr 1993.
5. T. Tsukamoto, K. Fujii, H. Murayama, M. Yamaguchi, and Y. Okada, *Phys. Rev.* D **51**, 3153 (1995).
6. J.L. Feng, M.E. Peskin, H. Murayama, and X. Tata, *Phys. Rev.* D **52**, 1418 (1995).
7. K. Fujii, M. Nojiri and T. Tsukamoto, *Phys. Rev.* D **54**, 6756 (1996).
8. Y. Kawamura, H. Murayama, and M. Yamaguchi, *Phys. Lett.* B **324**, 52 (1994).
9. J. Hisano, H. Murayama, and T. Goto, *Phys. Rev.* D **49**, 1446 (1994).
10. J.F. Gunion, *Phys. Rev. Lett.* **72**, 199 (1994); Debajyoti Choudhury and D.P. Roy, *Phys. Lett.* B **322**, 368 (1994); S.G. Frederiksen, N. Johnson, G. Kane, and J. Reid, *Phys. Rev.* D **50**, 4244 (1994).
11. V. Barger, Kingman Cheung, T. Han, R.J. Phillips, *Phys. Rev.* D **52**, 3815 (1995).

12. A. Miyamoto, K. Hikasa, and T. Izubuchi, KEK Preprint 94-203, "Heavy vector resonance effect on the $e^+e^- \to W^+W^-$ process at JLC-I."
13. T. Barklow, in the Albuquerque DPF94 Meeting, ed. by Sally Seidel, p. 1236; and references therein.
14. T.L. Barklow and C.R. Schmidt, in *DPF '94: The Albuquerque Meeting*, ed. by S. Seidel, World Scientific, Singapore, 1995.
15. Uma Mahanta, H. Murayama, M.V. Ramana, in preparation.
16. T. Appelquist and J. Terning, *Phys. Lett.* B **315**, 139 (1993).
17. L. Randall, *Nucl. Phys.* B **403**, 122 (1993); H. Georgi, *Nucl. Phys.* B **416**, 699 (1994).

Physics at LHC and NLC

Frank E. Paige

Physics Department
Brookhaven National Laboratory
Upton, NY 11973 USA

Abstract. The Standard Model describes everything that we know about particle physics today, but it is probably incomplete, with new physics below the 1 TeV scale. The Large Hadron Collider at CERN will be the first accelerator to probe this scale. The Next Lepton Collider could provide valuable additional information.

I INTRODUCTION

Particle physics from the discovery of charm to the precision measurements of the Z boson at LEP and the SLC and the discovery of the top quark at the Tevatron has been a triumph for the gauge sector of the Standard Model [1]. We are now confident that the strong, electromagnetic, and weak interactions are described by gauge theories with an $SU(3)_{\text{color}} \otimes SU(2)_L \otimes U(1)$ gauge group. But $SU(2)_L \otimes U(1)$ gauge invariance requires massless gauge bosons and fermions. Simply adding mass terms for these would break the gauge symmetry, destroying the Standard Model. Hence we must introduce some new dynamics to break the gauge symmetry spontaneously. We know the mass scale for this symmetry breaking,

$$v = 2^{-1/4} G_F^{-1/2} = 246\,\text{GeV}\,,$$

but we do not know the mechanism.

Electroweak symmetry is broken in the Standard Model by adding an elementary scalar field, the Higgs boson, which gets vacuum expectation value v. Interactions with this Higgs boson give masses both to gauge bosons and to quarks and leptons. But elementary scalars cause theoretical problems. The first problem is *triviality*: the Higgs self-coupling λ is not asymptotically free and so grows with Q,

$$\frac{d\lambda}{d\log Q} = +b\lambda^2 + \ldots, \qquad b > 0\,.$$

The solution is
$$\lambda(Q) \sim \frac{\lambda_0}{1 - \lambda_0 b \log Q^2/\mu_0^2},$$
which diverges at some Q unless $\lambda_0 = 0$. This implies that we must treat the Standard Model as an effective theory, *i.e.*, as the leading term in an expansion in the scale Λ of new physics. To include gravity, the new scale must be no greater than the Planck mass, approximately 10^{19} GeV.

A second and closely related problem is *naturalness:* the Higgs mass diverges quadratically, so radiative corrections give corrections to it proportional to the new physics scale,
$$\delta M_H^2 \sim \Lambda^2.$$
While it is formally possible to renormalize the mass, subtracting a fine-tuned counterterm of order Λ from a bare mass also of order Λ to obtain a physical mass of order 100 GeV, this seems very unnatural for large Λ.

Only a few possible solutions have been suggested for the triviality and naturalness/fine-tuning problems:

- Accept the Standard Model with fine tuning, assuming that this somehow emerges from the unknown physics at the scale Λ.

- Replace the elementary Higgs boson with a dynamical condensate at the TeV mass scale such as technicolor.

- Embed the Standard Model in supersymmetry (SUSY). Then the contribution of each normal particle to the quadratic divergence is canceled by that of its SUSY partner, so the correction to the Higgs mass is $\sim M_{SUSY}^2$ rather than $\sim \Lambda^2$.

Deciding which of these solutions — or perhaps something else as yet not considered — is correct has been the central question in particle physics for many years [2].

It is certainly possible that we may resolve the question with existing accelerators. For example, LEP-II might find a Higgs boson, either a Standard Model or a SUSY one, or the Tevatron might find SUSY particles, most likely from the process [3]
$$q + \bar{q} \to \tilde{\chi}_1^{\pm} + \tilde{\chi}_2^0 \to \tilde{\chi}_1^0 \ell^{\pm} \nu + \tilde{\chi}_1^0 \ell^+ \ell^-.$$
But is is likely that the answer requires study of the TeV mass scale. This was the primary motivation of the SSC and remains the motivation for the LHC at CERN.

Supersymmetry (SUSY) [4] is the most radical of the suggested solutions but perhaps also the most appealing. SUSY requires doubling the known particle spectrum — every particle must have a partner differing by 1/2 unit of angular momentum. This is not a small price to pay for explaining how the Higgs boson can be light while shedding no light on fermion masses and mixings. But SUSY at some mass scale has strong theoretical motivations as the maximal extension of the Lorentz group and as an integral part of attempts to build a successful quantum theory of gravity. Furthermore, SUSY at the TeV scale leads to a (nearly) successful grand unification [5] of the $SU(3)$, $SU(2)$, and $U(1)$ gauge couplings, thus explaining the peculiar fermion representations appearing in the Standard Model. Furthermore, the unification mass scale is high enough to be consistent with present bounds on proton decay. SUSY corrections to precision electroweak data are small if the masses are naturally small if the SUSY masses are large.

The discussion that follows will be restricted to Higgs and SUSY physics at two future machines: the Large Hadron Collider (LHC) at CERN with

$$pp: \sqrt{s} = 14\,\text{TeV}, \mathcal{L} = 10^{33}\text{--}10^{34}\,\text{cm}^{-2}\text{s}^{-1}.$$

and the NLC with

$$\ell^+\ell^-: \sqrt{s} = 0.5\text{--}2\,\text{TeV}, \mathcal{L} \sim 10^{34}\,\text{cm}^{-2}\text{s}^{-1}.$$

The "N" in NLC should be taken as "Next" without prejudice as to the location or design of the machine. The "L" may be taken as "Linear" (by e^+e^- advocates) or as "Lepton" (by $\mu^+\mu^-$ advocates) although all of the subsequent plots have been taken from e^+e^- studies without inclusion of the large backgrounds intrinsic to $\mu^+\mu^-$ colliders. Finally, it will be implicitly assumed that the NLC will turn on substantially later than the LHC.

II HIGGS BOSONS

The discussion of triviality in the previous section implies that Higgs bosons must be treated as effective field theories, valid below some scale Q. Certainly this scale must be above the mass of the Higgs itself. Lattice calculations lead to a bound $M_H \lesssim 700\,\text{GeV}$ for the Standard Model Higgs [6]. A bound is obtained by requiring that the perturbative $W_L W_L$ scattering amplitude is less than the unitarity bound [7]. If the theory is to be perturbatively valid up to the grand unification scale, $\sim 10^{16}\,\text{GeV}$, then the Higgs mass must satisfy $M_H \lesssim 200\,\text{GeV}$ [8] Under the same assumption, there is also a lower bound, $M_H \gtrsim 130\,\text{GeV}$, from the requirement that the Higgs potential be bounded from below [9]. SUSY models favor even lighter Higgs bosons; in the Minimal Supersymmetric Standard Model (MSSM), the mass of the lightest Higgs boson must be lighter than about 130 GeV [10].

Production cross sections for Higgs bosons at LHC are relatively large, 1–100 pb, but only rare decay modes can be detected. The search for the Standard Model Higgs can be divided into three mass regions [11]:

1. $M_H > 2M_Z$: the signal is $H \to ZZ \to \ell^+\ell^-\ell^+\ell^-$.

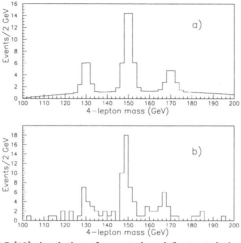

FIGURE 1. ATLAS [12] simulation of expected and fluctuated signals with $30\,\text{fb}^{-1}$ for $M_H = 130$, 150, and 170 GeV.

2. $130\,\text{GeV} \lesssim M_H < 2M_Z$: the signal is $H \to ZZ^* \to \ell^+\ell^-\ell^+\ell^-$.

3. $M_H \lesssim 130\,\text{GeV}$: the signal is $H \to \gamma\gamma$, either inclusively or with a lepton tag for WH and $t\bar{t}H$ production.

ZZ mode: For $M_H > 2M_Z$, the dominant Higgs decay modes are W^+W^- and ZZ. The coupling becomes strong for large M_H, so that the Higgs is wide,

$$\Gamma_H \sim 0.5\,\text{TeV}\left(\frac{M_H}{1\,\text{TeV}}\right)^3$$

For $ZZ \to \ell^+\ell^-\ell^+\ell^-$ only the real ZZ background important [12,13], so calculations are simple. The S/B is good, so the only limit comes from the rate and the fact that a heavy Higgs is so broad. A larger rate can in principle be obtained from $ZZ \to \ell^+\ell^- q\bar{q}$. In practice severe cuts are needed to reject the background from Z plus QCD jets, so that the gain is marginal.

ZZ^* mode: Again one must use the decay to $\ell^+\ell^-\ell^+\ell^-$, but with only one Z is on shell. Hence there are potential backgrounds from processes such as $Zb\bar{b}$, where each b decays leptonically. These can be rejected with isolation cuts and good mass resolution [12,13]. Detailed studies such as the one shown in Figure 1 find a low rate but a good S/B ratio. Discovery of a Higgs in this mass range should be straightforward, although it may take some time to accumulate sufficient luminosity, especially for $M_H \sim 170\,\text{GeV}$.

$\gamma\gamma$ mode: The $H \to ZZ^*$ branching ratio is too small for $M_H \lesssim 130\,\text{GeV}$. Decays into the dominant modes, $b\bar{b}$ and $\tau\tau$, have far too much background, so one must rely on $H \to \gamma\gamma$. There is still a large background from the $\gamma\gamma$ continuum. Since a Higgs in this mass range is very narrow, the background

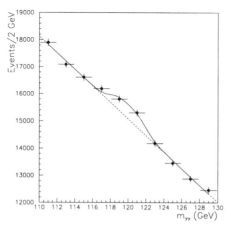

FIGURE 2. Signal and $\gamma\gamma$ continuum background for $M_H = 120$ GeV with the ATLAS detector [12].

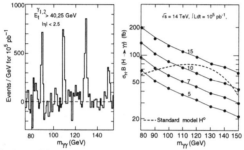

FIGURE 3. Simulation of background-subtracted signals for $M_H = 90, 110, 130,$ and 150 GeV with the CMS detector [13].

can be suppressed by excellent mass resolution, which requires both energy resolution and angular resolution in the electromagnetic calorimeter. The calorimeters of both ATLAS and CMS have been designed primarily with the goal of observing this process. Nevertheless, the S/B is marginal even for the most favorable case; see Figure 2.

In addition to the QCD $\gamma\gamma$ continuum, there are also huge potential jet-jet and γ-jet backgrounds. A rejection of 10^4 for jets while keeping 80%–90% γ efficiency is required. Detailed simulations [12,13] have found that this is achievable using calorimeter and tracking isolation plus electromagnetic shower shape to reduce $\pi^0 \to \gamma\gamma$. Background-subtracted mass peaks for the CMS detector, which plans to use $PbWO_4$ crystals to obtain extraordinary energy resolution, are shown in Figure 3. The signal appears to be observable

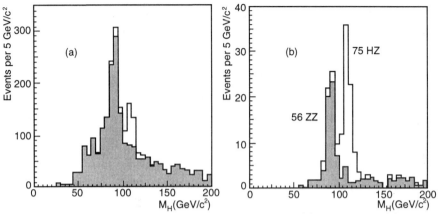

FIGURE 4. Simulation of Higgs signal and backgrounds (a) without and (b) with b tagging at NLC [15].

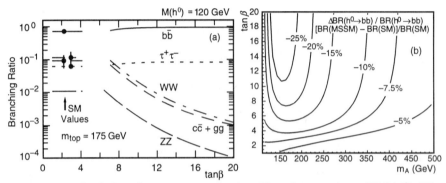

FIGURE 5. Simulation of measurement of various Higgs decay modes at NLC [15]. (The $\tan\beta$ variation shown in the left-hand figure is not that expected in the MSSM.)

in the Standard Model, although the significance is marginal at the lower masses. Since the signal is model dependent, it is important that LEP-2 run at $\sqrt{s} = 190\,\text{GeV}$ to cover $M_H \lesssim 95\,\text{GeV}$ [14].

Detection of low-mass Higgs bosons is much more robust at the NLC. The process $e^+e^- \to Zh$ can be observed with $10\,\text{fb}^{-1}$ just by selecting 4 jets, and a very good S/B can be obtained with modest b tagging, as can be seen from Figure 4 [15]. An energy $\sqrt{s} = 0.5\,\text{TeV}$ and a luminosity $\mathcal{L} \gtrsim 10\,\text{fb}^{-1}$ should be sufficient to find or to exclude any Higgs boson with $m_H < 200\,\text{GeV}$ [15].

The real strength of the NLC for Higgs physics is its ability to study many modes, not just rare ones [15,16]. Having identified the Z, one can then observe $h \to b\bar{b}$ by using b tagging; $h \to WW^*$ by requiring 6 jets and $M_{jj} \approx M_W$; $h \to \tau\tau$ by requiring low multiplicity jets; and $h \to c\bar{c}, gg$ by requiring a negative b tag. It may even be possible to use the vertex detector to separate $c\bar{c}$ and gg.

TABLE 1. Estimated relative branching fraction errors with $50\,\text{fb}^{-1}$ assuming Standard Model Higgs couplings [17].

Mode	$m_H = 140\,\text{GeV}$	$m_H = 120\,\text{GeV}$
$b\bar{b}$	±12%	±7%
WW^*	±24%	±48%
$c\bar{c} + gg$	±116%	±39%
$\tau\tau$	±22%	±14%

All of these can be measured with reasonable precision [17]. Unfortunately, the light Higgs boson in the MSSM can be quite similar to the Standard Model Higgs, as is indicated by Figure 5, so the errors listed in Table 1 may not be sufficient to distinguish them.

III SUPERSYMMETRY

In the Standard Model, the Higgs mass gets corrections of order any new physics scale Λ, so it unnatural for it to be light. SUSY [4] has a fermion for each boson with identical properties. These give opposite signs in loops and so cancel, producing instead $\Delta M_h^2 \sim M_{\text{SUSY}}^2$. Thus SUSY solves the naturalness and fine-tuning problems of the Standard Model provided that the SUSY masses are $\lesssim 1\,\text{TeV}$. Squarks and gluinos with such masses should easily be observable at the LHC [12,13].

Most of the detailed studies have been done for the minimal supergravity (SUGRA) model [18]. (The connection to supergravity is tenuous.) In the SUGRA model R parity is exact, so SUSY particles are produced in pairs and decay to the lightest SUSY particle $\tilde{\chi}_1^0$, which is stable, neutral, and weakly interacting. The SUGRA model assumes the minimal particle content, two Higgs doublets, grand unification, and universal soft SUSY breaking terms at the GUT scale. In particular, this means that squarks and Higgs bosons both have positive M^2. But when the renormalization group equations are run down to the weak scale, the large top Yukawa coupling drives a Higgs M^2 negative, breaking electroweak symmetry but not color or charge [3]. There remain just four parameters and a sign:

- m_0: the common scalar mass at M_{GUT}.

- $m_{1/2}$: the common gaugino mass at M_{GUT}.

- A_0: the common trilinear term at M_{GUT}.

- $\tan\beta$: the ratio of Higgs vev's at M_Z.

- $\text{sgn}\,\mu = \pm 1$.

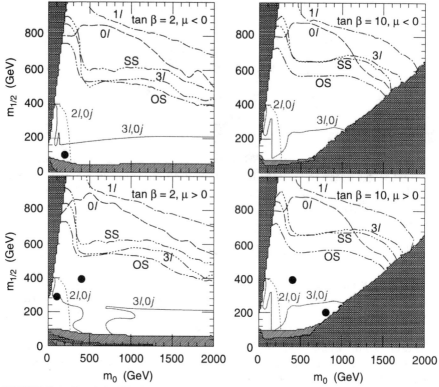

FIGURE 6. Reach in SUGRA parameter space at LHC. The shaded regions are excluded by theoretical constraints or experiment. The dots mark the five LHC points.

At the LHC gluino and squark production dominate. Typically they decay to the $\tilde{\chi}_1^0$ through several steps, e.g.

$$\tilde{g} \to \tilde{q}_L \bar{q}$$
$$\tilde{q}_L \to \tilde{\chi}_1^\pm q'$$
$$\tilde{\chi}_1^\pm \to \tilde{\chi}_1^0 W^\pm$$

These cascade decays give rise to a complex pattern of signatures, including

- 0ℓ: Multijets + $\not{E}_T$ + 0ℓ.
- 1ℓ: Multijets + $\not{E}_T$ + 1ℓ
- SS: Multijets + $\not{E}_T$ + $\ell^\pm \ell^\pm$.
- OS: Multijets + $\not{E}_T$ + $\ell^\pm \ell^\mp$.
- 3ℓ: Multijets + $\not{E}_T$ + 3ℓ.

TABLE 2. Representative masses and decay modes for LHC Point 5.

Particle	Mass(GeV)	Typical Decays
$\tilde{g}$	767	$\tilde{q}\bar{q}$, $\tilde{t}\bar{t}$
$\tilde{u}_L$	686	$\tilde{\chi}_2^0 q$, $\tilde{\chi}_1^\pm q$
$\tilde{u}_R$	664	$\tilde{\chi}_1^0 q$
$\tilde{\chi}_2^0$	231	$\tilde{\chi}_1^0 h$, $\tilde{\ell}^\pm \ell^\mp$
$\tilde{e}_R$	157	$\tilde{\chi}_1^0 e$
h	104	$b\bar{b}$

- $2\ell, 0j$: $\not{E}_T + 2\ell$ + jet veto, e.g., $\tilde{\ell}\tilde{\ell}$.
- $3\ell, 0j$: $\not{E}_T + 3\ell$ + jet veto, e.g., $\tilde{\chi}_1^\pm \tilde{\chi}_2^0$.

The reach in the SUGRA parameter space to observe a 5σ signal in each of these channels is summarized in Figure 6 [19]. The missing energy channels, with or without a lepton veto, can be used to find squarks and gluinos up to $\sim 2\,\text{TeV}$ with just $10\,\text{fb}^{-1}$. Many modes can be studied for masses up to $\sim 1\,\text{TeV}$, the upper limit suggested by naturalness.

If R-parity is conserved, every SUSY event contains two $\tilde{\chi}_1^0$, which are neutral and escape the detector. Hence no masses can be reconstructed directly. However, it is still possible to make precision measurements of combinations of masses [20]. The strategy for doing so is to start with characteristic decay at bottom of the decay chain and work backwards. What measurements are available is very model dependent, so the analysis has to be carried out separately for each point in parameter space.

Some examples of precision measurements at one of the LHC points were described by Hinchliffe [21] in these Proceedings. As a second example, consider LHC Point 5, a SUGRA point with $m_0 = 100\,\text{GeV}$, $m_{1/2} = 300\,\text{GeV}$, $A_0 = 300\,\text{GeV}$, $\tan\beta = 2.1$ and $\text{sgn}\,\mu = +$. These parameters were chosen so that the $\tilde{\chi}_1^0$ provides cosmologically acceptable cold dark matter. Some representative masses and decay modes are listed in Table 2.

For this analysis signal and background events were generated with ISAJET 7.22 [22]. Events were selected with four jets, one with $p_T > 100\,\text{GeV}$ and the rest with $p_T > 50\,\text{GeV}$, an effective mass

$$M_{\text{eff}} \equiv \not{E}_T + p_{T1} + p_{T2} + p_{T3} + p_{T4} > 800\,\text{GeV}$$

and $\not{E}_T > 0.2 M_{\text{eff}}$. These cuts are sufficient to make the signal dominate all Standard Model backgrounds. Then the mass of pairs of jets tagged as b's with $p_T > 25\,\text{GeV}$ and $\eta < 2$ was plotted, assuming a tagging efficiency of 60%. The result, Figure 7, shows a peak at the Higgs mass with a substantial background from SUSY and a small Standard Model background. This signal is much easier to detect than the $h \to \gamma\gamma$ signal discussed in the previous section.

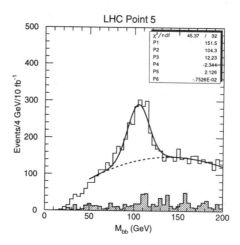

FIGURE 7. Masses of b-jet pairs at Point 5 with cuts described in the text plus a Gaussian plus polynomial fit. The Standard Model background is shown by the shaded histogram [20].

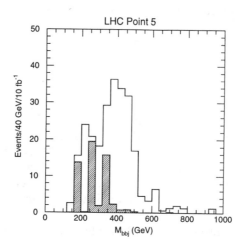

FIGURE 8. Smaller of two masses formed from $b\bar{b}$ pair from Figure 7 plus additional jet [20].

Gluino pair production is large at this point. If one gluino decays via $\tilde{g} \to \tilde{q}_L \bar{q} \to \tilde{\chi}_2^0 q \bar{q} \to \tilde{\chi}_1^0 h q \bar{q}$ and the other via $\tilde{g} \to \tilde{q}_R \bar{q} \to \tilde{\chi}_1^0 q \bar{q}$, the event will contain an $h \to b\bar{b}$, two additional hard jets, and some soft jets. The smaller of the two $b\bar{b}q$ masses should be less than the kinematic limit for the q_L decay,

$$(M_{hq}^{\max})^2 = M_h^2 + \left(M_{\tilde{q}}^2 - M_{\tilde{\chi}_2^0}^2\right) \times$$

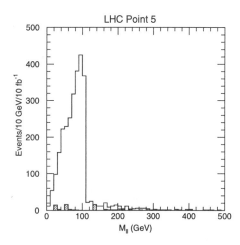

FIGURE 9. Opposite-sign, same-flavor dilepton mass distribution at Point 5 [20].

$$\left[\frac{M_{\tilde{\chi}_2^0}^2 + M_h^2 - M_{\tilde{\chi}_1^0}^2 + \sqrt{(M_{\tilde{\chi}_2^0}^2 - M_h^2 - M_{\tilde{\chi}_1^0}^2)^2 - 4M_h^2 M_{\tilde{\chi}_1^0}^2}}{2M_{\tilde{\chi}_2^0}}\right],$$

namely $M_{hq}^{max} = 506\,\text{GeV}$. This distribution is shown in Figure 8 and has an endpoint at about the right point. This endpoint could be measured to about $\pm 40\,\text{GeV}$, and would constrain the combination of the squark mass and other masses to that error.

The decay $\tilde{\chi}_2^0 \to \tilde{\ell}_R^\pm \ell^\mp$ produces dileptons at this point. Events were selected with four jets, two leptons, and $\not{E}_T$. The opposite-sign, same-flavor dilepton mass distribution is shown in Figure 9. This sharp endpoint cannot result from direct $\tilde{\chi}_2^0 \to \tilde{\chi}_1^0 \ell^+ \ell^-$, since this would have to compete with 2-body $\tilde{\chi}_2^0 \to \tilde{\chi}_1^0 h$. Hence it must be from $\tilde{\chi}_2^0 \to \tilde{\ell}\ell$. Therefore, the endpoint measures

$$M_{\ell\ell}^{max} = M_{\tilde{\chi}_2^0}\sqrt{1 - \frac{M_{\tilde{\ell}}^2}{M_{\tilde{\chi}_2^0}^2}}\sqrt{1 - \frac{M_{\tilde{\chi}_1^0}^2}{M_{\tilde{\ell}}^2}}.$$

The error would be of order 1 GeV. Additional information on these masses can be obtained by looking at the ratio p_{T2}/p_{T1} of the two lepton transverse momenta.

If a SUSY signal is observed at LHC, one will clearly fit a wide range of SUSY models to all available distributions to determine the allowed models and parameters. As a first attempt at this program, the combinations of masses determined from precision measurements of the sort described above have been fit to the SUGRA model. For Point 5, the inputs to this fit were

- $M_{\tilde{\chi}_2}\sqrt{1 - M_{\tilde{\ell}}^2/M_{\tilde{\chi}_1^0}^2}\sqrt{1 - M_{\tilde{\chi}_1^0}^2/M_{\tilde{\ell}}^2} = 108.6 \pm 1$ GeV,

TABLE 3. Fits for Point 5 SUGRA parameters.

	Input (GeV)	Fit 1 (GeV)	Fit 2 (GeV)
m_0	100	100.5^{+12}_{-8}	91 ± 3
$m_{1/2}$	300	298^{+16}_{-9}	288 ± 18
$\tan\beta$	2.1	$1.8^{+0.3}_{-0.5}$	3.1 ± 0.2
sgn μ	+	+	−

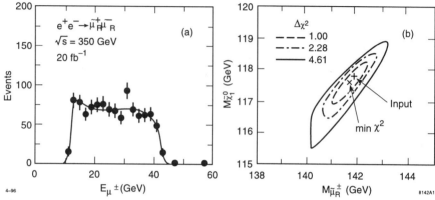

FIGURE 10. μ energy distribution from $\tilde{\mu}$ events and Standard Model background at NLC [16].

- $\tilde{g} \to \tilde{t}t$ is allowed,
- $M_{hj}^{\max} = 506 \pm 40$ GeV,
- $M_h = 104.15 \pm 3$ GeV.

A brute force scan found two solutions, shown in Table 3. A_0 poorly constrained and is not listed. Many masses relevant to LHC physics are quite similar for the two solutions, but the $\tilde{t}$, heavy gaugino, and heavy Higgs sectors are different. The different $\tilde{t}_1$ mass changes the $\tilde{g} \to \tilde{t}_1 t$ branching ratio substantially. It is possible to reconstruct $t \to q\bar{q}b$ decays in these events, determining the background using sidebands and mixed events. The number of t decays selected the sgn $\mu = +$ solution.

The study of SUSY at the NLC is much simpler: one can generally find everything whose production is kinematically allowed. The only question is whether the NLC will have sufficient energy. It needs 1–1.5 TeV to have the same discovery reach in the SUGRA parameter space as the LHC.

To offset its more limited reach, the NLC has several additional handles that are very useful for studying SUSY [15]. First, SUSY particles will be pair produced with the beam energy, providing an additional kinematic constraint that allows determination of the $\tilde{\chi}_1^0$ mass without model assumptions. Second, the beam energy can be varied to study each SUSY threshold sep-

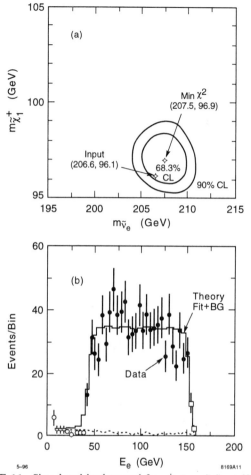

FIGURE 11. Signal and background for $e^+e^- \to \tilde{\nu}_e\tilde{\nu}_e$ at NLC [15].

arately. It is much easier to separate different SUSY processes than at the LHC, where the main background for SUSY is SUSY. Finally, the polarized electron beam is valuable both for suppressing Standard Model backgrounds and for understanding the nature of the signal.

The beam energy constraint allows on to select a particular two-body decay and then use both endpoints of the energy distribution to determine two masses. The simplest example of this is

$$e^+e^- \to \tilde{\mu}^+\tilde{\mu}^- \to \tilde{\chi}_1^0\mu^+\tilde{\chi}_1^0\mu^-.$$

Events with two acollinear muons were selected, and the E_μ distribution was plotted in Figure 10. The two endpoints determine both $M_{\tilde{\mu}}$ and $M_{\tilde{\chi}_1^0}$ with very good precision.

As a more complex example of the same strategy, assume 95% left polarization to enhance $\tilde{\nu}_e$ and select events consistent with

$$e^+e^- \to \tilde{\nu}_e\tilde{\nu}_e \to e^-\tilde{\chi}_1^+ e^+ \tilde{\chi}_1^- \to e^-\mu^+\nu\tilde{\chi}_1^0 e^+ q\bar{q}\tilde{\chi}_1^0$$

There is very little Standard Model background. The distribution of E_e determines both $M(\tilde{\nu}_e)$ and $M(\tilde{\chi}_1^+)$, as can be seen in Figure 11. Many other similar examples have been studied.

The greatest strength of the NLC is probably in the study of sleptons, which have small direct production cross sections at LHC. If SUGRA is to account for the cold dark matter needed for cosmology, then sleptons probably are fairly light. Studies of sleptons at NLC can be used [23], e.g., to determine $\tan\beta$ directly from the model independent relation

$$M_{\tilde{\nu}_L}^2 - M_{\tilde{e}_L}^2 = M_W^2 \cos(2\beta).$$

They can also be used to check the unification of masses via

$$M_{\tilde{l}_L}^2 - M_{\tilde{l}_R}^2 = (0.5 M_2)^2.$$

In summary, the LHC should discover SUSY if it exists at the electroweak scale. It can make many measurements, including some precision measurements of combinations of masses that may in favorable cases be sufficient to determine the model completely. The NLC could provide additional precision measurements of the lighter SUSY particles. To be optimally useful, however, the NLC would need to have enough energy to produce the heavy gauginos and Higgs bosons, since these are generally difficult to study at LHC. This requires pair producing particles with masses comparable to the squarks and gluino.

REFERENCES

1. For a review see W.J. Marciano, these proceedings.
2. Compare with E. Eichten, I. Hinchliffe, K. Lane, and C. Quigg, Rev. Mod. Phys. **56**, 579 (1984).
3. For a review see H. Baer, et al., hep-ph/9503479 (1995).
4. For reviews see, H.P. Nilles, Phys. Rep. **111**, 1 (1984);
   H.E. Haber and G.L. Kane, Phys. Rep. **117**, 75 (1985).
5. U. Amaldi, et al., Phys. Rev. **D36**, 1385 (1987)
6. A. Hasenfratz, *Quantum Fields on the Computer*, (World Scientific, Singapore, 1992), ed. M. Creutz.
7. B.Lee, C. Quigg, and H. Thacker, Phys. Rev. **D16** (1977) 1519;
   D. Dicus and V. Mathur, Phys. Rev. **D7** (1973) 3111.
8. N.Cabibbo, L. Maiani, G. Parisi, and R. Petronzio, Nucl. Phys. **B158** (1979) 295.

9. M. Sher, Phys. Lett. **B317** (1993) 159; **B331** (1994) 448.
10. P. Chankowski, S. Pokorski, and J. Rosiek, Phys. Lett. **B274** (1992) 191;
    J. Espinosa and M. Quiros, Phys. Lett. **BB267** (1991) 27;
    H. Haber and R. Hempfling, Phys. Rev. **D48** (1993) 4280;
    J. Ellis, G. Ridolfi, and F. Zwirner, Phys. Lett. **B257** (1991) 83.
11. J.F. Gunion, H.E. Haber, G. Kane, and S. Dawson, *The Higgs Hunter's Guide*, ISBN 0-201-50935-0 (1990).
12. ATLAS Collaboration, *Technical Proposal*, LHCC/P2 (1994).
13. CMS Collaboration, *Technical Proposal*, LHCC/P1 (1994).
14. M. Carena, P. Zerwas, *et al.*, *Higgs Physics*, hep-ph/9602250.
15. S. Kuhlman, et al., *Physics and Technology of the Next Linear Collider*, BNL-52502 (1996).
16. H. Murayama and M. Peskin, Ann. Rev. Nucl. Part. Sci. **46**, 533 (1996).
17. M.D. Hildreth, T.L. Barklow, D.L. Burke Phys. Rev. **D49**, 3441 (1994).
18. L. Alvarez-Gaume, J. Polchinski and M.B. Wise, Nucl. Phys. **B221**, 495 (1983);
    L. Ibañez, Phys. Lett. **118B**, 73 (1982);
    J.Ellis, D.V. Nanopolous and K. Tamvakis, Phys. Lett. **121B**, 123 (1983);
    K. Inoue *et al.* Prog. Theor. Phys. **68**, 927 (1982);
    A.H. Chamseddine, R. Arnowitt, and P. Nath, Phys. Rev. Lett., **49**, 970 (1982).
19. H. Baer, C.-H. Chen, F. Paige, and X. Tata, Phys. Rev. **D53**, 6241 (1996).
20. I. Hinchliffe, F.E. Paige, M.D. Shapiro, J. Soderqvist, and W. Yao, LBL-39412, hep-ph/9610544 (1996).
21. I. Hinchliffe, these proceedings.
22. Howard Baer, F.E. Paige, S.D. Protopopescu, and X. Tata, hep-ph/9305342 (1993).
23. M. Peskin, Prog. Theor. Phys. Suppl. **123**, 507 (1996).

The TESLA Superconducting Linear Collider

R. Brinkmann for the TESLA Collaboration
DESY, Notkestr. 85, D-22603 Hamburg, Germany

Abstract

This paper summarizes the present status of the studies for a superconducting Linear Collider (TESLA).

1 INTRODUCTION

As it has been discussed at many conferences, workshops and symposia in the past years, studying electron-positron collisions at energies well beyond the reach of LEP ($E_{cm} \approx 200\,\text{GeV}$) has a high potential for the future development of Particle Physics, in many respects complementary to the LHC (see the contributions [1, 2] to this Symposium). There is broad agreement in the High Energy Physics community that a linear e^+e^- collider with an initial center of mass energy of 350...500 GeV and a luminosity above $10^{33}\,\text{cm}^{-2}\text{s}^{-1}$ should be built as the next accelerator facility.

The feasibility of a linear collider has been demonstrated by the successful operation of the SLC, the only existing facility of this kind. Nevertheless, meeting the requirements for a next generation linear collider is by no means an easy task. In particular, high beam powers and very small spot sizes at the collision point are needed in order to obtain a sufficiently high luminosity. Several groups worldwide are pursuing different linear collider design efforts [3]. The fundamental difference of the TESLA approach compared to other designs is the choice of superconducting accelerating structures. The challenge of pushing the superconducting linac technology to a high accelerating gradient and at the same time reducing the cost per unit length, both necessary in order to be competitive with conventional approaches, is considerable, but the advantages connected with this technology (as summarized below) are significant and we are convinced that the potential for the machine performance is unrivaled by other concepts.

TESLA uses 9-cell Niobium cavities (Fig. 1) cooled by superfluid Helium to T=2K and operating at L-band frequency (1.3 GHz). The design gradient for a 500 GeV collider is g=25 MV/m with an unloaded quality factor of $Q_0 =$

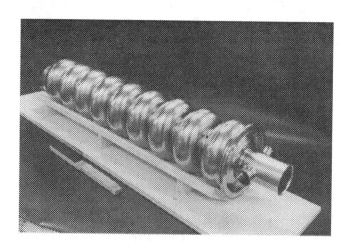

Figure 1: The 9-cell Niobium cavity for TESLA.

Table 1: Parameters for different 500 GeV linear collider designs. A more detailed discussion of the TESLA parameters is given in section 2.

	S-band	X-band	Two-beam	TESLA
Accelerating gradient [MV/m]	17	29...55	78	25
RF-frequency [GHz]	3	11.4	30	1.3
RF-peak power [MW/m]	12	50...100	144	0.2
Av. beam power [MW]	7.3	4	4	8.2
Beam pulse length [μs]	2	0.1	0.006	800
Bunch spacing [ns]	6	1.4	0.67	708
Vertical spot size at IP [nm]	15	5	7	19
Luminosity [10^{33}cm^{-2}s^{-1}]	5	6	5	6

$5 \cdot 10^9$. This technology provides several important advantages for the design of a linear collider. The power dissipation in the cavity walls is extremely small which allows to produce the accelerating field with long, low peak power RF-pulses and yields a high transfer efficiency of RF-power to the beam. With a high average beam power, the required luminosity can be achieved with a spot size at the interaction point (IP for short) only moderately (about a factor of 3.5) smaller than what has been achieved at the Final Focus Test Beam experiment performed at the SLAC linac [4]. At the same time the AC power consumption remains within acceptable limits (100 MW). The long RF-pulse allows for a large bunch spacing (see Table 1), making it easy for the experiment to resolve single bunch crossings. In addition, a fast bunch-to-bunch feedback can be used to stabilise the orbit within one beam pulse, which makes TESLA practically immune to mechanical vibrations which could otherwise lead to serious luminosity reduction via dilution of the spot size and separation of the beams at the IP. Further benefits of the long pulse are the possibility to use a head-on collision scheme with large-aperture superconducting quadrupoles in the interaction region and to employ a safety system which can "turn off" the beam within one pulse in case an emergency is indicated by enhanced loss rates.

The choice of a low drive frequency for TESLA results in very small transverse and longitudinal wakefields in the accelerating structures. This leads to relatively relaxed alignment tolerances for the linac components required for the transportation of a low emittance beam. It is instructive here to compare the different linear collider design concepts on a basis of simple scaling arguments [5]. One of the most essential contributions to emittance dilution results from short-range transverse wakefields due to random offsets of the accelerating structures with respect to the beam orbit. The emittance dilution from this effect can be written as

$$\frac{\Delta \epsilon}{\epsilon} \propto F \cdot \bar{\beta} \cdot \delta y_c^2 \qquad (1)$$

where $\bar{\beta}$ denotes the average β-function in the linac (the stronger the focussing, the smaller $\bar{\beta}$), δy_c the rms-offset of the structures and F the dilution factor which depends on the beam parameters and very strongly on the linac frequency f_{RF}:

$$F \propto \frac{N_e^2 \sigma_z f_{RF}^6}{g^2 \epsilon_y} \qquad (2)$$

The considerable variation of F for different linear collider designs is shown in Fig. 2. It becomes clear that TESLA can afford very much relaxed requirements for the alignment tolerances (e.g. 0.5 mm (rms) for the accelerating structures compared to 50 μm for the S-Band and 10 μm for the X-Band designs, respectively) and for the beam optics. We consider it a crucial point to have a large safety margin concerning beam stability in order to be able to guarantee efficient and stable operation of such a future facility. In addition, part of the safety margin can be used to upgrade the machine performance beyond the "standard" requirements for the next generation linear collider.

In order to demonstrate the feasibility of the s.c. cavity technology, the TESLA collaboration decided to start an R&D program and to build the TESLA Test Facility (TTF) several years ago [6]. In the formative stages of the TESLA collaboration, three 5-cell L-band cavities were built and tested to reach accelerating fields of 26-28 MV/m [7]. The TTF [8] includes the infrastructure for applying different processing techniques to the Niobium cavities obtained from industrial series production. A test linac is under construction incorporating 32 cavities in order to demonstrate acceleration of a beam up to 800 MeV. Operation with an RF- and beam-pulse structure

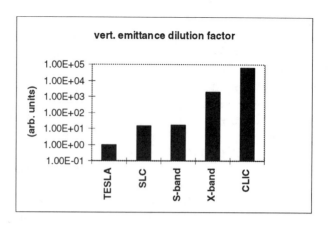

Figure 2: Wakefield emittance dilution factor for different 500 GeV linear collider designs. The SLC is included for comparison.

as in the linear collider design is foreseen, so that a full integrated system test relevant for TESLA-500 is possible. While construction work on the TTF linac and commissioning of its first stage is in progress, several important results have already been obtained:

- Several of the cavities have reached or even surpassed the TESLA goal, see Fig. 3 for an example.

- Field emission has been overcome by handling the cavities in a dust-free environment and by proper treatment.

- Gradients and quality factors achieved in horizontal tests were similar to those achieved in vertical tests.

While demonstration of a successfully working superconducting linac system is the primary goal at the TTF, its construction will also provide a sound basis for a cost analysis. Studies of cost effective component design are under way and the preparation of a detailed cost model for all major components and sub-systems of TESLA will be the next step of our design work. There are, however, already important ingredients for a substantial cost reduction in the design of the systems. One measure is to build cavities with a larger number of cells than done before (9 cells instead of typically 4-5 cells). Longer structures with higher number of cells lead to a reduced number of input couplers, HOM-couplers, tuning devices, RF-waveguides, etc. Building long cryostats with many cavities (there are 8 cavities per cryostat in the TTF design) also contributes to a reduction of the system costs. A large number of cryostats is grouped together in a 2.5 km long cryogenic unit

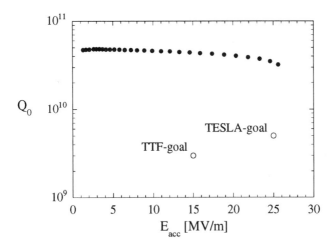

Figure 3: Quality factor vs. accelerating gradient for one of the 9-cell TESLA cavities tested with CW-RF in a vertical cryostat. The initial goal for the TTF and the design goal for TESLA are indicated.

for the TESLA linac. The helium distribution system is incorporated in the cryostats with only one feedbox every 2.5 km. The number of cold-to-warm transitions, which are costly because of the penetration of the heat shields and the vacuum vessel and which also contribute to the heat load, is reduced to a minimum.

X-ray FEL and Further Options

Due to its ability to sustain highest beam quality during acceleration, the TESLA linac is the ideal driver for an X-ray Free Electron Laser (FEL). Such a device is now considered by the major part of the synchrotron radiation community as a truely new generation of X-ray sources. It is explained in [9] how an X-ray FEL can be integrated into TESLA, allowing to operate this facility in parallel with the linear collider.

The X-ray FEL concept represents a considerable extrapolation of present day FEL technology. It has therefore been decided to perform a proof-of-principle experiment at the TTF [10]. After a successful test at about 50 nm wavelength, an energy upgrade of the TTF will open up the possibility to build a soft X-ray FEL user facility. Thus operational and scientific experience can be gained which seems indispensable for the construction of a large-scale X-ray FEL laboratory. Furthermore, a long-term adequate usage of the TTF will be provided.

If the linear collider is constructed close to an existing large proton storage ring (as the HERA-p ring or the TEVATRON), a linac-ring electron-proton collider with a center-of-mass energy of about 1 TeV is a possible option. There is also the possibility to convert the electron beam into a γ-beam by compton scattering, thus obtaining γ-p collisions. The most serious issue here concerns the achievable luminosity (see e.g. refs. [13, 14, 15]). More detailed studies are required before a conclusion concerning the feasibility of a linac-ring e-p collider with reasonably high luminosity can be drawn.

In case TESLA is constructed at the DESY site, an additional option becomes attractive. The HERA electron ring can be used as a pulse stretcher to deliver a continuous electron beam of 15...25 GeV for Nuclear Physics experiments [11]. The ring is filled by additional beam pulses generated in the lower energy part of one of the TESLA linacs. The additional investment required for this option would be minimal, still the quality of the beam delivered from the stretcher ring being comparable to the original ELFE proposal [12] for a continuous beam facility.

2 PARAMETERS AT 500 GeV

The second key parameter for a linear collider, besides the center-of-mass energy of the colliding beams, is the luminosity L. It is given by

$$L = \frac{n_b N_e^2 f_{rep}}{4\pi \sigma_x^* \sigma_y^*} \times H_D \qquad (3)$$

n_b number of bunches per pulse
N_e number of electrons (positrons) per bunch
f_{rep} pulse repetition frequency
$\sigma_{x,y}^*$ horizontal (vertical) beam size at interaction point
H_D disruption enhancement factor (typically $H_D \approx 1.5$)

Introducing the average beam power $P_b = E_{cm} n_b N_e f_{rep}$, the luminosity can be written as

$$L = \frac{P_b}{E_{cm}} \times \frac{N_e}{4\pi \sigma_x^* \sigma_y^*} \times H_D \qquad (4)$$

An important constraint on the choice of interaction parameters is due to the effect of beamstrahlung: the particles emit hard synchrotron radiation in the strong space-charge field of the opposing bunch. The average fractional beam energy loss from beamstrahlung is approximately given by [18]:

$$\delta_E \approx 0.86 \frac{r_e^3 N_e^2 \gamma}{\sigma_z (\sigma_x^* + \sigma_y^*)^2} \qquad (5)$$

r_e classical electron radius
γ relativistic factor $E_{beam}/m_0 c^2$

Beamstrahlung causes a reduction and a spread of the collision energy and can lead to undesirable experimental background. The energy loss δ_E therefore has to be limited to typically a few percent. By choosing a large aspect ratio $R = \sigma_x^*/\sigma_y^* \gg 1$, δ_E becomes independent of the vertical beam size and the luminosity can be increased by making σ_y^* as small as possible. Since $\sigma_y^* = (\epsilon_{y,N}\beta_y^*/\gamma)^{1/2}$, this is achieved by both a small vertical beta function at the IP and a small normalized vertical emittance. The lower limit on β_y^* is given by the bunch length ("hourglass effect"). Setting $\beta_y^* = \sigma_z$, the luminosity can be expressed as:

$$L \approx 5.74 \cdot 10^{20} \mathrm{m}^{-3/2} \times \frac{P_b}{E_{cm}} \times \left(\frac{\delta_E}{\epsilon_{y,N}}\right)^{1/2} \times H_D \qquad (6)$$

Due to the small wakefields, the superconducting linac is ideal for transporting a beam with an extremely small emittance $\epsilon_{y,N}$. In order to define a standard parameter set for the 500 GeV collider, we do not have to make full use of this potential, though. The high AC-to-beam power conversion efficiency of TESLA of about 16% allows to produce a high beam power and to obtain the required luminosity with a total AC-power consumption of 99 MW and a normalized emittance achievable with tolerances (between 0.1 and 0.5 mm) in the linac and in the damping rings which can realistically be obtained at the time of machine installation. The additional use of beam-based correction and alignment methods will then allow to push $\epsilon_{y,N}$ to much smaller values, opening up options for upgrading the machine. The other basic parameters, not directly entering into eq. 6, have been chosen such that freedom in parameter space is maintained, not being at a technical limit in any of the collider subsystems. For instance, the large bunch spacing (708 ns) leaves a considerable safety margin for technical components like the damping ring injection and extraction devices, bunch-to-bunch orbit measurement and correction, and for resolving single bunch crossings in the experiment. As another example, the relatively large horizontal emittance ($\epsilon_{x,N} = 1.4 \cdot 10^{-5}$ m) relaxes the beam optics requirements in the damping rings. An overview of the TESLA standard parameters is given in Table 2.

In Table 2 we also show a possible low-ϵ_y parameter set, in order to give an example for the large flexibility in parameter space available for TESLA. The version shown here combines a very low level of beamstrahlung with a somewhat higher luminosity and reduced AC-power. Such a parameter set with low δ_E would e.g. be desirable for operation around the top quark threshold (at $E_{cm} \approx 350$ GeV) or for a polarized positron source. Obtaining the small emittance will require to apply beam-based techniques, especially to improve the accuracy of the beam position monitor alignment w.r.t. the focussing magnets. The necessary accuracy (several 10's of μm, both in the linac and the damping rings) is easy to achieve and, once established,

Table 2: TESLA parameters for E_{cm}=500 GeV. The machine length includes a 2% overhead for energy management. The klystron power and AC power quoted include a 10% regulation reserve. The standard parameter set is the basis for the design study. The additional parameter set (right-hand column) is given as an example for the potential of TESLA operating with a lower beam emittance.

	Standard	Reduced ϵ_y
Accelerating gradient g [MV/m]	25	25
RF-frequency f_{RF} [GHz]	1.3	1.3
Total site length L_{tot} [km]	33	33
Active length [km]	20.4	20.4
# of acc. structures	19712	19712
# of klystrons	616	616
Repetition rate f_{rep} [Hz]	5	4
Beam pulse length T_P [μs]	800	800
RF-pulse length T_{RF} [μs]	1330	1330
# of bunches p. pulse n_b	1130	2260
Bunch spacing Δt_b [ns]	708	354
Charge p. bunch N_e [10^{10}]	3.63	1.82
Emittance at IP $\gamma\epsilon_{x,y}$ [10^{-6}m]	14, 0.25	12, 0.025
Beta at IP $\beta^*_{x,y}$ [mm]	25, 0.7	25, 0.5
Beam size at IP $\sigma^*_{x,y}$ [nm]	845, 19	783, 5.1
Bunch length at IP σ_z [mm]	0.7	0.5
Beamstrahlung δ_E [%]	2.5	1.0
Luminosity L [10^{33}cm^{-2}s^{-1}]	6	8.4
Beam power P_b [MW]	16.3	13
AC power P_{AC} [MW]	99	83

expected to be stable over long periods of time.

3 GENERAL LAYOUT

A sketch of the overall layout of the TESLA linear collider is given in Fig. 4. We start the description of the various subsystems of the machine at the beam sources.

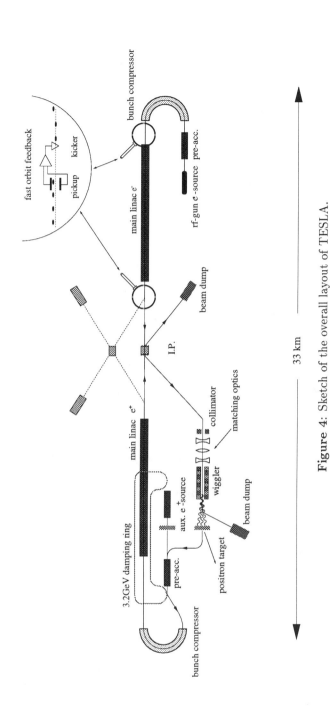

Figure 4: Sketch of the overall layout of TESLA.

3.1 Injection System

With the relatively large design beam emittance, it may be possible to obtain the required electron beam quality directly from a laser-driven RF gun. This design aspect is not fundamental but considered as a possible low-cost version of the electron injection system for an initial stage of operation. An additional damping ring, similar to the one on the positron side (see below), is likely to be required in order to exploit the full potential of the collider with smaller beam emittance. A polarized electron source with a larger emittance is used in that case. The requirements for longitudinal bunch compression are rather moderate and can be met with a single stage compressor. After the bunch compressor, injection into the main linear accelerator takes place at a beam energy of 3.2 GeV.

The positron injection system has to provide a total charge of about $4 \cdot 10^{13} e^+$ per beam pulse, which does not seem feasible with a conventional source. The method considered here is to produce the positrons from γ-conversion in a thin target [16, 17]. The photons are produced by passing the spent high-energy electron beam after the IP through a wiggler. This requires a special beam-optical system and various collimators in order to deliver an electron beam of sufficiently small size to the wiggler. After the wiggler the electron beam is deflected, bypasses the photon target and is sent on a dump. The positrons are preaccelerated in a conventional 200 MeV L-band linac, followed by a 3 GeV superconducting accelerator. The method proposed here has a lower limit for the electron beam energy required to generate photons of sufficiently high energy. This limit is at about 150 GeV, which puts a lower boundary on the center-of-mass energy accessible with TESLA at maximum luminosity. In case lower energy running at high luminosity becomes important, solutions with additional electron beam pulses and bypass-beamlines are conceivable.

A low-intensity auxiliary e^+ source will be needed for commissioning and machine study purposes. We assume that the auxiliary source should be capable of generating single bunches at full design intensity and bunch trains at a few per cent of the design intensity.

Changes of the beam parameters at the IP can introduce fluctuations of the positron production efficiency, which in turn feed back to the spent beam properties via the collision at the next pulse. It has been checked that the coupled dynamical system is damped and stable and that the expected fluctuations are acceptable.

Besides providing a sufficiently high positron beam intensity, this concept offers additional advantages. With a thin target, the positron beam tends to have a smaller transverse emittance than from a conventional source. The considerable investment and operating costs for a high-power electron linac needed in a conventional scheme are avoided. Furthermore, it is conceivable to obtain a polarized positron beam by using a helical undulator instead of a wiggler. This option is more ambitious concerning the required quality of the spent beam passing through the undulator.

The positron beam is injected into the damping ring at an energy of 3.2 GeV. The bunch train is stored in the ring in a compressed mode, with the bunch spacing reduced by about a factor of 14. Still a large ring circumference of about 17 km is required. We choose a damping ring design (see Fig. 5) here which avoids having to build an additional ring tunnel of this size. The layout has two 8 km straight sections, entirely placed in the main linac

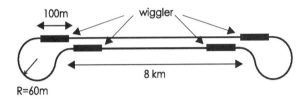

Figure 5: General layout of the "dogbone"-shaped damping ring

tunnel. Additional tunnels are only required for the small loops at the ends. Despite its unconventional shape (it has been named the "dogbone ring"), no particular difficulties concerning single particle and collective beam dynamics were found. A 400 m long wiggler section is foreseen to achieve sufficient damping. Special kicker devices are required for compression and decompression of the bunch train at injection and extraction, respectively.

3.2 Main Linac

The two main linear accelerators comprise about ten thousand one meter long superconducting cavities each. Groups of 8 cavities are installed in a common cryostat, a cost-efficient modular solution which is already applied at the TTF linac. Superconducting magnets for beam focussing and steering, beam position monitors and absorbers for higher order modes induced by the beam are integrated in the modules.

The RF-power is generated by some 300 klystrons per linac, each feeding 32 9-cell cavities. The required peak power per klystron is 8 MW, which includes a margin for correcting phase errors which occur during the beam pulse due to mechanical deformations of the cavities (Lorentz force detuning, microphonics). A multi-beam klystron with low perveance is presently under construction, which will be able to deliver up to 10 MW peak power at a high efficiency of 70%. The high-voltage pulses for the klystrons are provided by conventional type modulators, or, as an alternative option presently under study, in "SMES" devices where the pulse energy is stored as magnetic field energy in superconducting solenoids.

The cryogenic system of the TESLA linac is divided into 2.5 km long subsections, each supplied by a cryogenic plant. The total cooling power required is 27 kW at 2 K, 34 kW at 4 K and 260 kW at 70 K, respectively.

3.3 Beam Delivery System

The beam transport between the linac and the IP (so-called beam delivery system) consists of collimation, beam diagnostics and correction, and final focus sections (see Fig. 6).

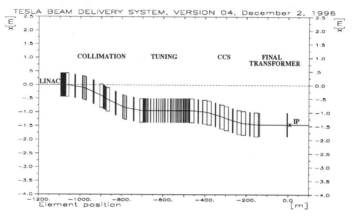

Figure 6: Horizontal survey of the beam delivery system.

In the interaction region, no crossing angle is required, because the beams can be separated by an electrostatic deflector well before the first parasitic interaction with the subsequent bunch would occur (at 150 m from the IP). Large aperture superconducting quadrupoles can be used in the IR, with one benefit being that collimation requirements upstream for protection of the experiment from background are rather relaxed. With gas scattering practically absent in the TESLA linac and wakefields being very small, the expected amount of beam halo which must be collimated is small so that background from muons originating at the collimators is unlikely to be a problem. In case the loss rates at the collimators exceed an acceptable limit (e.g. due to mis-steering or possible failures upstream), the large bunch spacing allows to trigger a safety system which sends the remaining bunch train to a beamdump.

The concept of the final focus sytem is essentially the same as for the successfully tested FFTB system at SLAC. Beam size demagnification and chromatic corrections for the TESLA design parameters are even somewhat less ambitious than at the FFTB. The beams can be kept in collision at the IP with high accuracy by using a fast bunch-to-bunch feedback which measures the beam-beam deflection and steers the beams back into head-on collision within a time small compared to the beam pulse length. A similar system is foreseen at the entrance to the main linac and at its end, removing possible pulse-to-pulse orbit jitter generated in the injection system or in the linac.

The layout of the beam delivery system chosen here is optimized for a single interaction point. It can be entirely accomodated in a straight tunnel and is kept as simple as possible, with benefits for the required tunnel length (about 2.3 km for the total system between the two linacs) and for facilitating commissioning and operation of the machine. There is no fundamental problem arranging a 2nd IP by adding another beam line as sketched in Fig. 4. The detailed geometry and beam optics will depend on whether the 2nd experiment is to study e^+e^- collisions as well, or is aiming at the investigation of $\gamma\gamma$ and $e\gamma$ collisions.

3.4 Tunnel Layout

The two linear accelerators as well as the beam delivery system will be installed in an underground tunnel of 5 m diameter, see Fig. 7. A $30 \times 80 m^2$ experimental hall is foreseen to accomodate the detector, with an option to be extended in case a 2nd separate experiment is to be installed. Six additional surface halls are required at a distance of about 5 km along the linacs, each housing two cryogenic plants. These halls are connected to the underground tunnel by access shafts. They will also contain the modulators which generate the HV pulses for the klystrons. The pulse transformers are placed in the tunnel close to the klystrons and are connected to the modulators by cables. The long cables contribute to a few % of power losses, but it is advantageous that maintenance of the modulators will be possible while the machine is running without the need of access to the tunnel. Exchange of klystrons requires to interrupt operation of the machine. With an energy surplus of 2% foreseen in the design, such maintenance breaks of one day are necessary only every few weeks, assuming an average klystron life time of 40,000 h.

4 UPGRADE TO HIGHER ENERGY

The center-of-mass energy reach of TESLA is clearly limited by the maximum

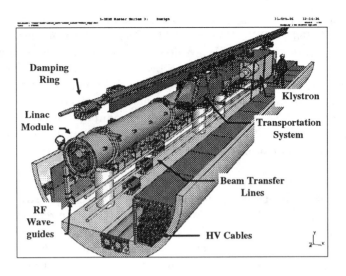

Figure 7: Sketch of the 5 m diameter TESLA linac tunnel.

gradient achievable with the superconducting cavities and cannot, unlike designs using conventional accelerating structures, be increased by upgrading the RF-pulse power. In order to go to energies beyond $E_{cm} \approx 1\,\text{TeV}$, the length of the machine would therefore have to be increased. We do believe, though, that a significant energy upgrade of TESLA will be possible within the site length for the 500 GeV design by operating the cavities at a gradient above 25 MV/m. The main reasons which justify this optimism are:

- The fundamental limit for the gradient in a Nb-structure at 2 K is well above 50 MV/m.

- Cavity tests at the TTF have shown that field emission can be very effectively suppressed and a quality factor far in excess of the design value ($Q_0 = 5 \cdot 10^9$) at 25 MV/m is feasible.

- Accelerating gradients of 30...40 MV/m [19, 20, 21] have been achieved with single-cell L-band cavities.

We are aware that a number of questions have to be adressed before a definite conclusion can be drawn (e.g. stronger Lorentz-force detuning, potentially more severe dark current problems), but with further R&D it seems conceivable that a maximum gradient of about 40 MV/m at $Q_0 = 5 \cdot 10^9$ can be reached with the 9-cell cavities. The average gradient for the entire linac may be lower initially, so that a possible scenario for an energy upgrade in steps

Table 3: TESLA parameters for an upgrade to 800 GeV and for a 2nd stage at 1.6 TeV with doubled linac length (see text).

	800 GeV	800 GeV low ϵ_y	1.6 TeV
Accelerating gradient g [MV/m]	40	40	40
total site length L_{tot} [km]	32	32	≈ 62
# of acc. structures	19712	19712	39424
# of klystrons	1232	1232	2464
repetition rate f_{rep} [Hz]	3	3	3
beam pulse length T_P [μs]	640	640	640
RF-pulse length T_P [μs]	1330	1330	1330
# of bunches p. pulse n_b	1130	2260	2260
bunch spacing Δt_b [ns]	566	283	283
charge p. bunch N_e [10^{10}]	3.63	1.82	1.82
emittance at IP $\gamma\epsilon_{x,y}$ [10^{-6}m]	14, 0.25	12, 0.025	12, 0.025
beta at IP $\beta^*_{x,y}$ [mm]	25, 0.7	25, 0.5	25, 0.5
beam size at IP $\sigma^*_{x,y}$ [nm]	669, 15	618, 4.0	436, 2.8
bunch length at IP σ_z [mm]	0.7	0.5	0.5
beamstrahlung δ_E [%]	5.2	2.2	6.7
luminosity L [10^{33}cm^{-2}s^{-1}]	5.7	11	20
beam power P_b [MW]	15.6	15.6	31.2
AC power P_{AC} [MW]	115	115	230

could be to exchange groups of modules containing the "weakest" (i.e. lowest gradient) cavities with new, high-performance modules. This upgrade path leads to a maximum energy of E_{cm}=800 GeV. The subsystems for TESLA (in particular the beam delivery system) are designed such that the energy upgrade can be accomodated without further hardware modifications. In order to obtain optimum luminosity at higher energy, an upgrade of the RF-system is also necessary, though. At higher gradient, either the RF-pulse power or pulse length must be increased to maintain a reasonable ratio of beam-on time to RF-on time. We follow the first approach, which has the advantage that the modulators providing the pulse power for the klystrons do not have to be modified. The 800 GeV version of TESLA would then have twice as many klystrons and modulators as the initial stage at 500 GeV. The pulse repetition rate is reduced so that the dynamic load for the cryogenic system remains almost constant and the AC-power consumption is only slightly increased. As in the previous section, we show two sets of beam parameters in Table 3. The alignment tolerances in the damping rings and in the main linac are for both versions similar as for the respective parameter sets of TESLA-500.

The potential for operating with a very small vertical beam emittance becomes particulary important when we consider a linear collider at energies beyond 1 TeV. According to eq. 6, scaling the luminosity as $L \propto E^2_{cm}$, as desirable, becomes difficult unless a high level of beamstrahlung is accepted. Assuming that the same vertical emittance as for the 800 GeV low-ϵ_y version can be maintained in a machine doubled in length, we arrive at the parameter set for a 1.6 TeV collider shown in the last column of Table 3. It demonstrates that the TESLA approach has the capability of maintaining reasonable interaction parameters at an energy scale well above 1 TeV. We have not yet investigated this 2nd stage upgrade in detail, but several aspects are already included in the initial stage design. The tunnel is kept straight in the beam delivery section, so that it can later be used for the installation of linac structures. The two linacs of the 1st stage would be used as one linac of the 2nd stage (note that the superconducting cavities allow to accelerate a beam in both directions) and the machine length be extended to one side. Thus the part of the linac being used for the FEL facility and potentially other options would remain unchanged. A new beam delivery system and interaction region would have to be constructed. The entire injection system can be kept basically unchanged, except for modifications concerning the spent beam capture system and an additional transfer line from one of the damping rings to the beginning of the new 800 GeV linac.

References

[1] H. Murayama, e^+e^- *Physics*, contribution to this Symposium.

[2] F. Paige, *Complementarity of Lepton and Hadron Colliders*, contribution to this Symposium.

[3] G. A. Loew (ed.), *International Linear Collider Technical Review Committee Report*, SLAC-R-95-471, 1995.

[4] D. Burke for the FFTB Collaboration, *Results from the Final Focus Test Beam*, Proc. of the IVth European Particle Accelerator Conf., London 1994, Vol. I, p. 23.

[5] R. Brinkmann, *Beam Dynamics in Linear Colliders- What Are the Choices ?*, DESY-M-95-10, 1995.

[6] *Proposal for a TESLA Test Facility*, DESY-TESLA-93-01, 1992.

[7] C. Crawford et al., Particle Accelerators 49, p. 1, 1995.

[8] D. A. Edwards (ed.), *TESLA Test Facility Linac - Design Report*, DESY-TESLA-95-01, 1995.

[9] R. Brinkmann, G. Materlik, J. Roßbach, J. R. Schneider and B. H. Wiik, *An X-Ray FEL Laboratory as Part of a Linear Collider Design*, Proc. Int. Conf.

on Free Electron Lasers, Rome 1996, to be published in Nucl. Instr. Meth., 1996.

[10] J. Roßbach, ed., *A VUV Free Electron Laser at the TESLA Test Facility at DESY*, DESY-TESLA-FEL-95-03, 1995.

[11] R. Brinkmann, P. Bruinsma, J.-M. de Conto, J. Faure, B. Frois, M. Gentner, D. Husmann, R. Kose, J. Maidment, P. Nghiem, J. Payet, F. Tazzioli, A. Tkatchenko, Y. Y. Wu, *ELFE@DESY: An Electron Laboratory for Europe at DESY*, 1997, to be published.

[12] J.M. De Conto (ed.), *Electron Laboratory for Europe - Accelerator Technical Proposal*, 1993.

[13] M. Tigner, B. H. Wiik and F. Willeke, Proc. Particle Accelerator Conf., San Francisco 1991, Vol. 5, p. 2910.

[14] Z. Z. Aydin et al., *HERA+LC Based γp Collider: Luminosity and Physics*, DESY-95-014, 1995.

[15] R. Brinkmann and M. Dohlus, *A Method to Overcome the Bunch Length Limitation on β_p^* for Electron-Proton Colliders*, DESY-M-95-11, 1995.

[16] V. E. Balakin and A. A. Mikhailichenko, *The Conversion System for Obtaining High Polarized Electrons and Positrons*, Preprint INP 79-85, 1979.

[17] K. Flöttmann, *Investigations Toward the Development of Polarized and Unpolarized High Intensity Positron Sources for Linear Colliders*, DESY-93-161, 1993.

[18] P. Chen and K. Yokoya, *Beam-Beam Phenomena in Linear Colliders*, KEK-report 91-2, 1991.

[19] T. Higuchi et al, *Investigation of Barrel Polishing for Superconducting Niobium Cavities*, KEK-report 95-220, 1996.

[20] P. Kneisel, R. W. Röth and H.-G. Kürschner, *Results from a Nearly "Defect-Free" Niobium Cavity*, Proc. 7th Workshop on RF-Superconductivity, p. 449,Saclay 1995.

[21] M. Ono, S. Noguchi, K. Saito, E. Kako, T. Shishido, H. Inoue, Y. Funahashi, T. Fujino, S. Koizumi, T. Higuchi, T. Suzuki, H. Umezawa and K. Takeuchi, *Achivement of High Acceleration Field (40 MV/m) in L-Band Superconducting Cavity at KEK*, Proc. 21st Linear Accelerator Meeting, Tokyo 1996, NUP-A-96-10, p. 38.

Scaling Linear Colliders to 5 TeV and Above*

Perry B. Wilson

Stanford Linear Accelerator Center, Stanford University, P. O. Box 4349, Stanford, CA 94309

Abstract. Detailed designs exist at present for linear colliders in the 0.5–1.0 TeV center-of-mass energy range. For linear colliders driven by discrete rf sources (klystrons), the rf operating frequencies range from 1.3 GHz to 14 GHz, and the unloaded accelerating gradients from 21 MV/m to 100 MV/m. Except for the collider design at 1.3 GHz (TESLA) which uses superconducting accelerating structures, the accelerating gradients vary roughly linearly with the rf frequency. This correlation between gradient and frequency follows from the necessity to keep the ac "wall plug" power within reasonable bounds. For linear colliders at energies of 5 TeV and above, even higher accelerating gradients and rf operating frequencies will be required if both the total machine length and ac power are to be kept within reasonable limits. An rf system for a 5 TeV collider operating at 34 GHz is outlined, and it is shown that there are reasonable candidates for microwave tube sources which, together with rf pulse compression, are capable of supplying the required rf power. Some possibilities for a 15 TeV collider at 91GHz are briefly discussed.

1. INTRODUCTION

Detailed design parameters have been developed for linear colliders in the 0.5–1.0 TeV center-of-mass energy range (see, for example, the report of the International Linear Collider Technical Review Committee (1)). Some basic rf-related parameters for these machines at the 1 TeV design energy are given in Table 1. The designs listed are the SBLC (S-band Linear Collider) at DESY; the JLC (Japan Linear Collider) C-band and X-band designs developed at KEK, Japan; the NLC (Next Linear Collider) at SLAC; and the VLEPP (Russian acronym for colliding linear electron positron beams) at the Branch Institute for Nuclear Physics, Protvino. Parameters for the first (and as yet the only) operating linear collider, the 0.1 TeV SLC machine at SLAC, are shown for comparison. All the machines shown in the table use copper accelerating structures. Not shown is the TESLA linear collider at DESY, which is based on superconducting rf technology. Operating at a frequency of 1.3 GHz and a gradient of 25 MV/m, the TESLA technology is difficult to scale to a higher gradients and energy. Also not shown in Table I are two proposed linear colliders which are based on the two-beam accelerator approach: CLIC at CERN and the Two-Beam NLC proposed by LBNL, Berkeley, and LLNL, Livermore. Although the drive beam of a two-beam accelerator is capable of producing copious amounts of rf power at good efficiency, for the purpose of scaling linear colliders with frequency we consider here only colliders powered by discrete rf sources.

---

*Work supported by the Department of Energy, contract DE-AC03-76SF00515.

Table 1
Linear Collider RF Parameters at 1 TeV Center-of-Mass Energy

	SBLC (DESY)	JLC-C (KEK)	JLC-X (KEK)	NLC (SLAC)	VLEPP (BINP)	SLC (SLAC)
RF Frequency GHz	3.0	5.7	11.4	11.4	14.0	2.856
Accelerating Gradient Unloaded/Loaded (MV/m)	42/36	58/47	73/58	77/64	100/91	21
Dark Current Capture Gradient (MV/m)	16	31	61	61	75	15
Peak Power per Meter at Structure Input (MV/m)	49	97	100	96	120	12
Klystron Peak Power (MW)	150	100	67	75	150	65
Klystron Pulse Length (μs)	2.8	2.4	0.75	1.44	0.5	3.5
Repetition Rate (Hz)	50	50	150	120	300	120
Pulse Compression Type	SLED	SLED	DLDS	BPC	VPM	SLED
Compression Ratio	≈3.5	5.0	3.0	4.0	4.6	4.2
Compression Power Gain	2.0	3.5	2.9	3.5	3.3	2.6
Compression Efficiency (%)	≈60	70	97	87	73	62
RF System Efficiency (%)	23	26	31	40	40	14
Number of Klystrons	4900	6200	9200	6500	2800	240
Active Length (km)	29	22	18	18	12	2×2.85
Wall Plug Power (MW)	285	200	220	180	115	25

The Technical Review Committee report is primarily concerned with the design of linear colliders at 500 GeV c.m. energy; upgrade paths to 1 TeV are only briefly considered. However, we have selected the 1 TeV parameters because particle physicists strongly urge that collider designs should include the potential to reach this energy. Also, it is more likely that the higher-energy designs push the limitations of rf sources at each frequency. If the unloaded accelerating gradients listed in Table 1 are plotted as a function of rf frequency, it will be seen that the gradients vary approximately as $G_0 \sim \omega_{rf}^{0.55}$. For the more detailed designs at 0.5 TeV, it is found that the unloaded gradients vary approximately linearly with rf frequency. There is, of course, a simple physical reason for this correlation with frequency; the stored energy per meter of accelerating structure varies approximately as $(G_0\lambda)^2$, where λ is the rf wavelength. Thus, for a linac which is pulsed at a fixed repetition rate, a higher operating frequency makes it possible to achieve a given final energy with both a shorter accelerator length and a lower ac power. Further details on frequency scaling are given in the following section.

Starting with the NLC 1.0 TeV parameters as a base design, and following (at least approximately) the scaling relations developed in Section 2, rough rf parameters for linear collider designs at 5, 10 and 15 TeV are presented in Section 3. It will be found that the peak rf power per meter of structure varies roughly as $\omega_{rf}^{3/2}$. Because the peak power that can be obtained from discrete rf sources such as klystrons is limited and tends to decrease with increasing rf frequency, there will be some cross-over frequency (not sharply defined) above which rf pulse compression is required to enhance the power available from practical microwave tubes. From the designs as proposed in the Technical Review Committee Report (1), the frequency at which pulse compression is required seems to be at about C-band (5 GHz) and above. Some background on pulse compression methods is given in Section 4, along with a rough design for a 34 GHz pulse compression system for a 5 TeV collider.

Even with the peak power enhancement provided by pulse compression, the power required from a 34 GHz rf source will be in the 100–150 MW range. In Section 5 some limitations are given on the power that can be reached by round-beam klystrons, and how this maximum power output scales with frequency. It will be found that a conventional round-beam klystron operating at a reasonable beam voltage cannot provide 100 MW of power at 34 GHz. Other possible microwave power sources which can provide power at this frequency and this power level are discussed.

The conclusion of Section 5 is that there is a reasonable expectation that, together with rf pulse compression, a microwave tube source can be built which can provide adequate peak power for driving a 34 GHz linear collider at a gradient on the order of 200 MV/m. However, it does not seem reasonable that discrete rf sources can provide the 2400 MW/m required to drive a 91 GHz, 15 TeV collider at 600 MV/m, unless a new scheme for pulse compression or active power switching is developed. Perhaps at this frequency and gradient the two-beam accelerator approach is the best route to rf power production. A possible two-beam, 91 GHz scheme, due to H. Henke (2), is outlined in Section 6. Finally, we note that this paper is concerned with scaling rf parameters in order to obtain higher gradients and hence a higher energy in a reasonable linac length. How the beam parameters must be scaled in order to reach the 5–15 TeV energy range is another story. A number of papers have been written which provide parameter lists for these, and even higher, collider energies. A set of beam parameters specific to the 34 GHz, 5 TeV collider described here is given in (3).

2. SCALING WITH FREQUENCY

Several factors must be taken into account in scaling the rf system of a linear collider to a higher frequency. One of the most important is the threshold for the capture of an electron at rest by a traveling wave with a phase velocity equal to the velocity of light. This "dark current" capture threshold gradient is given by

$$G_{th} = \pi\, mc^2/e\lambda = 1.605 \text{ MV}/\lambda \,. \tag{1}$$

Some linear colliders are designed at gradients that exceed this threshold by a modest factor (30% or so for the 0.5 TeV c.m. designs in the Technical Review Committee report. However, it would be unwise to plan on an operational gradient which is considerably in excess of G_{th} without strong experimental support from measurements at a test facility showing that the dark current level is acceptable.

The peak power required per meter of structure is given by the energy stored per unit length, $u \sim G_0^2\, \lambda^2$, divided by the filling time, $T_f = \tau T_0$, where τ is the attenuation parameter and T_0 is the decrement time, $T_0 = 2Q/\omega \sim \lambda^{3/2}$:

$$P_m \sim G_0^2 \lambda^2/T_f \sim G_0^2 \lambda^{1/2} \quad . \quad (2)$$

The above scaling assumes a constant group velocity and iris aperture, a/λ, relative to the wavelength. If the normalized aperture varies, the above expression should be multiplied by a function $F(a/\lambda)$, where F is approximately proportional to $(v_g/c)^{1/3}$
in the group velocity range of interest for linear colliders, $v_g/c \approx .03$ to 0.10.

Ignoring dark current considerations, the accelerating gradient will eventually be limited by rf breakdown. Some evidence indicates (this is a long story) that the breakdown gradient scales as $\omega^{1/2}/(T_p)^{1/4}$, where T_p is the rf pulse length. Since the pulse length tends to scale in proportion to the structure filling time, $T_p \sim \omega^{-3/2}$, then $G_{br} \sim \omega^{7/8}$. Loew and Wang(4) obtained a breakdown threshold for the peak surface field of 330 Mv/m at 2.856 GHz for a pulse length of several times a typical structure filling time. Using this as a calibration point, the breakdown field limit becomes

$$G_b(Mv/m) \approx 130 \, [f(GHz)]^{7/8} \quad . \quad (3)$$

Using the fact that the peak surface field is on the order of 2.3 times the average accelerating gradient in a typical accelerating structure, the breakdown threshold accelerating gradient will be about 140 MV/m at 2.856 GHz. This is about 9 times the dark current capture gradient. At 91 GHz this ratio is about 6. Thus for any projected linear collider rf frequency, the accelerator will always operate well below the breakdown threshold and operation will be limited to effects related to dark current capture and rf processing.

In addition to the obvious limitation on accelerating gradient imposed by the intense electric field on the copper surface near the iris opening, there may also be a problem due to the pulse heating by the magnetic surface field at the cell walls. The temperature rise at the end of a pulse of duration T_p is

$$\Delta T = (R_s/K) \, (DT_p/\pi)^{1/2} (G/Z_H)^2 \quad (4)$$

Here R_s is the surface resistance, K is the thermal conductivity and D is the thermal diffusivity, given by $D = K/C_v$ where C_v is the specific heat per unit volume. Z_H is an impedance defined as $Z_H \equiv G/H_s$ where H_s is the maximum surface magnetic field anywhere in the structure for a given average (not local) accelerating gradient. For the NLC structure, the minimum value of Z_H is 307 ohms (the maximum value is about 400 ohms). Using $D = 1.15$ cm^2/sec and $K = 3.95$ W/cm/^{0}C for copper, Eq. (5) gives a temperature rise of 16^{0}C for the current 1 TeV NLC parameters ($G_0 = 77$ MV/m and $T_p = 360$ ns). Assuming the pulse length scales as $\omega^{-3/2}$, the temperature rise scales as $\Delta T \sim \lambda^{1/4} G^2$.

There is considerable controversy over the detailed mechanism for surface damage due to pulsed heating, and the resultant effect on structure Q. Rough estimates indicate that the yield strength in copper is exceeded at a pulse temperature rise on the order of 40^0 C, resulting in surface roughening and fatigue cracks (5). However, the extent to which this surface damage might degrade the rf surface resistance is not clear. To help answer these questions, an experiment is underway at SLAC in which a pulse temperature rise of several hundred degrees is applied to a demountable surface in an X-band cavity (6). After measuring possible changes in surface resistance through changes in cavity Q, the test surface can be removed for microscopic inspection of surface damage.

Finally, we consider the effects of these frequency scaling relations on accelerator length and an ac wall plug power. In the next section, we take some linear collider examples in which the accelerating gradient is scaled in proportion to the dark current capture threshold. The active structure length then scales as $L_A \sim E_{c.m.} \lambda$. In scaling the ac power, we consider a rf pulse repetition rate which is fixed at 120 Hz. The ac power (and luminosity) can always be decreased by lowering the repetition rate. However, as the repetition rate decreases, pulse-to-pulse feedback against the effects of the high frequency component of ground motion becomes more difficult. For a fixed repetition rate, the ac power scales as

$$P_{ac} \sim G \lambda^2 E_{c.m.} / \eta_{rf}, \quad (5a)$$

where η_{rf} is the efficiency for conversion of power from the ac line to the structure input. Again, if the iris aperture is opened up at shorter wave lengths to reduce wakefields and to increase the length of the individual accelerating structures, the above expression must be multiplied by $F(a/\lambda) \sim (v_g/c)^{1/3}$. In the next section we will consider a collider scaling which follows approximately $v_g/c \sim \omega^{1/3}$ and $G \sim \omega$. In this case,

$$P_{ac} \sim \lambda^{8/9} E_{c.m.} / \eta_{rf} \quad (5b)$$

3. COLLIDER DESIGNS AT THREE FREQUENCIES

Table 2 shows some linear collider designs scaled in frequency and energy, based on the 1 TeV NLC design shown in the first row. In this design the parameters of the NLC damped, detuned structure are assumed: $v_g/c = 0.05$, structure length = 1.8 m, $T_f = 120$ ns, $T_0 = 220$ ns, $\tau = 0.545$ and an effective shunt impedance per unit length of 94 MΩ/m. Effective shunt impedance means that, using this value in the standard expression for acceleration in a constant gradient (CG) structure, an unloaded gradient of 77.4 MV/m is obtained in the actual NLC structure (which differs somewhat from a true CG design because of the Gaussian detuning) for a peak power per meter of 96 MW. A beam loading gradient of 13.6 MV/m is obtained for bunches with 1.1×10^{10} electrons spaced 2.8 ns (32λ) apart (an effective shunt impedance of 96.5 MΩ/m must be used in standard CG expression to obtain the beam loading gradient for the actual NLC detuned structure). The loaded gradient is then 63.8 MV/m. Assuming an overhead factor of 1.15 (to account for off-crest operation for BNS damping, feedback overhead etc.), and allowing for an injection energy of 10 GeV per linac, the total active structure length is 17.7 km. The rf pulse length of 360 ns allows for a train of 81 bunches (225 ns), the structure filling time (120 ns) and a switching and rise time allowance of 15 ns. The beam size at the IP is 5 nm × 250 nm with a pinch enhancement factor of 1.4, giving a luminosity of 1.0×10^{34}/cm^2/sec. More complete beam parameters are given in the NLC ZDR Design Report (7). The required klystron power is obtained assuming a binary pulse compression system with a factor of 4 compression ratio and an efficiency of 0.875 (power gain = 3.5). Assuming four klystrons and six 1.8 m accelerating structure per BPC system, the klystron power is (96 MW/m)(10.8m)/(4)(3.5) = 74 MW. The ac power assumes an rf system efficiency of 40%: klystron efficiency = 61%, modulator efficiency = 75%, pulse compression and rf power transmission efficiency = 87.5%. The rf energy per pulse at the structure input is (360 ns)(96 MW/m) = 34.5 J/m. Multiplying this by the active length and repetition rate, and dividing by the rf efficiency gives P_{ac} = 183 MW. The number of pulse compression systems is 17.66 km/10.8 m = 1635. The number of klystrons is four times this or about 6500.

The next row in Table 2 shows how the 11.4 GHz, 1 TeV collider might be scaled to 1.5 TeV with only a modest increase in length by employing a higher power, higher efficiency rf system. The unloaded gradient is increased to 100 MW/m by increasing the klystron power to 124 MW. This implies an extended beam or multiple beam device (see Sec. 5), or the use of two 62 MW tubes in place of each 74 MW klystron. The ac power is computed on the assumption that the klystron efficiency has been raised to 66% (efficiencies of this order are already being obtained in simulations). Also, it is assumed that the modulator has been eliminated and that the klystron beam is switched by a grid from a dc supply with a switching efficiency of 95%. The net rf system efficiency is then 55%.

In the third row parameters are given for a 5 TeV collider with a frequency and unloaded gradient which are both three times the 1 TeV X-band values. The group

Table 2. Linear Collider Designs Scaled in Frequency and Energy

$E_{c.m}$ (TEV)	Frequency (GHz)	G_0 (MV/m)	G_0/G_{th}	P_m (MW/m)	P_{ac} (MW)	L_A (km)	Max ΔT (°C)
1.0	11.4	77	1.26	96	180	17.7	16
1.5	11.4	100	1.64	160	260	20.6	27
5.0	34.3	225	1.23	530	280	30.4	112
10	91.4	500	1.02	1700	160	27.4	405
15	91.4	600	1.23	2400	290	34.3	585

velocity has been increased to .072c, giving a structure length and filling time of 0.5 m and 23 ns. The decrement time scales to $T_0 = 220/3^{3/2} = 42$ ns, giving $\tau = 0.54$. Taking into account the larger iris aperture, the shunt impedance scales to 144 MΩ/m. This gives a peak power per meter of 530 MW/m for an unloaded gradient of 225 MV/m. The beam loading gradient is based on a charge of 2.4×10^9 electrons per bunch and a bunch spacing of $12\lambda = 0.350$ ns. The loaded gradient then works out to be 188.5 MV/m, and the active length to 30.4 km, assuming again a 15% overhead in length. The rf pulse length of 80 ns is based on a train of 150 bunches (52 ns), plus the 23 ns filling time, plus a 5 ns allowance for phase switching and rise time. The ac power then works out to be 280 MW. Assuming a transverse beam size of 0.3×30 nm at the IP, a luminosity of 1.5×10^{35}/cm^2/sec is obtained with a pinch enhancement factor of 1.7. More detailed beam parameters are given in (3). To calculate the required klystron power a pulse compression system design must be assumed. This is described in Sec. 4.

The 91 GHz (8×11.42 4 GHz) parameters listed in Table 2 are obtained assuming a group velocity of 0.10c and a structure length 0.18 m, with $T_f = 6$ ns, $T_0 = 10$ ns, and $\tau = 0.60$. The beam loaded gradient is assumed to be 84% of the unloaded gradient, as in the 5 TeV example, giving a total active structure length of about 34 km. An rf pulse length of 16 ns is chosen, allowing for a train of 100 bunches spaced 8λ apart. For the assumed beam loading gradient, the bunch charge is about 1×10^9 electrons. The beam cross section will have to be incredibly small (on the order of $.04 \times 4.0$ nm) to achieve a luminosity of 1×10^{36}/cm^2/sec.

The pulse heating is marginally high, but within the realm of possibility for the 5 TeV collider. Perhaps same sort of surface treatment or surface coating can ameliorate the effects of a pulse temperature rise of this magnitude. However, it does not seem reasonable that pulse temeprature rises of 400–600°C can be tolerated. Also, even with pulse compression, a very large number of very high power rf sources will be needed for the 91 GHz machines because of the high peak power per unit length. It is also questionable whether the required peak power flow can be accommodated without breakdown in the power transmission components. For these reasons, a 5 TeV, 34 GHz linear collider may be the end of the line for a machine based on conventional rf technology.

4. RF PULSE COMPRESSION

An rf pulse compression system can enhance the peak power output from a microwave tube by trading increased peak power for reduced pulsewidth. The power gain is given by the compression ratio in pulsewidth, R, times a compression efficiency which takes into account intrinsic losses (e.g., reflected power in a SLED pulse compression system), resistive copper losses, and the efficiency reduction due to a non-flat output pulse. Pulse compression reduces the burden on the rf power source and helps to match the modulator pulse length to the accelerating structure filling time. This is especially important at higher frequencies where the filling time is short and the production of peak power is more difficult. A pulse compression system always involves an energy storage element to delay or transfer energy from the early portion of the rf pulse into the compression output pulse.

The first large-scale pulse compression system for an accelerator application was the SLED scheme, implemented on the SLAC linac in the late 1970's. Using a pair of TE_{015} cylindrical cavities ($Q_0 \approx 1 \times 10^5$) as energy storage elements, SLED produces a power gain of about 2.7 with a compression efficiency of 62% (R = 4.4). A characteristic feature of the SLED compression method is a 180^0 phase reversal in the klystron drive, which triggers the release of the energy stored in the high Q cavities. Two cavities and a 3 db coupler are used so that the energy reflected and emitted from the cavities will not return to the klystron but will be directed into the transmission line to the accelerator.

Because the SLED output pulse has a shape which is dominated by the exponential decay of energy in the storage cavities, it is poorly adapted for powering a linear collider with long bunch trains. The pulse shape problem can be solved by replacing the two storage cavities with shorted delay lines. In this scheme, called SLED-II, the delay line length (in travel time) is equal to one-half the desired output pulse length. The power gain is optimized by adjusting the reflection coefficient of an iris at the entrance to the delay lines. Assuming lossless components, the power gains (and intrinsic efficiencies) for a SLED-II system with compression ratios of 4, 5 and 6 are 3.44 (86%), 4.0 (80%) and 4.5 (75%). The 0.5 TeV NLC design uses a SLED-II pulse compression (PC) system with R = 5. Losses in the delay lines and other components reduce the efficiency from 80% to 76%. Power transmission losses from the klystron to the PC system and from the PC system to the accelerator are usually included in calculating the PC system efficiency and power gain. Including transmission losses, the net efficiency of the NLC system falls to 72% and the power gain to 3.6.

If the 0.5 TeV NLC design is scaled to 1 TeV, the power loss due to the intrinsic inefficiency of the SLED-II system translates to an intolerable 30–40 MW increase in ac power. There are several schemes for pulse compression which eliminate or reduce this loss. In the so-called Binary Pulse Compression (BPC) scheme, the input rf pulse is divided into 2^n time bins, where n is the number of compression stages and $R = 2^n$ is the net compression ratio. At the input to the first stage, the power from two klystrons is combined by a 3–db directional coupler (hybrid). The phase relation between the waves at the input ports of the hybrid is such that, during the first half of the pulse, the power from both klystrons is directed into a delay line with a time length equal to half the pulse length. Half way through the pulse the relative phase at the input ports of the hybrid is switched by 180^0 (by reversing the drive phase of one of the klystrons), causing the combined klystron power to come out of the second output port of the hybrid, in time coincidence with the power emerging from the delay line. A clever phase switching pattern devised by Z. D. Farkas allows compression stages to be chained in series, the peak power being doubled and the pulse width cut in half at each stage. The total length of delay line needed for a BPC system is $(R-1)T_p$, where T_p is the output pulse width. For the NLC 1 TeV design, a two-stage BPC system is envisional with an efficiency (due mostly to copper losses) of 0.875 and a power gain of 3.5. The total delay length is 3×360 ns = 1.08 μs.

If cylindrical pipe is used to achieve this delay, the total length of pipe would need to be about 300 m. In principle, the pipe can be replaced by a sequence of high Q TE_{01n}-mode cylindrical cavities having a much shorter total length. An analysis using an equivalent circuit model (a chain of inductively-coupled resonant circuits) has shown that a relatively small number of coupled cavities can produce a reasonably flat output pulse. The transmission efficiency of such a system can be quite high. For example, a cavity 20 cm in diameter and 2.5 m long would have a Q of 1.5×10^6 and a decrement time of 42 μs at 11.4 GHz. A two stage BPC system employing such cavities would have a transmission efficiency of 95% (neglecting pulse shape effects) for an output pulse length of 360 ns.

This concept for pulse compression can be applied to produce the 530 MW/m required by the 5 TeV, 34 GHz linear collider example listed in Table 2. Assume eight 0.5 m accelerating structures per 3-stage BPC system. A TE_{01n}-mode cavity 1 m long by 10 cm in diameter would have a Q of 1.25×10^6 and a decrement time of 11.7 μs. For an output pulse length of 80 ns the transmission efficiency due to cavity losses would be 91% (there would be an additional efficiency loss due to pulse-top ripple and rise time degradation). Assuming additional losses of 6.5%, the net compression efficiency would be 85% and the power gain would be 6.8. The required klystron power is (530 MW/m)(4.0 m)/(2)(6.8) = 155 MW. In the next section, we discuss how this source power might be achieved at 34 GHz. Note also that the peak power flow at the output of the BPC system is about 1100 MW. It will be difficult to design the hybrid and other waveguide components to accommodate this power flow without breakdown, but carefully designed oversized components could do the job.

5. RF POWER GENERATION FOR 34 GHz

There are two basic limitations on the power that can be generated by a conventional round-beam klystron. First of all, it is well known that the electronic efficiency of a klystron depends on the microperveance, defined as $K_\mu \equiv (I_b/V_b^{3/2}) \times 10^6$. The perveance sets the scale for space charge forces, which in turn limit the compactness of the electron bunches and hence the rf component of the beam current. An expression which fits recent simulations on X-band klystrons at SLAC is:

$$\eta \approx 0.75 - 0.17\, K_\mu. \qquad (6)$$

To achieve the desired efficiency of 68%, $K_\mu \approx 0.41$. At a beam voltage of 500 kV, the output power would be $P_k = \eta K V_b^{5/2} \approx 50$ MW. This limitation is independent of rf frequency.

A second limitation on klystron power is related to the area of the electron beam, which does depend on wavelength. To achieve good coupling to the longitudinal rf fields in the output gap, the radius of the beam should not be larger than about $\lambda/8$. If the beam radius exceed this, then electrons on the beam axis and electrons at the edge of the beam will see a substantially different rf gap voltage, and efficiency will suffer. Next, the beam area at the cathode can be larger than the beam area in the drift region by a factor A_c, the area convergence ratio. This ratio is limited by aberrations in the gun optics, transverse emittance, alignment tolerances, etc. A good measure of these effects is the convergence half-angle, which is related to the f ratio of conventional optics. In practice, it is found that the convergence half-angle is limited to about 40^0, corresponding roughly to f 0.6. Because of the dynamics of space-charge-limited electron flow in the gun region, the gun focal length and hence the area convergence ratio depends on the perveance (8). By plotting A_c vs. K_μ for a variety of gun designs with a convergence half angles of 35^0–40^0, the points can be crudely fit (within a factor of two or so) by

$$A_c \approx 150/K_\mu^2 \ . \qquad (7)$$

A further limitation is the acceptable cathode loading current per square centimeter, I_A. Putting these factors together, the maximum output power is

$$P_k \approx \eta\, V_b\, I_A A_c \pi\, (\lambda/8)^2, \qquad (8)$$

where both η and I_A are functions of perveance. If Eq. (7) is equated to $P_k = \eta K v_b^{5/2}$, an expression is obtained for the maximum perveance allowable which is consistent with the limitations imposed by Eq. (7):

$$K_\mu = 194\, I_A^{1/3}\, \lambda^{2/3}/V_b^{1/2} \ . \qquad (9)$$

Taking $I_A = 10 A/cm^2$ and $V_b = 500$ kV, Eq. (9) gives K_μ (max) = 0.54, η = 0.66 and $P_k = 63$ MW at 34 GHz. A lower perveance can, of course, be chosen for better efficiency at lower output power.

The bottom line is that it should be possible to build a klystron at 34 GHz which has good efficiency (65–70%) and an output power on the order of 50 MW. To obtain the 150 MW desired to drive the 34 GHz pulse compression system described in the previous section, there are several possibilities. For example, three or four 40–50 MW beams can be packaged together in the same vacuum envelope. Such a multibeam klystron having common rf cavities but separate PPM-focused beams has indeed been proposed (8). Klystrons using a sheet beam, which is essentially equivalent to many round beams in parallel, are also capable (in simulations) of producing 150 MW at 34 GHz with good efficiency (9). It is well known that gyroklystrons are also capable of producing high power at high frequency. At the University of Maryland, a coaxial-circuit gyroklystron frequency doubled from 17 to 34 GHz has been designed which produces an output power of 150 MW at a simulated efficiency of 42% (10). A single-stage depressed collector can increase the efficiency to 56%.

Another annular-beam device capable of delivering high power output at high frequencies is the Ubitron (FEL) proposed by McDermott et. al (11). Using a TE_{01}-mode coaxial cavity and a PPM wiggler, it produce a simulated output power of 250 MW at 11.4 GHz with an efficiency of 50%. It should be capable of producing a high output power when scaled to 34 GHz.

6. TWO-BEAM 15 TeV COLLIDER CONCEPT

As shown in Table 2, a high gradient, 91 GHz collider runs into trouble from pulse heating, in spite of a very short rf pulse length of 16 ns. As a possible way around this difficulty, H. Henke (2) has proposed a two-beam accelerator scheme which uses a beat-wave transformer to extract rf power from the drive beam in extremely short bursts. Very briefly, the transformer consists of two 91 GHz side-by-side coupled cavities with an aperture for the drive beam in one cavity and an aperture for the main accelerated beam in the other. The coupling is adjusted to produce a beat frequency period which is equal to on-half of the period of the rf for the drive beam re-acceleration cavities. A drive beam bunch, which has a bunch length of about $\lambda/2$ so that higher modes are not significantly excited, deposits energy in the fundamental mode in the first cavity. This energy then passes into the accelerating cavity in a time given by one-half the beat frequency period. At this time the bunch (or short train of bunches) to be accelerated passes through the second cavity, removing a fraction of the stored energy. The remaining energy now couples back into the first cavity during the second half of the beat frequency cycle. At this time a scavenger bunch, riding at the decelerating phase of the drive beam rf, passes through the first cavity and removes a major portion of the unused energy and returns it to the drive beam.

As an example, assume that the 1.3 GHz TESLA superconducting accelerator provides the drive beam. The beat frequency period would then be $(2.6$ GHz$)^{-1}$ = 385 ps. The effective time the energy spreads in the accelerating cavity is about a fourth of this, or about 100 ps. The temperature rise at a gradient of 600 MV/m is then only 45°C.

Finally, we note that, even with efficient pulse compression by a factor of eight, a 15 TeV accelerator powered by conventional rf sources would require five 70 MW tubes per meter of active length. The two-beam accelerator scheme outlined above seems to be a much cleaner and efficient way to provide the rf power. It could be viewed at as a 15 TeV upgrade to the TESLA machine.

REFERENCES

1. Loew, G. A., ed., *International Linear Collider Technical Review Report: 1995*, SLAC-R-95-471, SLAC, Stanford University (1995).

2. H. Henke, "An Energy Recuperation Scheme for a Two-Beam Accelerator", CERN-LEP-RF/88-55, CERN, Geneva, Switzerland (1988).

3. Chen, P. et al., "A 5-TeV-c.m. Linear Collider on the NLC Site", to be published in the Proceedings of the 1996 Workshop on New Directions for High Energy Physics (Snowmass 96), Snowmass, Colorado, June 25–July 12, 1996.

4. Loew, G. A., and Wang, J. W., "RF Breakdown Studies in Room Temperature Electron Linac Structures", SLAC-PUB-4676, SLAC, Stanford University (1988).

5. Siemann, R., Internal Report ARDB-12, SLAC, Stanford University (1996).

6. Siemann, R., private communication.

7. A Report submitted to Snowmass'96: *Physics and Technology of the Next Linear Collider*, SLAC Report 485, SLAC, Stanford University (1996), p. 11.

8. Phillips, R., SLAC, private communication.

9. Yu, D.U.L., Kim, J. S., and Wilson, P. B., "Design of a High Power Sheet Beam Klystron" in *Proceedings of the Workshop on Advance Accelerator Concpets, Port Jefferson, New York 1992*: AIP Conference Proceedings 279, 1993, pp. 85-102. Also Kim, J. S., private communication.

10. Saraph, G., University of Maryland, private communication.

11. McDermott, D. B., et al., "Periodic Permanent Magnet Focusing of an Annular Electron Beam and its Application to a 250 MW Ubitron FEL", Submitted to *Physics of Plasmas* (1996).

The Problems and Physics Prospects for a $\mu^+\mu^-$ Collider

David B. Cline

*University of California, Los Angeles
Department of Physics and Astronomy, Box 951547
Los Angeles, CA 90095-1547 USA*

Abstract. In 1992, a new study of $\mu^+\mu^-$ colliders was started at a workshop at Napa, California. Subsequently, many problelms have been solved except for two: (1) The best cooling system to use for the machine and (2) the real particle physics needs of such a machine. We argue here that these two issues are related and that the only compelling scientific argument today is for a $\mu^+\mu^-$ collider Higgs factory. We show that a 4-TeV collider may not even be the correct high-energy range based on the possible future observations at the LHC and the NLC and the Higgs factory. Such a collider requires very cold $\mu^\pm$ beams and will select the cooling method.

INTRODUCTION

A $\mu^+\mu^-$ collider offers some compelling particle physics opportunities at a tradeoff with immense technical problems (1,2). This is very similar to the situation for the $\bar{p}p$ collider where the scientific goals were clearly the W,Z discovery by Rubbia, McIntyre, and the author (3). The technical problem was to cool, store, reinject, accelerate, and collide the rare antiprotons, and this was a tour de force of accelerator physics carried out by the CERN and FNAL accelerator teams.

What are the similar scientific goals for a $\mu^+\mu^-$ collider? Some would say that we should blindly go to the highest energy (*i.e.*, 4 TeV). However, unlike 1976, there is no clear physics advantage of this energy, and the LHC will already have 14 TeV in the center of mass with a very high luminosity and years of detector operation before any 4-TeV $\mu^+\mu^-$ collider could be built. Current theory is very restrictive, suggesting that all new physics will occur below or at 1 TeV and, thus, this could be observed or totally rejected by 2010. In which case one may need 40 TeV or 400 TeV to find new physics. Thus our viewpoint is and has been that the only useful $\mu^+\mu^-$ collider is to explore the electroweak Higgs sector (a $\mu^+\mu^-$ Higgs factory). We will show that the machine requires extremely cold $\mu^\pm$ beams and, therefore, will be directly related to the μ cooling scheme. The μ polarization view will also be very important.

In the next section, we discuss the Higgs collider, followed by our scheme to produce ultracold $\mu^\pm$ beams for the Higgs collider. Finally we give our conclusions.

A $\mu^+\mu^-$ COLLIDER HIGGS FACTORY

Recently there has been a great deal of activity concerning $\mu^+\mu^-$ colliders, starting with the Napa workshop in 1992 (4,5). We have proposed that such a collider is very useful to study the scalar sector of the electroweak interaction (2). In this brief report, we discuss the arguments for a Higgs factory (see Table 1).

The strongest argument for the low-energy 250 × 250-GeV collider comes from the growing evidence that the Higgs should exist in this low-mass range from:

1. The original works of Cabibbo and colleagues (6), which shows that, when $m_t > M_Z$ and assuming a grand unification theory (GUT), $M_H < 2 M_Z$ (6);

2. Fits to LEP data imply that a low mass h^0 could be consistent with $m_t > 150$ GeV (7);

3. The extrapolation to the GUT scale that is consistent with SUSY also implies that one of the Higgs should have a low mass, perhaps below 130–150 GeV (7).

This evidence implies the exciting possibility that the Higgs mass is just beyond the reach of LEP II and in a range that is very difficult for the LHC to detect (1).

We expect the supercollider LHC to extract the signal from background (*i.e.*, seeing either $h^0 \to \gamma\gamma$ or the very rare $h^0 \to \mu\mu\mu\mu$ in this mass range, since $h \to b\bar{b}$ is swamped by hadronic background). However, detectors for the LHC are designed to extract this signal. Figure 1 gives a picture of the various physics thresholds that may be of interest for a $\mu^+\mu^-$ collider. In this low mass region, the Higgs is also expected to be a fairly narrow resonance and, thus, the signal should stand out clearly from the background from

$$\mu^+\mu^- \to \gamma \to b\bar{b} \to Z_{tail} \to b\bar{b} \quad . \tag{1}$$

For masses above 180 GeV, the dominant Higgs decay is

$$h^0 \to W^+W^- \quad \text{or} \quad Z^0Z^0 \quad , \tag{2}$$

and the LHC should easily detect this Higgs particle. Thus the $\mu\mu$ collider is better adapted for the low mass region. The report of Barger *et al.* is very illuminating regarding the physics potential of a $\mu\mu$ collider (see Table 2) (1,7,8).

TABLE 1. Arguments for a $\mu^+\mu^-$ Collider Higgs Factory

1. The m_μ/m_e ratio gives coupling 40,000 times greater to the Higgs particle. In the SUSY model, one Higgs m_h < 120 GeV!!
2. The low radiation of the beams makes precision energy scans possible.
3. The cost of a "custom" collider ring is a small fraction of the $\mu^\pm$ source.
4. Feasibility report to Snowmass established that $\mathcal{L} \sim 10^{33}$ cm^{-2} s^{-1} is feasible.

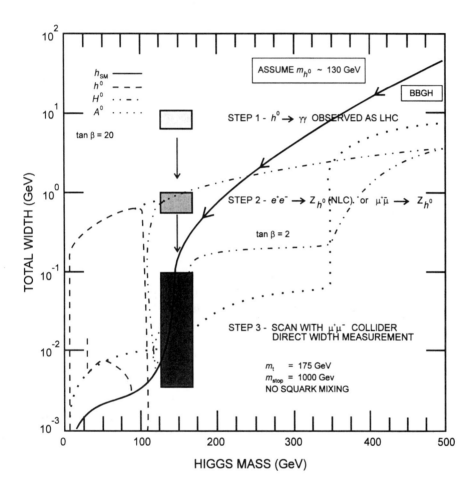

FIGURE 1. A $\mu^+\mu^-$ collider Higgs-factory concept. The Higgs is discovered at the LHC (CMS) and the width further reduced at the NLC or at a $\mu^+\mu^-$ collider. The final stage is to scan for the Higgs at the $\mu^+\mu^-$ collider. Existing models can be distinguished by their widths. [Adapted from (7) (BBGH = Barger, Berger, Gunion, Han) and (9).]

TABLE 2. The Scalar Sector

With a high-mass *t* quark, precision LEP/SLD data and the theorists' dreams of a SUSY world, the scalar (pseudo-scalar sector) is possibly very complex and may require several types of colliders.[8] Consider:

- If the low-mass Higgs has *m* > 130 GeV, MSSM is not allowed.
- If *m* > 200 GeV, there are constraints from the requirement that perturbation theory be useful up to very high energy and from the stability of the vacuum.
- If *m* < 130 GeV, MSSM is possibly ok, but we may expect other particles (*H*, *A*), and the width of the low mass Higgs may change.
- The scalar sector may be extremely complex, requiring *pp* (LHC) and $\mu^+\mu^-$ colliders (and possibly NLC and $\gamma\gamma$ colliders).
- In high energy collisions, vector states are allowed unless a special method is used. Consider $\mu^+\mu^-$ colliders with polarized $\mu^\pm$:

$\mu^+\mu^-$ ⟨ (100–500) GeV - scalars (*H*, *A*, ...) W^+W^-
        ≥ 2 + TeV Z^0Z^0 production in scalars

This cannot be done for *pp* or e^+e^- colliders.

- A $\mu^+\mu^-$ collider is complimentary to the LHC/CMS detector.

Scan for Higgs Mass and Width: SUSY or Not

In this section, we assume for the sake of argument that the CMS detector at the LHC has barely detected a signal at $m \sim 130$ GeV ($h^0 \rightarrow \gamma\gamma$) and at an experimental width of ~8 GeV (Step 1, illustrated in Fig. 1). The question will now be

1. Is this a Higgs boson or not?
2. Is it the standard model Higgs or a SUSY Higgs?

We envision the next step would be to construct the $\mu^+\mu^-$ collider operating between the energies of $E_{\mu^+\mu^-} \sim m_{h^0}$ (CMS) and $E_{\mu^+\mu^-} \sim m_z + m_{h^0}$ (CMS) or the use of the NLC to observe $e^+e^- \rightarrow Z^0 h^0$ (10). We build the $\mu^+\mu^-$ collider (after already having built a $\mu^\pm$ source), and for Step 2 operate near the $Z^0 + h^0$ (CMS) threshold to determine m_{h^0} and Γ_{h^0} to ~ 1 GeV. (See Fig. 2 for the cross sections.) For Step 3, we envision an energy scan of the mass region by varying the $\mu^+\mu^-$ energy (1,9). At some point, the mass and width are determined and then used to distinguish between the standard model Higgs and a SUSY Higgs (Fig. 3).

The final step is to measure the branching fractions for different decay modes (6,7). Figure 4 shows the expectations for the standard model Higgs.

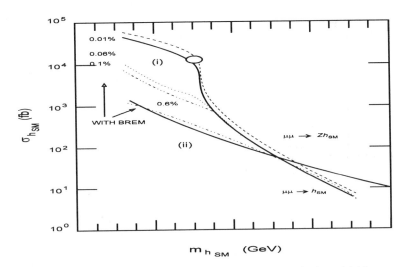

FIGURE 2. Cross sections versus $m_{h_{SM}}$ for inclusive standard-model Higgs production: (i) the s-channel $\bar{\sigma}_h$ for $\mu^+\mu^- \to h_{SM}$ with $R = 0.01\%$, 0.06%, 0.1%, and 0.6%; and (ii) $\sigma(\mu^+\mu^- \to Zh_{SM})$ at $\sqrt{s} = m_Z + \sqrt{2}\, m_{h_{SM}}$. Also shown is the result for $R = 0.01\%$ if bremsstrahlung effects are not included. [Adapted from (7).]

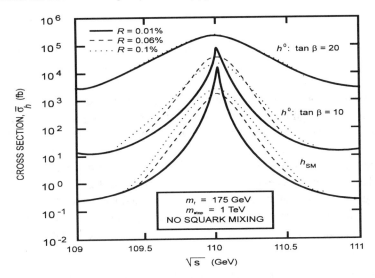

FIGURE 3. The effective cross section, $\bar{\sigma}_h$, obtained after convoluting σ_h with the Gaussian distributions for $R = 0.01\%$, 0.06%, and 0.1%, is plotted as a function of $\sqrt{s}$ taking $m_h = 110$ GeV. Results are displayed in the cases h_{SM}, h^0 with $\tan\beta = 10$ and $= 20$. In the MSSM h^0 cases, two-loop/RGE-improved radiative corrections have been included for Higgs masses, mixing angles, and self-couplings assuming $= 1$ TeV and neglecting squark mixing. The effects of bremsstrahlung are not included in this figure. [Adapted from (7).]

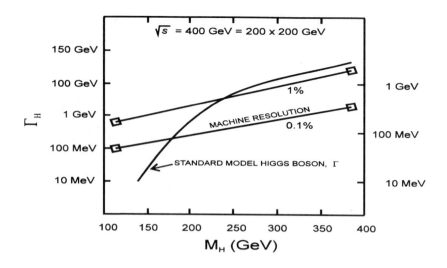

FIGURE 4. Machine requirements for Higgs scan (11).

Polarized Collider

The most interesting question now in particle physics is associated with the origin of mass. It is generally assumed that the exchange of fundamental scalar particles, called the "scalar sector" is somehow responsible for this. For super-symmetry modes, this scalar sector is even more complex and interesting (see Table 2) (11,12).

In this section, we highlight one of the most interesting goals of a $\mu^+\mu^-$ collider: the discovery of a Higgs boson in the mass range beyond that to be covered by LEP I & II (~ 80 - 90 GeV) and the natural range of the supercolliders.

There are several ways to determine the approximate mass of the Higgs boson in the future (10). Suppose it is expected to be at a mass of 135 ± 2 GeV, the energy spread of a $\mu^+\mu^-$ collider can be matched to the expected width (see Fig. 5). An energy scan could yield a strong signal to background especially with polarized $\mu^+\mu^-$ in the scalar configuration [11,12]. Once the Higgs is found, the following could be carried out:

1. Measurement of width, to separate standard-model Higgs from SUSY or other Higgs models (6,7),

2. Measurement of the Branching fractions, the rare decay will involve loop effects that can sample very high energies.

Polarization will play an essential role for any $\mu^+\mu^-$ collider (12,13)!

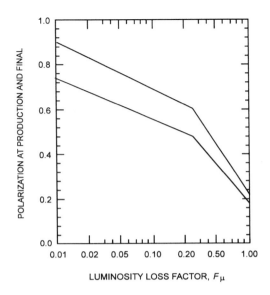

FIGURE 5. Polarization vs the fraction, F_μ, of muons accepted (solid line: polarization at source; dashed line: after cooling) (8).

Polarization is natural for $\mu^\pm$ since they are produced in weak decays and are initially fully polarized because of this V-A interaction. There are three proposed methods for producing intense polarized $\mu^\pm$ beams:

- Accelerate polarization and cool the $\pi^\pm$ (A. Skrinsky *et al.*) (14),
- Use $K^\pm$ decays and "narrow-band neutrino-like beam," and
- Use pion decays and a short proton bunch (9).

Figure 5 shows the tradeoff between intensity and polarization in one of these schemes (9,12,14). This is one of the major areas of research for $\mu^+\mu^-$ colliders.

Some Examples of Possible $\mu^+\mu^-$ Colliders for Higgs Factories

At the Snowmass '96 meeting, the U.S. $\mu^+\mu^-$ collider consortium presented a feasibility design of a $\mu^+\mu^-$ collider (9). The important point is that $\mathcal{L} \sim 10^{33}$ cm^{-2} s^{-1} was shown to be possible for this collider. We consider this an existence of proof of sorts. This collider is complex, the simplest part being the actual storage ring for the $\mu^+\mu^-$ collisions. It is important to note that this collider ring is likely a minor part of the cost of the overall complex.

There are other possible $\mu^+\mu^-$ collider designs that may serve as a Higgs factory. These designs differ by either the assumptions about the $\mu^\pm$ cooling method or the type of overall collider. Figure 6(A) shows a schematic design for a $\mu^+\mu^-$ collider in Japan that uses the high-current 50-GeV accelerator now being designed for KEK (15). The cooling method is by frictional cooling of low-energy $\mu^\pm$ beams.

Figure 6(B) shows a scheme worked out by the author and A. Bogacz, which uses crystal channeling for both the cooling and the collisions (16). In the latter case, if the $\mu^\pm$ can be confined to a crystal channel ($\sim$ 10 - 30 Å) then high luminosity can be achieved using modest $\mu^\pm$ intensities, greatly reducing the background and possible cost of the Higgs factory.

A hypothetical schedule for a Higgs-factory $\mu^+\mu^-$ collider is given in Table 3, which is of course entirely the author's own viewpoint.

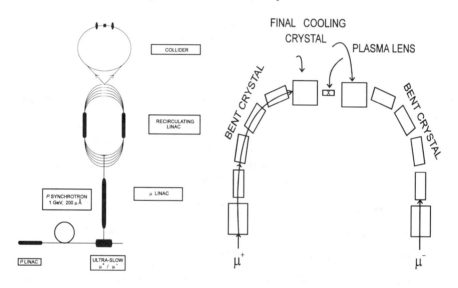

FIGURE 6. Collider concepts: (A) Japanese $\mu^+\mu^-$ collider (15); (B) Crystal quantum collider (16).

TABLE 3. Possible Scheme for a $\mu^+\mu^-$ Collider Higgs Factory

$\sim$ 2003:	Start construction of $\mu^\pm$ source.
$\sim$ 2006:	First observation of h^0 in CMS (ATLAS).
$\sim$ 2007:	Design final collider; start construction.
$\sim$ 2009:	Higgs factory operates; scan for h^0.
$\sim$ 2010:	$\rightarrow 10^5$ h^0 in direct channel.

FINAL CRYSTAL COOLING TO REDUCE THE BEAM CURRENT AND DETECTOR BACKGROUNDS FOR A $\mu^+\mu^-$ COLLIDER

Over the past four years the possibility of a real $\mu^+\mu^-$ collider has gained interest (1,17,18). With the *Proceedings* of the 1995 San Francisco conference (8) and the "Snowmass Book" (9), the design goals have gained even more credibility. However, there are several serious problems still remaining:

1. The high backgrounds in the detector from $\mu^\pm$ decay products (1,17,18),

2. The very high $\mu^\pm$ content in the final collider ($\sim 5 \times 10^{12}$ $\mu^\pm$ per bunch),

3. The relatively poor reduction of the phase-space using medium energy ionization cooling (18), and

4. The current high cost of the source due to the large yield of $\mu^\pm$ required to reach high luminosity.

Most of these problems can be partially cured if the number of $\mu^\pm$ required to produce high luminosity can be reduced.

Over the past few years we have studied the use of crystal channels for cooling (19,20) and even for a $\mu^+\mu^-$ collider that would use bent-crystal beam confinement. Here, we show how crystal cooling for the high-energy $\mu^\pm$ beams could result in a dramatic decrease in the beam emittance and, therefore, provide high luminosity with a substantially reduced beam intensity and backgrounds of $\mu^\pm$ in the storage ring (1,17). Also, reducing the beam emittance at high energy helps the relatively poor low-energy beam cooling (12). Finally, the lower yield of $\mu^\pm$ reduces the required proton current in the $\mu^\pm$ source, thus possibly reducing considerably the cost of the overall $\mu^+\mu^-$ collider (8).

We now turn to a discussion of the cooling concept. Our model calculation, presented here, shows that one can decrease the normalized emittance to less than $\epsilon_N = 10^{-8}$ mrad by passing the muon beam through a cascade of many cooling modules.

Crystal-Channel Beam Cooling

We consider motion of planar channeled particles in a crystal, which is bent elasticity in a direction perpendicular to the particle velocity and to the channeling planes. The effect of bending introduces a centripetal force to the equation of transverse motion (21) (by adding a linear piece to the crystal potential), which is equivalent to lowering one side of the continuum potential well and raising the other. The equilibrium planar trajectory moves away from the midpoint of the planar channel toward the plane on the convex side of the curved planar channel. However, such a shift would cause some fraction of the channeled particles to leave the potential well (dechannel) (22). The curvature at which no particle can remain channeled is reached when the equilibrium point of planar channeled motion is shifted to the position of the planar wall on the outside of the curve. This critical radius of curvature, known as the Tsyganov radius (23), ρ_T, is

$$\rho_T = \frac{2E_\mu}{\phi a}, \qquad (3)$$

where $\phi = 6 \times 10^{12}$ GeV m^{-2} is a material constant, related to the curvature of the potential well (24), and a is the distance between adjacent atomic planes (≈ 2 Å).

Using a simple formula linking equivalent magnetic bending field, B, with the trajectory's curvature, ρ, namely

$$B[\text{tesla}] \times \rho_T[\text{m}] = 3.34 \times E_\mu[\text{GeV}], \qquad (4)$$

one can calculate the maximum available equivalent bending field corresponding to the Tsyganov curvature. From Eqs. (3) and (4), this field is given by

$$B_T[\text{tesla}] = 3.34 \times (1/2)\phi a. \qquad (5)$$

Its numerical value for silicon is evaluated as $B_T = 2 \times 10^3$ tesla. We note in passing, that the maximum bending field is energy independent.

Here we propose a fast muon-cooling scheme based on the ionization energy loss (25) experienced by high-energy muons (25 GeV) channeling through an Si crystal. Applying classical theory of ionization energy loss (4(A)), a relativistic (γ) charged particle passing through an Si crystal of length, ΔL, loses total energy of ΔE[MeV] $= 4 \times 10^2 \times \Delta L$[m]. One can introduce a characteristic damping length, Λ,

$$\frac{1}{\Lambda} = \frac{1}{E_\mu} \frac{\Delta E}{\Delta L}, \qquad (6)$$

over which the particle loses all its energy. Relativistic muons passing though the crystal lose energy uniformly in both the transverse and longitudinal directions according to Eq. (6). After passing through a short section of a crystal ($\Delta L \ll \Lambda$), muons are re-accelerated longitudinally to compensate for the lost longitudinal energy. This leads to the transverse emittance shrinkage.

Introducing normalized transverse emittance, $\epsilon_N = \gamma \sigma_x \sigma_{x'}$, one can write the normalized emittance budget in the form of the following cooling/heating equation:

$$\frac{d\epsilon_N}{dL} = -\frac{\epsilon_N}{\Lambda} + \left(\frac{\Delta \epsilon_N}{\Delta L}\right)_{\text{scatt}} \qquad (7)$$

The last term in the above equation accounts for transverse heating processes contributing to the beam divergence increase according to the following relationship (24):

$$\left(\frac{\Delta \epsilon_N}{\Delta L}\right)_{scatt} = \frac{1}{2}\gamma\beta \frac{\Delta \langle\theta\rangle^2_{scatt}}{\Delta L} . \qquad (8)$$

Here β is the beta function of a focusing crystal channel, which has an enormously small value ($\beta = 2 \times 10^{-6}$ m, for 25-GeV muons channeling through a silicon crystal).

For muon channeling in a dielectric crystal, the dominant scattering process comes from elastic (Rutherford) muon scattering off the conduction electrons present in the channel. One can integrate the Rutherford cross section over the solid angle, which yields the following formula:

$$\alpha = \left(\frac{\Delta \epsilon_N}{\Delta L}\right)_{scatt} = 40\pi n \frac{r_\mu^2}{\gamma}\beta . \qquad (9)$$

Here $r_\mu = 1.4 \times 10^{-17}$ m is the classical muon radius and $n = 6 \times 10^{29}$ m^{-3} is the average concentration of the conduction electron gas in silicon crystal.

Integrating the cooling equation, Eq. (7), one obtains the following compact solution in terms of the normalized transverse emittance evolution:

$$\epsilon_N = \epsilon_N^0 e^{-(L/\Lambda)} + \Lambda\alpha\left[1 - e^{-(L/\Lambda)}\right] . \qquad (10)$$

The last term in Eq. (10) sets the equilibrium cooling limit of

$$\epsilon_N^{min} = \Lambda\alpha , \quad L \to \infty . \qquad (11)$$

Assuming 25-GeV muons, one gets $\Lambda = 62.5$ m and the equilibrium limit of the normalized emittance of

$$\epsilon_N^{min} = 8 \times 10^{-9} \text{ mrad} . \qquad (12)$$

This value of the normalized emittance will be used in our achievable luminosity estimate.

Bent-Crystal Cooling Ring

Here we employ previously discussed properties of the planar channeling of high energy muons in silicon to design components of a storage ring. Particularly, we are interested in a section of bent crystal followed by two straight pieces providing alternating horizontal–vertical focusing. A basic guiding cell is depicted schematically in Fig. 7. It can be noticed that the induced configuration of guiding fields in this element is equivalent to a powerful alternating gradient achromat.

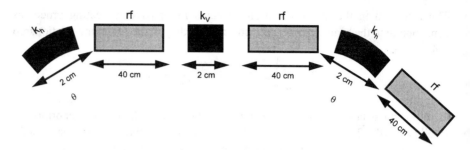

FIGURE 7. Layout of a cooling ring consisting of 50 bending-focusing-acceleration multi-functional cells. A straight piece of Si crystal rotated by 90° separating two sections of bent crystals provides vertical focusing, which maintains betatron phase stability in the proposed lattice. Conventional rf, 40-cm-long inserts (20 MeV/m) follow every 2-cm-long section of Si crystal absorber. (E_μ = 25 GeV, k_h = k_v = 180 m^{-2}, θ = 2π × 10^{-2} rad, rf grad = 20 MeV/m)

Relativistic muons channeling through an Si crystal are confined between two neighboring atomic planes – they experience strong focusing electrostatic crystal potential in the direction perpendicular to these planes, while there is virtually no confinement in the direction parallel to the planes (no focusing or defocusing). The focusing gradient, $k = 1/\beta$, is equivalent to the magnetic quadrupole strength, k_1 (magnetic gradient), where

$$k_1 = \frac{1}{B\rho}\frac{\partial B_y}{\partial x} \quad , \tag{13}$$

and

$$k = \frac{\phi}{E_\mu} \quad . \tag{14}$$

E_μ is the total muon energy and $\phi = 6 \times 10^{12}$ GeV m^{-2} is a material constant, related to the curvature of the potential well. Assuming 25-GeV muons, crystal focusing gradient, k, yields an enormous value of 180 m^{-2}, exceeding conventional quadrupole strength by four or five orders of magnitude.

As discussed previously in "Crystal Channel Beam Cooling," the crystal can be bent slightly, so that channeling muons follow the curvature of the guiding field, which results in bending of muon trajectories similar to the effect of a bending magnetic field. Projecting experimental results for proton channeling in a bent silicon crystal, one can assume that 25-GeV muons channeling through a 2-cm-long crystal should follow (without significant dechanneling effects) a bend of θ = 2π × 10^{-2} rad (compared with the critical bending angle θ_T = 2 × 10^{-1} rad). The lattice design presented here is based on these two numbers, k and θ.

Figure 7 illustrates a functional bending-focusing cell, where alternating sections of the horizontal and vertical continuous focusing channels are combined with sections of horizontally bent Si crystals. The following is the sequence of crystal elements: a horizontally focusing bent crystal (2-cm long) – a short drift space – a conventional rf re-acceleration section (40-cm long) – a short (2-cm long) vertically focusing straight crystal – a short drift space – another conventional rf re-acceleration section (40-cm long) – and finally, a horizontally focusing bent crystal (2-cm long) followed by a conventional rf re-acceleration section (40-cm long) completes the proposed elementary cell. At 25 GeV, one could close the entire collider ring using 50 of the above $F_hOF_vOF_hO$ cells. For a sequence of the above described cells, one can find periodic betatron trajectories in both the horizontal and vertical planes – the betatron phase stability is provided by the proposed lattice configuration (alternating horizontal/vertical focusing). By virtue of [110] planar channeling, discussed in detail in the previous section, a crystal channel provides an ultra-strong electrostatic focusing gradient in the [110] direction with practically no confinement or defocusing in the plane perpendicular to the [110] direction. This fact guarantees both local and global decoupling of the horizontal and vertical betatron motions for the proposed collider lattice.

Practical realization of muon cooling at 25 GeV could be done in a compact "cooling ring." Assuming characteristic damping length, Λ, of 62.5 m, the energy loss suffered by the muon beam after passing through a 2-cm-long section of an Si crystal is equal to 8 MeV. In principle, conventional high-gradient (20-MeV/m) acceleration inserts (a 40-cm-long rf insert following every 2-cm-long crystal absorber) could be used to replenish the suffered energy loss (0.4 m × 20 MeV/m = 8 MeV). The proposed cooling ring of 50-fold symmetry would have a nominal circumference of 63 meters!

Our goal is to start with the initial muon phase-space of the normalized emittance of 2.5×10^{-7} mrad and cool it down to the final emittance of 2.5×10^{-9} mrad. One can see from Eq. (10) that to achieve this goal, muons have to pass through the total silicon crystal length of

$$L = 2 \log 10 \times \Lambda = 280 \text{ m} \ . \qquad (15)$$

In the proposed cooling cell architecture, the total cooling medium (silicon) length of $L = 280$ m is equivalent to about 90 turns of the beam circulation in the ring. The lost energy is replenished every $\Delta L = 2$ cm, which satisfies the adiabatic re-acceleration condition ($\Delta L \ll \Lambda = 62.5$ m).

To go beyond the above simple analytic calculation, we are planning to carry out realistic computer simulations of planar channeling in bent crystals. One should track a charged particle through the distorted crystal lattice with the use of a realistic continuous-potential approximation and take into account the processes of both single and multiple scattering of electrons and nuclei, as well as on various defects and imperfections of the crystal lattice.

Application To High Energy Cooling: Conclusions

In this section, we show how crystal cooling for the high energy $\mu^\pm$ beams could result in a dramatic decrease in the beam emittance and, therefore, would provide high luminosity with a substantially reduced beam intensity and backgrounds of $\mu^\pm$ in the storage ring. Also, reducing the beam emittance at high energy helps the relatively poor low-energy beam cooling. Finally, the lower yield of $\mu^\pm$ reduces the required proton current in the $\mu^\pm$ source, thus possibly considerably reducing the cost of the overall $\mu^+\mu^-$ collider.

We suggest employing ionization energy loss in an alternating focusing crystal channel as a cooling mechanism, since initially small muon phase-space allows for efficient channeling through long sections of silicon crystal. Ultra-strong focusing in a crystal channel combined with alternating bending makes it a powerful focusing cell with ultra-small beta function. The cooling equation derived here shows that it is quite feasible to reduce transverse emittance by two orders of magnitude. Our model calculation done for 25 GeV muons shows that final emittances as low as 10^{-9} mrad are readily achievable, limited only by multiple scattering off the valence electrons in the crystal.

We conclude our study with the following observation: The proposed ionization crystal cooling could be used at some later stages of the collider scheme (*e.g.*, for the final cooling), because of 'favorable' energy scaling of the relevant cooling characteristics, α, Λ, and ϵ. They can be summarized as follows:

$$\alpha \sim \gamma^{-3/2} , \qquad (16)$$

$$\Lambda \sim \log \gamma , \qquad (17)$$

and

$$\epsilon_N^{min} = \Lambda\alpha \sim \left(\gamma^{-3/2}\right)\log\gamma . \qquad (18)$$

Therefore, the proposed cooling mechanism scaled to higher energies looks even more attractive.

CONCLUSIONS

We have discussed the importance of a low-energy $\mu^+\mu^-$ collider that will be a Higgs factory. We do not believe that there will be any compelling need for a higher energy $\mu^+\mu^-$ collider until the results of the experiments at the LHC are obtained. A crucial requirement is to cool the $\mu^\pm$ to a very cold beam temperature. There are several ways one might do this, *e.g.*, starting with a friction-cooled, low-energy beam. However, we have discussed another method here that uses crystal cooling. While this is a long shot, it may be the key to reducing the beam phase-space for the precision Higgs collider application.

ACKNOWLEDGMENTS

I wish to thank members of the CMS collaboration, the U.S. $\mu^+\mu^-$ consortium, and V. Barger, J. Gunion, and T. Han for helpful comments. I also wish to thank J. Wurtle and R. Fernow for discussions, and S. A. Bogacz and D. S. Sanders for their work on ultracold $\mu^\pm$ beams reported here. Stimulating comments from S. Geer and P. Chen are also acknowledged.

REFERENCES

1. *Physics Potential and Development of $\mu^+\mu^-$ Colliders* (Proc., 2nd workshop, Sausalito, CA, 1994), ed. D. Cline, New York: AIP Conference Proceedings 352, 1995.
2. Cline, D., *Nucl. Instrum./Meth A* **350**, 24 (1994), and the following 4 papers [*Nucl. Instrum./Meth. A* **350**, 27-56 (1994)] constitute a mini-proceedings of the Napa (Dec. 1992) meeting.
3. Rubbia, C., McIntyre, P., and Cline, D., in *Proceedings of the Int. Neutrino Conf.* (Aachen, 1976), eds. H. Faissner, H. Reithler, and P. Zerwas, Braunschweig: Vieweg & Sohn, 1977, p. 683.
4. Early references for $\mu\mu$ colliders are: (A) Perevedentsev, E. A. and Skrinsky, A. N., in *Proceedings of the 12th Int. Conf. on High Energy Accelerators* (Fermilab, 1983), eds. R. T. Cole and R. Donaldson, p. 481; (B) Neuffer, D., *Part. Accel.* **14**, 75 (1984) 75; (C) Neuffer, D., in *Advanced Accelerator Concepts*, New York: AIP Conference Proceedings 156, 1987, p. 201.
5. Neuffer, D. V., *Nucl. Instrum./Meth. A* **350**, 27 (1994).
6. For important references, see Dawson, S., Gunion, J. F., Haber, H. E., and Kane, G. L., *The Physics of the Higgs Bosons: Higgs Hunter's Guide*, Menlo Park, CA: Addison Wesley, 1989.
7. Barger, V., et al., in *Physics Potential & Development of $\mu^+\mu^-$ Colliders* (Proc., 3rd Int. Conf., San Francisco, 1995), ed. D. Cline, *Nucl. Phys. B (PS)* **51A**, 13 (1996).
8. See papers in *Physics Potential & Development of $\mu^+\mu^-$ Colliders* (Proc., 3rd Int. Conf., San Francisco, 1995), ed. D. Cline, *Nucl. Phys. B (PS)* **51A** (1996).
9. "$\mu^+\mu^-$ Collider A Feasibility Study" ("Snowmass Book"), report nos. BNL-52503, Fermi Lab-conf.-96/092, LBNL-38946 (July 1996).
10. CMS Proposal for the LHC, CERN (1994), unpublished.
11. Cline, D., "Physics Potential and Development of $\mu^+\mu^-$ Colliders," UCLA preprint no. CAA-0115-12/94 (1994).
12. Cline, D., in *Beam Dynamics and Technology Issues for $\mu^+\mu^-$ Colliders* (Proc., 9th Advanced ICFA Beam Dynamics Wksp., Montauk, LI, NY, 1995), ed. J. C. Gallardo, New York: AIP Conference Proceedings 372, 1996, p. 279.
13. Norum, B. and Rossmanith, R., in *Physics Potential & Development of $\mu^+\mu^-$ Colliders* (Proc., 3rd Int. Conf., San Francisco, 1995), ed. D. Cline, *Nucl. Phys. B (PS)* **51A**, 191 (1996).
14. Skrinsky, A., in *Physics Potential & Development of $\mu^+\mu^-$ Colliders* (Proc., 3rd Int. Conf., San Francisco, 1995), ed. D. Cline, *Nucl. Phys. B (PS)* **51A**, 201 (1996).

15. Nagamine, K., in *Physics Potential & Development of $\mu^+\mu^-$ Colliders* (Proc., 3rd Int. Conf., San Francisco, 1995), ed. D. Cline, *Nucl. Phys. B (PS)* **51A**, 115 (1996).
16. Bogacz A., and Cline, D. B., in *Physics Potential & Development of $\mu^+\mu^-$ Colliders* (Proc., 3rd Int. Conf., San Francisco, 1995), ed. D. Cline, *Nucl. Phys. B (PS)* **51A**, 90 (1996).
17. Cline, D. B, in *Advanced Accelerator Concepts*, ed. P. Schoessow, New York: AIP Conference Proceedings 335, 1994, p. 659.
18. Fernow, R., Brookhaven National Laboratory (1995), private communication.
19. Bogacz, S. A., and Cline, D. B, *Int. J. Mod. Phys. A* **11**, 2613 (1996).
20. Bogacz, S. A., Cline, D. B., and Sanders, D. A., in *Proceedings Relativistic Channeling Wksp.* (Aarhus, Denmark, July 1995), ed. H. H. Andersen, *NIMB*, **119** (1,2), 206–209 (1996).
21. Gibson, W. M., in *Relativistic Channeling*, NATO ASI Series, vol. 165B, ed. R. A. Carrigan, Jr. and J. A. Ellison, New York: Plenum, 1986, p. 27.
22. Biryukov, V. M., *Phys. Rev. Let. E.* **51**, 3522 (1995).
23. Tsyganov, E. N., Fermilab internal report no. TM-682 (1976).
24. Bogacz, S. A., and Cline, D. B., "A Bent Crystal Undulator to Provide Gold Positron Beams," UCLA report no. CAA-0123, *NIMB* (1996), submitted.
25. Skrinsky, A. N., and Parakhomchuk, V. V., *Sov. J. Part. Nucl.* **12**, 223 (1981).

The Physics Capabilities of $\mu^+\mu^-$ Colliders*

The Muon Quartet Collaboration
V. Barger[a], M.S. Berger[b], J.F. Gunion[c], T. Han[c]

[a] *Physics Department, University of Wisconsin, Madison, WI 53706, USA*
[b] *Physics Department, Indiana University, Bloomington, IN 47405, USA*
[c] *Physics Department, University of California, Davis, CA 95616, USA*

Abstract. We summarize the potential of muon colliders to probe fundamental physics. W^+W^-, $\bar{t}t$, and Zh threshold measurements could determine masses to precisions $\Delta M_W = 6$ MeV, $\Delta m_t = 70$ MeV, and $\Delta m_h = 45$ MeV, to test electroweak radiative corrections. With s-channel Higgs production, unique to a muon collider, the Higgs mass could be pinpointed ($\Delta m_h < 1$ MeV) and its width measured. The other Higgs bosons of supersymmetry can be produced and studied by three methods. If instead the WW sector turns out to be strongly interacting, a 4 TeV muon collider is ideally suited to its study.

I INTRODUCTION

In this report we address the exciting physics that could be accomplished at muon colliders in the context of the central physics issue of our time: how is the electroweak symmetry broken, weakly or strongly? Higgs bosons and a low energy supersymmetry (SUSY) are the particles of interest in the weakly broken scenario and new resonances at the TeV scale of a new strong interaction dynamics are the alternatives.

Muon colliders would have decided advantages over other machines in providing (i) sharp beam energy for precision measurements of masses, widths and couplings of the Higgs, W, t and supersymmetry particles, and (ii) high energy / high luminosity for production of high mass particles and studies of a strongly interacting electroweak sector (SEWS).

In order to be able to do interesting physics at a muon collider, the minimum luminosity requirement is

---

* Talk presented by V. Barger

$$L > \frac{1000 \text{ events/year}}{\sigma_{\text{QED}}}, \qquad (1)$$

where σ_{QED} is the cross section for the process $\mu^+\mu^- \to \gamma \to e^+e^-$,

$$\sigma_{\text{QED}} \approx \frac{100 \text{ fb}}{s(\text{TeV})^2}. \qquad (2)$$

The prototype designs under consideration well exceed the figure of merit in (1).

- First Muon Collider (FMC)

  (250 GeV) × (250 GeV) $\mathcal{L} = 2 \times 10^{33} \text{cm}^{-2} \text{s}^{-1}$ (20 fb^{-1}/year),
  $N_{\text{QED}} \approx 8000$ events/year.

- Next Muon Collider (NMC)

  (2 TeV) × (2 TeV) $\mathcal{L} = 10^{35} \text{ cm}^{-2} \text{s}^{-1}$ (1000 fb^{-1}/year),
  $N_{\text{QED}} \approx 6000$ events/year.

Special purpose rings may be added to these designs at modest cost, to optimize luminosities at specific energies for study of s-channel resonances and thresholds.

Muon colliders offer several unique and highly advantageous features. First, s-channel Higgs boson production occurs at interesting rates. The Higgs coupling is proportional to the mass, so this process is highly suppressed at e^+e^- and pp colliders. Second, a fine energy resolution is an intrinsic property of muon colliders. A beam energy resolution $R = 0.04$ to 0.08% is natural and a resolution down to $R = 0.01\%$ is realizable.[1] By comparison, $R > 1\%$ is expected at an e^+e^- machine. The root mean square spread in center-of-mass energy $\sigma_{\sqrt{s}}$ is given in terms of R by

$$\sigma_{\sqrt{s}} = (7 \text{ MeV}) \left(\frac{R}{0.01\%}\right) \left(\frac{\sqrt{s}}{100 \text{ GeV}}\right). \qquad (3)$$

The monochromicity of the c.m. energy is *vital* for s-channel Higgs studies and *valuable* for threshold measurements. Furthermore, for mass measurements a c.m. calibration can be obtained of MeV accuracy, and the necessary energy calibration $\delta\sqrt{s} \sim 10^{-6}$ may be achieved with spin rotation measurements of polarized muons in the ring. A final, but no less significant aspect of muon colliders, is that their c.m. energy reach is extendable into the $\sqrt{s} > 1$ TeV range where new physics is expected (either SUSY or SEWS).

II THRESHOLD PHYSICS AT THE FMC

Precision measurements of M_W and m_t can provide important constraints on the Higgs mass in the SM, or other new physics beyond the SM, through the relation

$$M_W = M_Z \left[1 - \frac{\pi \alpha}{\sqrt{2}\, G_\mu\, M_W^2 (1 - \delta r)}\right]^{1/2}, \qquad (4)$$

where the loop contributions δr depend on m_t^2 and $\log m_h$ in the SM and also on sparticle masses in supersymmetry. The optimal relative precision of M_W and m_t measurements is

$$\Delta M_W \approx \frac{1}{140} \Delta m_t, \qquad (5)$$

for example $\Delta M_W \approx 2$ MeV for $\Delta m_t \approx 200$ MeV.

A WW and $t\bar{t}$ thresholds[2]

There is excellent potential to make very precise M_W, m_t and m_h measurements at the FMC because of the sharp beam energy, the suppression of initial state radiation and the absence of significant beamstrahlung. With 100 fb^{-1} luminosity just above the WW threshold ($\sqrt{s} \approx 161$ GeV) a M_W precision

$$\Delta M_W = 6 \text{ MeV} \qquad (6)$$

is attainable. With 100 fb^{-1} devoted to a 10 point scan ($\sqrt{s} = 1$ GeV) over the $t\bar{t}$ threshold, the top quark mass could be measured to an accuracy

$$\Delta m_t = 70 \text{ MeV}. \qquad (7)$$

Then, with improved determinations of α, α_s, and $\sin^2 \theta_w$, such M_W and m_t measurements would constrain the SM Higgs to a precision

$$m_h{}^{+0.13 m_h}_{-0.11 m_h}, \qquad (8)$$

(e.g. 100^{+13}_{-11} GeV). The limiting factor on Δm_h is the uncertainty on M_W. Correspondingly tight constraints would be placed on other new physics. The shape of the $t\bar{t}$ threshold also constrains $\alpha_s(M_Z)$, the top quark width Γ_t, and possibly also m_h (via h-exchange contributions).

B Zh threshold[3]

In the minimal standard model (MSSM), the parameters of the Higgs sector are

$$\tan\beta = \frac{v_2}{v_1} \quad \text{and} \quad m_A \tag{9}$$

at tree level, and also the top-quark and stop masses and stop mixing at the loop level. The light Higgs boson is the only supersymmetric particle with an iron-clad upper bound, which is[4]

$$m_h < 64\text{--}105 \text{ GeV} \quad \text{for } \tan\beta \simeq 1.8, \tag{10}$$
$$m_h < 98\text{--}125 \text{ GeV} \quad \text{for } \tan\beta \simeq 60. \tag{11}$$

Thus it is the first target for experimental searches.

A light Higgs boson will be produced at a lepton collider via the Z-bremsstrahlung process

$$\ell^+\ell^- \to Z^* \to Zh. \tag{12}$$

With high resolution energy measurements the recoil mass and h-mass reconstruction give high precision m_h. With 50 fb^{-1} luminosity at a $\sqrt{s} = 500$ GeV e^+e^- collider, the following precisions are anticipated

$$\Delta m_h = 180 \text{ MeV} \quad \text{(SLD type detector)}, \tag{13}$$
$$\Delta m_h = 20 \text{ MeV} \quad \text{(super-JLC detector)}. \tag{14}$$

A threshold measurement of the Zh production cross section provides a new approach to precisely determine m_h. Figure 1 shows the predicted energy dependence of the cross section. The $Zb\bar{b}$ background is very small, except in the case when $m_h \sim M_Z$. A measurement with 100 fb^{-1} luminosity at $\sqrt{s} = m_h - M_Z + 0.5$ GeV of the cross section at a muon collider would yield a SM Higgs mass to within

$$\Delta m_h = 45 \text{ MeV} \tag{15}$$

for $m_h = 100$ GeV.

III s-CHANNEL HIGGS PHYSICS[5]

A unique capability of a muon collider is the production of a Higgs resonance in the s-channel, $\mu^+\mu^- \to h \to b\bar{b}$. The light quark background can be rejected by b-tagging. The resonance cross section is

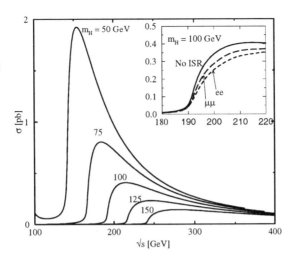

FIGURE 1. The cross section vs. $\sqrt{s}$ for the process $\ell^+\ell^- \to Z^*h \to f\bar{f}h$ for a range of Higgs masses. The inset figure shows the detailed structure for $m_h = 100$ GeV in the threshold region. Also shown in the inset figure are the effects of initial state radiation (ISR) and beam energy smearing assuming a Gaussian spread $R_e = 1\%$ for e^+e^- and $R_\mu = 0.1\%$ for $\mu^+\mu^-$.

$$\sigma_h = \frac{4\pi \Gamma(h \to \mu\bar{\mu}) \, \Gamma(h \to b\bar{b})}{\left(s - m_h^2\right)^2 + m_h^2 \Gamma_h^2}. \tag{16}$$

One needs to tune the Higgs energy to $\sqrt{s} = m_h$ by an energy scan in the vicinity of m_h. The signal is enhanced with polarized beams if the luminosity decrease with polarization[6] is less than $(1+P)^2/(1-P)^2$ which is 10 for $P = 0.84$.

The Higgs resonance profile depends on the total Higgs width, Γ_h, which is highly model dependent. Figure 2 shows expectations for Γ_h in the SM and the MSSM. A light ($\lesssim 125$ GeV) SM Higgs has a relatively narrow width, $\Gamma_h \sim$ few MeV. The width of the light Higgs in the MSSM is larger than that of the SM Higgs and scales up roughly with $(\tan\beta)^2$. Figure 3 shows the light Higgs resonance profile for the SM Higgs and the MSSM Higgs at $\tan\beta = 10$ and 20, for resolutions $R = 0.01\%$, 0.06% and 0.1%. For a resolution $\sigma \sim \Gamma_h$, the Breit-Wigner line shape can be measured and Γ_h determined. To be sensitive to Γ_h of a few MeV, a resolution $R = 0.01\%$ is needed.

Higgs Total Widths

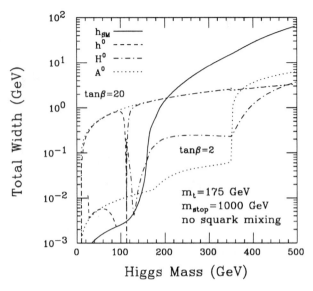

FIGURE 2. Total width versus mass of the SM and MSSM Higgs bosons for $m_t = 175$ GeV. In the case of the MSSM, we have plotted results for $\tan\beta = 2$ and 20, taking $m_{\tilde{t}} = 1$ TeV and including two-loop radiative corrections, neglecting squark mixing; SUSY decay channels are assumed to be absent.

A Energy scan

An energy scan can be made for a s-channel Higgs resonance in the mass range 63 GeV $< m_h < 2M_W$. For detection we require a significance $S/\sqrt{B} > 5$ and assume excellent resolution $R = 0.01\%$ ($\sigma_{\sqrt{s}} = 7$ MeV) to assure sensitivity to the narrow width of a SM Higgs. The necessary luminosity per scan point for representative m_h values is

L/point	m_h
0.01 fb^{-1}	110 GeV
0.1 fb^{-1}	except near M_Z
1 fb^{-1}	near M_Z

The number of scan points needed to zero in on $m_{h_{SM}}$ within 1 rms spread $\sigma_{\sqrt{s}} = 7$ MeV is

# scan points	L_{total}	δm_h
230	2.3 fb^{-1}	800 MeV (LHC)
3	0.03 fb^{-1}	20 MeV ($\ell^+\ell^- \to Zh$)

where the right-hand column is the assumed prior knowledge on δm_h.

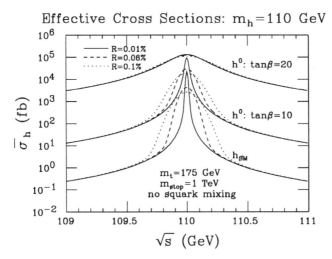

FIGURE 3. The effective cross section, $\bar{\sigma}_h$, obtained after convoluting σ_h with the Gaussian distributions for $R = 0.01\%$, $R = 0.06\%$, and $R = 0.1\%$, is plotted as a function of $\sqrt{s}$ taking $m_h = 110$ GeV. Results are displayed in the cases: h_{SM}, h^0 with $\tan\beta = 10$, and h^0 with $\tan\beta = 20$. In the MSSM h^0 cases, two-loop/RGE-improved radiative corrections have been included for Higgs masses, mixing angles, and self-couplings assuming $m_{\tilde{t}} = 1$ TeV and neglecting squark mixing. The effects of bremsstrahlung are not included in this figure.

B Fine Scan

Once a rough scan has determined the Higgs mass to an accuracy

$$\delta m_h = \sigma_{\sqrt{s}} d \quad \text{with } d \lesssim 0.3 \tag{17}$$

a three-point fine scan can pinpoint the Higgs mass to still higher precision. For this, the optimal distribution of luminosity is L on the peak location found in the rough scan and $2.5L$ on each of the resonance wings at $\pm 2\sigma_{\sqrt{s}}$ from the peak. The measurement of $\sigma_{\text{wings}}/\sigma_{\text{peak}}$ improves the m_h precision and measures Γ_h. As an illustration we consider $m_{h_{SM}} = 110$ GeV for which $\Gamma_{h_{SM}} = 3$ MeV. Then with a total luminosity $L_{\text{total}} = 3$ fb^{-1} and a resolution $R = 0.01\%$, a fine scan would yield

$$\delta m_h = 0.4 \text{ MeV} \qquad \delta\Gamma_h = 1 \text{ MeV}, \tag{18}$$

which is a 30% measurement of the SM Higgs width.

C h_{MSSM} or h_{SM}?

After the Higgs discovery, the next burning question is whether it is the SM Higgs or the MSSM Higgs boson. There are two ways to know. The first way is to measure Γ_h and $\Gamma(h \to \mu\mu) \times B(h \to \bar{b}b)$. The couplings of the MSSM Higgs to $\bar{b}b$ and $\mu^+\mu^-$ are substantially greater than the SM Higgs coupling to a heavy MSSM Higgs mass of $m_H \sim 400$ GeV. The two scenarios are thereby distinguishable at the 3σ level with $L = 50$ fb^{-1} and $R = 0.01\%$, except for m_h near M_Z.

The second way is to find the other heavier MSSM Higgs bosons. At the LHC there are some regions of $\tan\beta$ vs. m_A where only the lightest MSSM Higgs boson can be discovered. In the larger m_A limit of many supergravity models, the masses of H^0, A^0, and $H^\pm$ are approximately degenerate and h looks increasingly like h_{SM} in its properties. There are 3 possible H^0, A^0 search techniques at muon colliders:

1. Scan for s-channel Higgs

    With $L_{\text{total}} = 50$ fb^{-1} the H^0, A^0 discovery prospects are robust for 250 GeV $\leq m_{H^0,A^0} \lesssim \sqrt{s}$ and $\tan\beta \gtrsim 3$. Overlapping H^0, A^0 resonances can be separated by the scan; see Fig. 4. The H^0, A^0 widths ($\Gamma \sim 0.1$ to 0.6 GeV) are larger than resolution and can be measured by the scan.

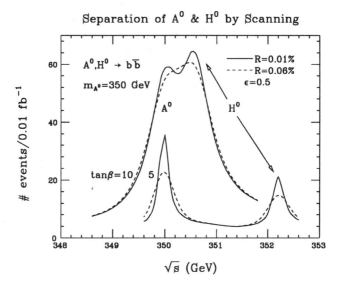

FIGURE 4. Plot of $b\bar{b}$ final state event rate as a function of $\sqrt{s}$ for $m_{A^0} = 350$ GeV, in the cases $\tan\beta = 5$ and 10, resulting from the H^0, A^0 resonances and the $b\bar{b}$ continuum background. We have taken $L = 0.01$ fb^{-1} (at any given $\sqrt{s}$), efficiency $\epsilon = 0.5$, $m_t = 175$ GeV, and included two-loop/RGE-improved radiative corrections to Higgs masses, mixing angles and self-couplings using $m_{\tilde{t}} = 1$ TeV and neglecting squark mixing. SUSY decays are assumed to be absent. Curves are given for two resolution choices: $R = 0.01\%$ and $R = 0.06\%$

2. Bremsstrahlung tail

When the muon collider is run at full energy, s-channel production of H^0, A^0 will result from the luminosity in the bremsstrahlung tail; see Fig. 5. This production is competitive with the scan search for $\tan\beta \gtrsim 5$–7 and invariant mass resolution $\Delta M_{b\bar{b}} = \pm 5$ GeV.

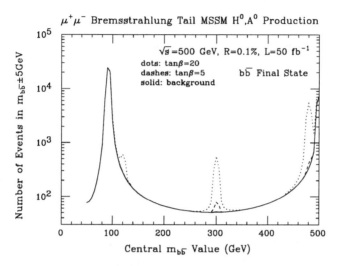

FIGURE 5. Taking $\sqrt{s} = 500$ GeV, integrated luminosity $L = 50$ fb^{-1}, and $R = 0.1\%$, we consider the $b\bar{b}$ final state and plot the number of events in the interval $[m_{b\bar{b}} - 5 \text{ GeV}, m_{b\bar{b}} + 5 \text{ GeV}]$, as a function of the location of the central $m_{b\bar{b}}$ value, resulting from the low $\sqrt{s}$ bremsstrahlung tail of the luminosity distribution. MSSM Higgs boson H^0 and A^0 resonances are present for the parameter choices of $m_{A^0} = 120, 300$ and 480 GeV, with $\tan\beta = 5$ and 20 in each case. Enhancements for $m_{A^0} = 120, 300$ and 480 GeV are visible for $\tan\beta = 20$; $\tan\beta = 5$ yields visible enhancements only for $m_{A^0} = 300$ and 480 GeV. Two-loop/RGE-improved radiative corrections are included, taking $m_t = 175$ GeV, $m_{\tilde{t}} = 1$ TeV and neglecting squark mixing. SUSY decay channels are assumed to be absent.

3. HA, H^+H^+ pair production

At the NMC (4 TeV) the discovery a very heavy Higgs boson is feasible via the the processes $\mu^+\mu^- \to Z^* \to HA, H^+H^+$. Cross sections are illustrated in Fig. 6. Once discovery is made, special storage rings can be constructed with c.m. energy $\sqrt{s} \sim m_A, m_H$ to measure the Higgs widths and partial widths.

Note that at the NMC the large event rates for production of the light Higgs boson may allow measurement of rare decay modes there, e.g. $h \to \gamma\gamma$.

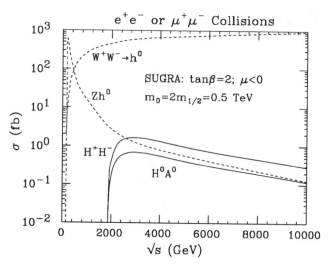

FIGURE 6. Pair production of heavy Higgs bosons at a high energy lepton collider. For comparison, cross sections for the lightest Higgs boson production via the Bjorken process $\mu^+\mu^- \to Z^* \to Zh^0$ and via the WW fusion process are also presented.

IV ADVANTAGES/NECESSITY OF A HIGH ENERGY MUON COLLIDER

A compelling case for building a 4 TeV NMC exists for both the weakly or strongly interacting electroweak symmetry breaking scenarios.

A Weakly interacting scenario[7]

Supersymmetry has many scalar particles (sleptons, squarks, Higgs bosons). Some or possibly many of these scalars may have TeV-scale masses. Since spin-0 pair production is p-wave suppressed at lepton colliders, energies well above the thresholds are *necessary* for sufficient production rates; see Fig. 7. Moreover, the single production mechanisms at lepton colliders and the excellent initial state energy resolution are *advantageous* in reconstructing sparticle mass spectra from their complex cascade decays.

B Strongly interacting electroweak scenarios (SEWS)[8]

If no Higgs boson exists with $m_h < 600$ GeV, then partial wave unitarity of $WW \to WW$ scattering requires that the scattering be strong at the 1–2 TeV energy scale. The $WW \to WW$ scattering amplitude is

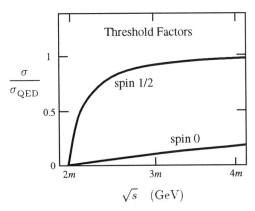

FIGURE 7. Comparison of kinematic suppression for fermion pairs and squark pair prodution at e^+e^- or $\mu^+\mu^-$ colliders.

$$A \sim m_H^2/v^2 \quad \text{if light Higgs}, \tag{19}$$
$$\sim s_{WW}/v^2 \quad \text{if no light Higgs}. \tag{20}$$

Then new physics must be manifest at high energies. Energy reach is a critical matter here with subprocess energies $\sqrt{s_{WW}} \gtrsim 1.5$ TeV needed to probe strong WW scattering. Since $E_\mu \sim (3\text{--}5)E_W$, this condition implies

$$\sqrt{s_{\mu\mu}} \sim (3\text{--}5)\sqrt{s_{WW}} \gtrsim 4 \text{ TeV}. \tag{21}$$

Thus the NMC would have sufficient energy for study of the SEWS.

The nature of the underlying physics will be revealed by the study of all possible vector boson–vector boson scattering channels, since the sizes of the signals depend on the resonant or non-resonant interactions in the different isospin channels; see Table 1.

TABLE 1. Sizes of SEWS signals in vector boson scattering channels: L (large), M (medium), S (small).

final state	resonant scalar (H^0)	resonant vector (ρ_{TC})	non-resonant (LET)
$W_L^+ W_L^-$	L	L	S
$Z_L Z_L$	L	S	M
$W_L^\pm W_L^\pm$	S	M	L
$W_L^\pm Z$	S	L	S

With 1000 fb^{-1} per year the NMC will allow comprehensive studies to be made of any SEWS signals. First, the vector resonance signals will be spectacular, as illustrated in Fig. 8. The production proceeds via vector meson dominance diagrams. On resonance, $\sqrt{s} \approx M_V$, the muon collider is a V-factory.

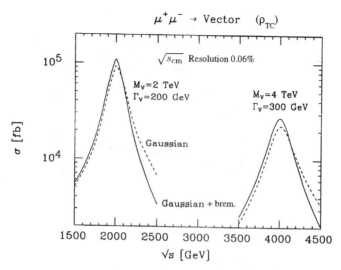

FIGURE 8. High event rates are possible if the muon collider energy is set equal to the vector resonance (Z' or ρ_{TC}) mass. Two examples are shown here with $R = 0.06\%$.

Similarly, Z' states would also give large signals.[7] Off-resonance production of a vector resonance ($\sqrt{s} > M_V$) can be detected via the bremsstrahlung luminosity. Second, the scalar particle (H) signals will be impressive. Figure 9 shows the signal

$$\Delta\sigma = \sigma(m_H = 1 \text{ TeV}) - \sigma(m_H = 0) \qquad (22)$$

expected from a 1 TeV scalar resonance. The signal cross sections are

$$\Delta\sigma(WW) = 70 \text{ fb} \quad \text{and} \quad \Delta\sigma(ZZ) = 40 \text{ fb}. \qquad (23)$$

Measurements could differentiate scalar models by measuring the resonance width to ± 30 GeV. Finally, there will be good low energy theorem (LET) signals too.

Angular distributions of the jets in the $WW \to 4$ jet final state will provide a powerful discrimination of SEWS from the light Higgs theory, as illustrated in Fig. 10. Here θ^* is the angles of q and $\bar{q}$ from W-decays in W-rest frames, relative to the W-boost direction in the WW c.m. averaged over all configurations.

The $W_L^+ W_L^- \to \bar{t}t$ channel is another valuable domain for SEWS studies, since $W_L^+ W_L^- \to \bar{t}t$ also violates unitarity at high energies. Figure 11 illustrates expected cross sections.

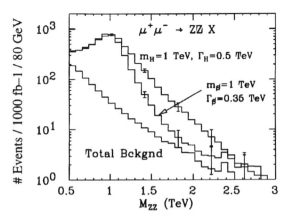

FIGURE 9. Events vs. M_{VV} for two SEWS models (including the combined backgrounds) and for the combined backgrounds alone in the ZZ final states. Signals shown are: (i) the SM Higgs with $m_{h_{SM}} = 1$ TeV, $\Gamma_H = 0.5$ TeV; (ii) the Scalar model with $M_S = 1$ TeV, $\Gamma_S = 0.35$ TeV. Results are for $L = 1000$ fb^{-1} and $\sqrt{s} = 4$ TeV. Sample error bars are shown at $M_{VV} = 1.02, 1.42, 1.82, 2.22$ and 2.62 TeV for the illustrated 80 GeV bins. Results are for $L = 1000$ fb^{-1} and $\sqrt{s} = 4$ TeV.

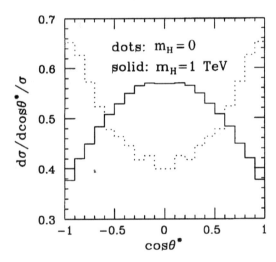

FIGURE 10. Plot of normalized cross section shapes and $dN/d\cos\theta^*$ (for $L = 200$ fb^{-1}) as a function of the $\cos\theta^*$ of the W^+ decays in the W^+W^+ final state. Error bars for a typical $dN/d\cos\theta^*$ bin are displayed. For this plot we require $M_{VV} \geq 500$ GeV, $p_T^V \geq 150$ GeV, $|\cos\theta_W^{\text{lab}}| < 0.8$ and $30 \leq p_T^{VV} \leq 300$ GeV.

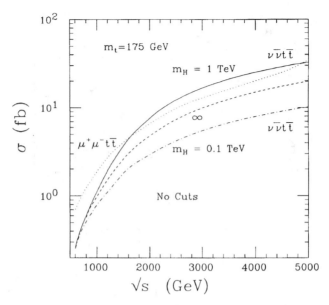

FIGURE 11. Cross section vs. $\sqrt{s}$ for $\mu^+\mu^- \to \nu\bar{\nu}t\bar{t}, \mu^+\mu^- t\bar{t}$ for Higgs masses $m_H = 0.1$ TeV, 1 TeV, and ∞.

V CONCLUSION

In summary, muon colliders are a natural match to the physics of electroweak symmetry breaking. The sharp muon beam energy allows

- precision threshold measurements of M_W, m_t, m_h to test electroweak radiative corrections.

- s-channel resonance scans to precisely determine m_h and measure Γ_h.

- discovery and study of the heavy MSSM Higgs bosons in 3 ways.

The NMC provides the c.m. energy and luminosity for

- heavy supersymmetry thresholds, or

- SEWS studies.

If the WW sector proves to be strongly interacting, the NMC is ideally suited to probe the nature of these new interactions.

ACKNOWLEDGEMENTS

This research was supported in part by the U.S. Department of Energy under Grant No. DE-FG02-95ER40896 and in part by the University of Wisconsin Research Committee with funds granted by the Wisconsin Alumni Research Foundation.

REFERENCES

1. R.B. Palmer, private communication; G.P. Jackson and D. Neuffer, private communication.
2. V. Barger, M.S. Berger, J.F. Gunion, and T. Han, Univ. of Wisconsin-Madison report MADPH-96-963 (1996).
3. V. Barger, M.S. Berger, J.F. Gunion, and T. Han, Univ. of Wisconsin-Madison preprint MADPH-96-979 (1996).
4. See e.g. M. Carena, J.R. Espinosa, M. Quiros, and C.E.M. Wagner, Phys. Lett. **B355**, 209 (1995).
5. V. Barger, M.S. Berger, J.F. Gunion, and T. Han, Phys. Rev. Lett. **75**, 1462 (1995); V. Barger, M.S. Berger, J.F. Gunion, and T. Han, Univ. of Wisconsin-Madison report MADPH-96-930 (1996), to appear in Physics Reports.
6. See, for example, Z. Parsa, $\mu^+\mu^-$ *Collider and Physics Possibilities*, unpublished.
7. V. Barger, M.S. Berger, J.F. Gunion, and T. Han, *Proceedings of the Symposium on Physics Potential and Development of $\mu^+\mu^-$ Colliders*, San Francisco, CA, 1995, ed. by D. Cline and D. Sanders, Nucl. Phys. B (Proc. Suppl.) **51A**, 13 (1996).
8. V. Barger, M.S. Berger, J.F. Gunion, and T. Han, Phys. Rev. **D55**, 142 (1997).

Linear Electron–Electron Colliders

Clemens A. Heusch

Institute for Particle Physics
University of California, Santa Cruz, CA 95064

I INTRODUCTION

In the framework of the discussion on what shape our future machine arsenal should take so as to maximize our chances of penetrating beyond the realm where our astonishingly successful Standard Model of Particle Interactions holds undisputed sway, the present contribution is somewhat unusual: I am not here to convince our community to build yet another machine. Instead, my task is to convince you that in the established choices that we are headed towards, it is of great importance that the *Electron Collider* of the next generation, i.e., in the 0.5 to 1.5 TeV energy range, should be configured just such, as an *Electron* collider, *NOT* dedicated to just one incoming charge state (say, e^+e^-).

Now that we have exceeded the energy range that can be reached with circular/recirculating machines, we are freed from the need to have oppositely charged electrons as projectiles and targets. The colliding linac configuration sets no preferential condition on the chosen net charge; in fact, this is the first time we have a machine that may well serve to collide a variety of initial states at full energy $2E_B$ and luminosity (e^+e^+, e^+e^-, e^-e^-), or at slightly reduced center-of-mass energy, but still full luminosity (γ–e, γ–γ).

I will not belabor the case for initial states including high-energy photons *beyond* mentioning, in Section IV, the intimate connection that a successful realization of these collisions has with the availability of a high-quality e^-e^- facility. Rather, I will attempt to show, briefly, that there is little if any problem in configuring an Electron Collider such that it can be run in either charge mode with comparable performance characteristics, excepting only the polarization parameter; and I will proceed to show you the very rich physics potential of the e^-e^- collision mode—some of it unique, some complementary to the promise of the more thoroughly discussed e^+e^- collision mode.

Before embarking on this enterprise, it is fair to remind you that the first electron collider was, in fact, built for the explicit purpose of testing the limits of precision to which the Standard Model of the 1950s, Quantum Electrodynamics, could be shown to follow its theoretically accepted pattern: Barber,

Gittelman, O'Neill, and Richter built their e^-e^- circular collider, with two rings, on the Stanford Campus, and were able to reach center-of-mass energies of 1012 MeV, at which they tested Møller scattering for possible cutoff or form factor effects. The first step toward testing the broader, emerging Standard Model that included the strong and weak interactions, showed the virtues of using e^+e^- annihilation, and led to the immensely successful operation of a slew of storage rings that would teach us a large fraction of our present state of knowledge, was initiated in Frascati by Bruno Touschek with his ADA ring [1]. Today's running of LEPII is, beyond any doubt, the last hurrah of the circular e^+e^- machines—and it would be disingenuous to suggest installation of a second ring in its tunnel for the purpose of running e^-e^- experiments.

Fortunately, the Linear Electron Colliders, the NLC version of which is described in detail in these proceedings, have no problem worth mentioning being configured in the e^-e^- (or, should that be of separate interest, the e^+e^+ initial state).

II MACHINE CONSIDERATIONS

Linear acceleration of electrons and positrons is identical once the phase difference with regard to the RF field is taken into account. What is not identical is the emittance of the beams entering the linac structure, and, as a result, the potential phase space effects due to wake fields building up in the accelerating structure. More differentiation needs to be considered for the interaction of the accelerated beams at the interaction point: the luminosity that can be reached with oppositely charged beams is enhanced by the electrostatic attraction of the two beams ("pinch effect"); conversely, like-charge beams repel each other and "blow up" the interaction area ("anti-pinch"). Also, there is the need for different handling of the "spent" beams beyond the interaction point—particularly in the case of non-zero crossing angle: like-sign beams need more attention because they do not automatically follow the optical path of the oppositely moving antiparticle, up to ejection into a beam dump. While all of these points are basically amenable to given technical solutions, there is one qualitative difference that cannot be made up for in any known way: electron guns can easily reach high degrees of polarization for the emerging e^- beams, and we do not believe there is any relevant limitation as to the available intensity. 80% polarized electrons are routinely used at SLAC, and there is no reason to believe that this value cannot be raised to above 90% in the intermediate time frame. It turns out that this capability is of immense value for the enhancement of a number of Beyond-the-Standard-Model effects, and for the suppression of backgrounds, as we will see below. For positron beams, there is strictly no way to reach similarly high polarization values; a number of schemes are being tested, but there is no hope of reaching anything beyond about 60%—which is not sufficient for precision work in a number of connotations.

Fairly detailed studies of the generalized luminosities $\mathcal{L}$ in terms of incident Gaussian bunches were made by J. E. Spencer [2]. He points out that while it would be nice to put the full current available from the gun, per RF pulse of the linac, into a single bunch for acceleration, this would be highly undesirable for reasons of emittance growth, energy spread, and beamstrahlen intensities. Rather, he plays a number of scenarios with multi-bunch operation ($n_B > 1$) and f_T, the number of bunch trains per second (which is the same as the RF rep rate). With N_B the number of electrons per bunch and a luminosity enhancement (or disruption) factor H_D, the luminosity can be expressed as

$$\mathcal{L} = \frac{f_T n_B N_B^2 H_D}{4\pi\sigma_z^*\sigma_y^*}\zeta \to \frac{f_T n_B N_B^2 \gamma H_D}{4\pi\epsilon_n \beta^*}\zeta = \frac{f_T n_B N_B^2 \gamma}{4\pi\hat{\sigma}^2}\zeta \propto \frac{P_b}{\hat{\sigma}^2}\left(\frac{\hat{N}_b^2}{N_b}\right). \quad (1)$$

Here, ζ is an efficiency factor which may well approach 1, and the σ are geometrical transverse spot sizes. Multibunch trains of $n_B = 100$ will help to distribute the total charge per RF pulse more evenly down the linac structure, which may well help to make electron currents easier to raise than positron currents that are injected into the linac from cooling rings. Many practical problems have to be addressed—the multibunch operation will necessitate a crab-crossing interaction geometry, and overall luminosity optimization may well make a plasma lens advisable for the compensation of the electrostatic beam-beam repulsion [3]—but overall there is an expectation that the implementation of a highly stable e^-e^- operation of the Next Linear Collider will have little trouble coming in with luminosities commensurate with what an e^+e^- version of the same machine can do.

Given the high demands on instrumentation that will be needed to produce efficient photon beams from laser photons backscattered within less than 1 cm of the IR off the incident electron beams, it will be very unwise to couple e^-e^- experimentation a priori with e-γ and/or $\gamma\gamma$ operation. Rather, the urgency of the physics program that can fruitfully be addressed by the e^-e^- mode argues powerfully for the implementation of the electron–electron version early on, at turn-on time of the colliding linacs. The physics motivations that will have to decide on the appropriate initial-state choice therefore are our paramount interest.

III PHYSICS PROMISE

In discussing the motivations for implementing the electron-electron version of the Next Linear Collider (or its equivalent), we will follow roughly the 1996 Snowmass Study organization. In an attempt to highlight both the uniqueness of the goals that lend themselves to experimental investigation from an e^-e^-

initial state and the complementarity of the different approaches, we will treat a few problems in more detail than others; this will serve to illustrate the strengths of this channel, but does not imply a lack of interest in the studies more cursorily advanced below. We hope to rectify any such impression by the concluding compilation of which physics problem will be accessible to which type of experimental approach.

A Weak Electroweak Symmetry Breaking: Higgs Bosons

Whether the LHC and/or the Tevatron manage to find credible evidence for Higgs boson production, it is almost certain that the Electron Collider will play a pivotal role in the investigation of the Higgs sector of electroweak symmetry breaking (EWSB). The much-touted discovery channel for the e^+e^- version of our collider is via the "Higgs radiation" graph of e^+e^- annihilation into a virtual Z boson [Fig. 1(a)]. While this graph provides a good signature (particularly for the fraction where the Z decays into muon pairs) and a sizeable cross-section close to threshold, its production rate drops with $1/s$. At higher energies, WW fusion and ZZ fusion [Fig. 1(b)] take over; but while the first of these has a factor of ten cross-section advantage over the latter, the undetectable neutrinos in the final state much decrease the discovery potential of this process, whereas the latter has the advantage of furnishing us with the scattered electron-positron pair for final-state reconstruction.

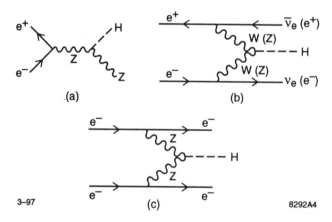

FIGURE 1. (a) "Higgs radiation": e^+e^- annihilation into a virtual Z boson followed by "Higgsstrahlung"; (b) WW fusion and ZZ fusion in e^+e^- collisions; (c) central production of Higgs boson in e^-e^- collisions. H^0 mass can be reconstructed if outgoing electron momenta are measured.

This is where the e^-e^- initial state takes over [Fig. 1(c)]: like-sign electron pairs are a rare background product, so that, for the kinematic region $\sqrt{s} > 2.5\, m_H$, the process $e^-e^- \to e^-e^-H$ provides the favored discovery and study channel. In a recent paper, Minkowski [4] investigated the usefulness of this method for a detailed study of Higgs boson production and decay by means of a good measurement of the tagged electron angles and energies: the central mass is well resolved by

$$M_{rec}^2 = (p_{rec})^2; \qquad (2)$$

its decay can be subsequently determined.

This procedure has several signal advantages:

a. In contrast to the Higgsstrahlung graph (in e^+e^- annihilation), the ZZ fusion cross-section saturates above the threshold region; it becomes proportional to m_Z^2, and it does not depend much on the scalar mass as long as this is well below the center-of-mass energy at which the measurement is being performed.

b. Once our detector imposes an angular cut (say, of a five degree cone in the forward and backward directions), the cross-section becomes geometric and decreases as s^{-1} (cf. Fig. 2), but remains roughly independent of the scalar mass.

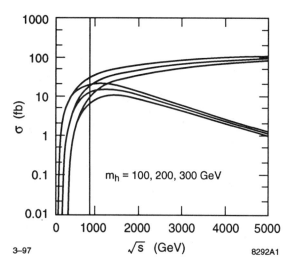

FIGURE 2. Cross-section for production of a standard scalar as a function of c.m. energy with and without angular cut (upper curves), for given H masses. Incoming electrons are pure left-handed.

c. There is essentially no influence on the event rates (shown, for typical parameters, in Fig. 3) due to the helicities of incoming electrons. Since the principal background is W^+W^- pair production via $\gamma\gamma$, and to a much reduced degree, ZZ fusion according to

$$e^-e^- \to e^-e^-W^+W^-, \tag{3}$$

the detector-imposed small-angle cut serves to diminish the background signal by virtue of the fact that many of the final-state electrons will exit at smaller angles.

d. The central mass thus studied may well decay "invisibly"—say, into gluino pairs—and it will not detract from our discovery potential; this is indicated in the mass plot shown in Fig. 4: the mass peak is strictly due to the final electrons' kinematics, while the background is due to added W pair production.

Obviously, this measurement is applicable to all CP-even scalars, such as either the basic H boson or its MSSM (minimal supersymmetric) cousin h; it changes only in angular distribution for the CP-odd A boson. If the LHC or the Tevatron discover any of these, our process may well help define the optimal running energy of an NLC, for a detailed and clean investigation of Higgs boson properties.

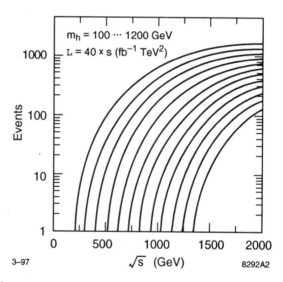

FIGURE 3. Event rates for production of neutral Higgs bosons by ZZ fusion from left-handed e^-e^- interactions, for given luminosity.

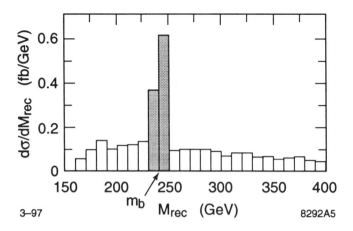

FIGURE 4. Differential cross-section for production of a 240 GeV/c² H^0 scalar at $\sqrt{s} = 850$ GeV, together with background due to the reaction $e^-e^- \to e^-e^-W^+W^-$. An angular cut between incoming and outgoing electrons is needed for recoil mass definition and background reduction.

B Extended Higgs Sector

Whereas neutral Higgs boson production will occur in minimal or extended models of mass generation, so that its observation will not *eo ipso* give definitive answers to the scenario in which they make their appearance, singly or doubly charged scalars are more specific to the models which contain them as telling components. Clearly, charge, helicity, and weak hypercharge conservation permit the e^-e^- initial state with adjustable helicities to fathom a broad class of extended models.

In the simplest tree-level realization of such extensions—most familiar to us from the MSSM framework—there is a minimal two-Higgs-doublet scenario. It lends itself to a detailed study by means of the reaction

$$e_L^- e_L^- \to W^- W^- \nu_e \nu_e \; (W^- W^- \to H^- H^-), \tag{4}$$

with two negative W bosons scattering into two negative Higgs bosons, and missing momentum taking the place of the unobserved neutrinos in the final state (cf. Fig. 5). Rizzo [5] studied this process in some detail and points out that, while the bare tree-level MSSM H^-H^- production cross-section is probably below detection level at an NLC, radiative corrections will raise the cross-section dramatically. Once we go beyond a basic two-doublet model, it may take some care to remain within the bounds of the $\rho = 1$ condition, but there are credible choices that add various basic triplet scalars (for a relatively simple version, see Ref. [6]). Spontaneous symmetry breaking within a custodial SU(2) framework leads to physical $\underset{\sim}{5}$ and $\underset{\sim}{3}$ scalar representations

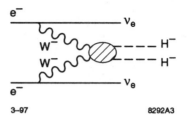

FIGURE 5. WW fusion graph for H^- pair production.

with singly and doubly charged Higgs bosons. Depending on parameter choices that include a possible resonance structure if the quintuplet mass is more than twice that of the triplets, such scenarios may well lead to wide variations in the resulting cross-sections, up to values in the 100 fb region.

Gunion [7] devoted a special study to the chances of discovering doubly charged Higgs bosons in e^-e^- scattering: given that we have no clear idea on the presumptive coupling strength c_{ee} for direct $H^{--}e^-e^-$ coupling [Fig. 6(b)], we have to resort to H^{--} production via W^-W^- fusion [shown in Fig. 6(a)]—which occurs to the tune of a full unit of R (the QED width for $e^+e^- \to \mu^+\mu^-$ at given energy), illustrated for various m_H values in Fig. 7. The H^{--} states may decay via W^-H^-, adding special spice to this field of inquiry.

Given that any putative effect can immediately be killed by appropriate manipulation of the incoming helicity, it becomes obvious that an electron-electron collider provides the ideal laboratory for the investigation of a large set of extended Higgs sector scenarios.

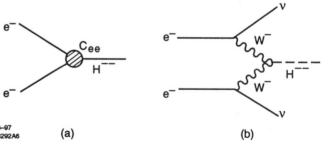

FIGURE 6. (a) Direct production of H^{--} from an e^- pair, with unknown coupling c_{ee}, and (b) WW fusion graph for H^{--} production.

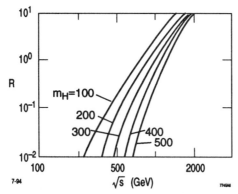

FIGURE 7. Production cross sections, by an NLC/TLC e^-e^- collider, of doubly charged H^{--} bosons in the mass range from 100 to 500 GeV/c^2, in units of R.

C Supersymmetry

In the coming years, the frenzied search for Supersymmetry signals will doubtlessly continue at LEPII and at the Tevatron. But it can confidently be anticipated that we will need higher-energy electron colliders for dedicated and focused studies of many relevant SUSY parameters—even though the first discoveries may well precede their turn-on. A detailed investigation of the sparticle mass spectrum, of decay distributions, and of all the relevant couplings will be much helped by the tight control we have over the energy and polarization of our incident beams. The electron-electron mode, simply by dint of its easily controlled helicities, is particularly suited to do selectron searches, to measure neutralino masses, and to determine the most important U(1) and SU(2) couplings.

The first of these concerns is well-illustrated in a study by Cuypers *et al.* [7], who studied the production of selectron pairs and chargino pairs in electron-electron collisions as in Figs. 8(a) and 8(b). As Peskin points out in his SUSY lecture notes [8], the preferential coupling of e_R^- to the selectron causes production cross-sections to be widely different (by three orders of magnitude) for the RR vs LL incoming helicity combinations, as shown in Fig. 9. That, in turn, has the considerable advantage of decoupling the right-handed incoming electrons, preferred for selectron pair production, from the W^- emitted in the lowest-order background graphs. The principal remaining background process in the final state e^-e^- plus missing transverse momentum (from unobserved neutralinos) would be due to the reaction

$$e^-e^- \to e^-e^- Z^0(Z^0 \to \nu\bar{\nu}). \tag{5}$$

It can fairly be excluded through judicious electron energy cuts, as indicated in Fig. 10(b). A purely kinematical evaluation of this plot can also yield a

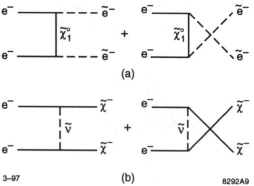

FIGURE 8. (a) $\tilde{e}$ pair production from e^-e^- collisions by $\tilde{\chi}_0$ exchange. (b) Like-sign $\tilde{\chi}^-$ pair production by $\tilde{\nu}$ exchange.

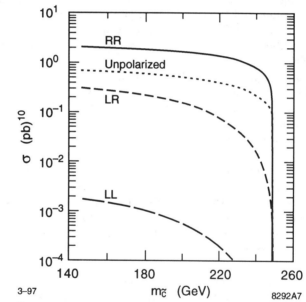

FIGURE 9. Cross-section for $\tilde{e}^-$ pair production from e^-e^- collisions as a function of $\tilde{e}$ mass, for an e^-e^- collider of $\sqrt{s} = 500$ GeV, with the incident electron helicities as indicated. For SUSY parameters see Ref. [7].

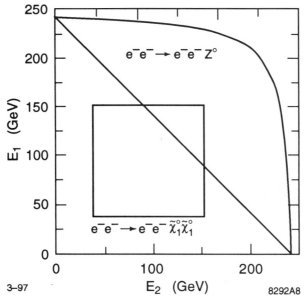

FIGURE 10. Permitted energy range for the two electrons observed in the two reactions mentioned in the figure. For $\tilde{e}$ decay, we assumed $m(\tilde{e}) = 200$ GeV, $m_{\tilde{\chi}^0} = 100$ GeV. The diagonal cut eliminates the background while losing little signal.

direct measurement of the masses both of the selectron and of the lightest neutralino.

The chargino pair production process (Fig. 8(b)) profits even more from the well-defined helicity content of the initial state: charginos couple only to left-handed leptons, so that only the LL combination will contribute. The fact that their masses as well as their couplings to selectrons (in the decay channel that leaves only a neutrino unobserved) are functions of the three SUSY parameters $\tan\beta$, μ, and M_2, further enhances the status of a clean search as an agent for a fuller exploration of our most elegant candidate for an extension of the Standard Model.

In the same vein, it is important to notice that for the case where the higgsino is the lightest SUSY particle, the masses of neutralino and chargino are closely spaced; this makes the like-sign selectron pair production process, the background of which is depressed by about an order of magnitude below the e^+e^- case, particularly significant.

D Strong EWSB: Strong WW Scattering

It is well-understood that, should we not find an elementary Higgs boson well below the TeV level, the Higgs mechanism will have to be impersonated

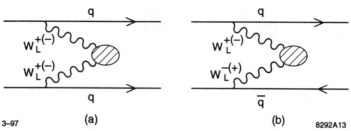

FIGURE 11. Strong WW scattering from (a) quarks in hadron collisions and (b) quarks-antiquarks in hadron collisions.

by a new strong interaction of longitudinal W bosons—dubbed the Fifth Force by M. Chanowitz [9]. This will lead to an entire field of new dynamics that will have to be fully explored—as basic in the TeV region as $\pi\pi$ scattering was below 1 GeV. This cannot possibly be exhausted at e^+e^- colliders, which are limited to the $J = 0, 1; I = 0, 1$ channels. The (possibly very distinctive) $I = 2$ channel can be uniquely well investigated in the e^-e^- collider at its upper energy reach, by means of the process

$$e^-e^- \to W^-W^-\nu\nu, \ W^-W^- \to \ell^-\ell^-\nu\nu. \tag{6}$$

Whereas hadron colliders in the TeV region will produce signals due to strong $W_L W_L$ scattering according to Figs. 11(a) and 11(b), these will be hard to separate from the expected backgrounds: the worst of these is tree-level

$$qq \to q'q' \ W^+W^+ (\text{or } W^-W^-), \tag{7}$$

visible in terms of the final states $\ell^+\ell^+$, (or $\ell^-\ell^-$) plus missing transverse momentum. Essentially all of the relevant W's are transverse, and effects of strong WW scattering will be very hard to isolate. The situation is more favorable in the e^-e^- initial state (Fig. 12), where the high polarization we can achieve for both incoming channels enhances the cross-section by a factor of four: any observable signal can therefore be strengthened by a set of judicious cuts [10,11]. If we are really lucky, it may well stick out prominently in the shape of a resonant scalar or vector state in the reconstructed WW or $b\bar{b}$ mass distributions. An example is shown in Fig. 13—which illustrates simultaneously the effect of an extended Higgs sector (cf. Fig. 6).

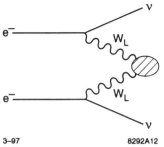

FIGURE 12. Strong $W_L W_L$ scattering in $e^- e^-$ collisions. This will work only for left-handed electrons.

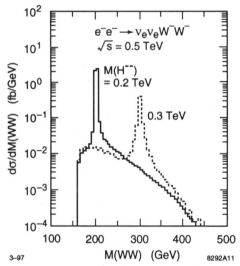

FIGURE 13. Distribution in the $W^- W^-$ invariant mass for $e^- e^- \to \nu_e \nu_e W^- W^-$, including the contribution of a doubly-charged Higgs boson of mass $M(H^{--}) = 0.2$ or 0.3 TeV.

E   Anomalous Gauge Couplings

One particular manifestation of an extended gauge group structure is the deviation of trilinear and/or quartic couplings from the Standard Model values. For the first set of these non-Abelian couplings, the relevant Lagrangean L_{eff}^{WWV} (see, e.g., Ref. [12]) contains the SM-specified coupling parameters

$$g_\gamma = e \qquad g'_\gamma = 1$$
$$g_Z = e\cot\theta_W \qquad g'_Z = 1$$

$$k_\gamma = 1 \qquad \lambda_\gamma = 0$$
$$k_Z = 1 \qquad \lambda_Z = 0, \qquad (8)$$

where electromagnetic gauge invariance fixes $g_\gamma^1 = 1$. To do a thorough investigation of all possible deviations, the Electron Collider should be used in all its charge modes for complementary information. In particular, the radiative corrections are quite different in the e^-e^- case, and should lead to added valuable information.

TABLE 1. A sampling of processes and associated gauge boson couplings measurable at e^-e^- colliders.

Process	Couplings probed
$e^-e^- \to e^-\nu W^-$	$WW\gamma, WWZ$
$e^-e^- \to e^-e^- Z$	$ZZ\gamma, Z\gamma\gamma$
$e^-e^- \to e^-\nu W^-\gamma$	$WW\gamma, WWZ$
$e^-e^- \to \nu\nu W^- W^-$	$WWWW$
$e^-e^- \to e^-\nu W^- Z$	$WWZZ$
$e^-e^- \to e^-e^- ZZ$	$ZZZZ$

The fact that the polarization of the incoming electron beams can easily be reversed, again makes the e^-e^- channel particularly useful, as the plots of the total cross-section sensitivity, for the process $e^-e^- \to e^-\nu W^-$, to small changes in the various indicated parameters illustrated in Figs. 14(a) and 14(b). Table 1 shows to which triple or quartic couplings individual e^-e^- reactions are sensitive; and Table 2 gives a comparison of the parameter space that is open to investigation by a 0.5 TeV Electron Collider with presently available coupling parameter limits. This comparison speaks for itself, but should be complemented by the capabilities of other incoming charge states.

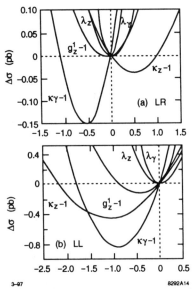

FIGURE 14. Contribution to the total cross section of each of the anomalous couplings, while all others are held to their Standard Model values, (a) for LR incoming electron helicities, (b) same, but for LL. The Standard Model cross sections are 3.19 pb for LL polarization and 0.348 pb for LR.

TABLE 2. Parameter values which can be tested by a particular experiment at 90% CL. The boldfaced numbers correspond to limits already set. For the generic 500 GeV linear collider LC500, we have assumed an integrated luminosity of 10 fb^{-1} for electron or photon beams. For the e^-e^- option we have used the combined information from LL and LR beam polarizations.

Machine/Experiment	g_z^1		K_γ		K_Z		λ_γ		λ_γ	
	min	max	min	max	min	max	min	max	min	max
UA2			-3.1	4.2			-3.6	3.5		
Tevatron			-2.4	3.7						
HERA			0.5	1.5			-2	2		
Tevatron			0.5	1.8	0.2		-0.2	0.2	-0.4	0.4
LHC			0.8	1.2	0.8		-0.02	0.02	-0.03	0.03
LEPII			0.86	1.87	0.76		-0.4	0.4	-0.4	0.4
LC500 e^+e^-			0.985	1.14			-0.02	0.04		
LC500 $e^+\gamma$			0.96	1.04			-0.05	0.05		
LC500 $\gamma\gamma$			0.98	1.015			-0.04	0.075		
LC500 e^-e^-	0.91	1.07	0.985	1.015	0.96	1.04	-0.045	0.075	-0.11	0.06

F Compositeness

As we are about to enter yet another energy regime of "point-like" particle interactions, it is only natural that we ask for the reach which this regime holds with respect to a potential energy scale where the "next layer of composite structure" might peel off. Møller scattering, it turns out, is more sensitive to the appearance of a new compositeness scale than Bhabha scattering, due to crossing term cancellations in the cross-section calculations. T. Barklow [13] investigated this issue in terms of a $1/\Lambda$ expansion of the relevant effective Lagrangean density operator, where the relevant four-fermion operators are written in terms of pure helicity fields.

By measuring the angular distribution of Møller scattering over the angular range $|\cos\theta| < 0.9$, 95% CL fits to the expected shape result in the sensitivities shown, assuming an integrated luminosity of 680 $pb^{-1} s/M_Z^2$ (for an easy comparison with existing PETRA limits), in Figs. 15 and 16. The first of these compares Bhabha with Møller scattering without giving the latter the benefit of polarization; clearly, e^-e^- must be the channel of choice at the energies covered in this plot. When we make the transition to polarized electrons, Møller scattering, at energies up to 1 TeV, is shown (Fig. 16) to be superior by a considerable stretch, reaching well beyond 150 TeV in its substructure sensitivity—obviously a highly topical capability of the TeV Electron Collider.

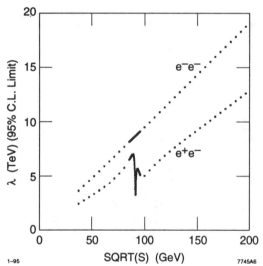

FIGURE 15. The 95% CL limits that can be obtained for the compositeness scale Λ_{LL} as a function of the e^-e^- or e^+e^- center-of-mass energy. The luminosity is given by $\mathcal{L} = 680\ pb^{-1} \times s/M_Z^2$. The beams are assumed to be unpolarized.

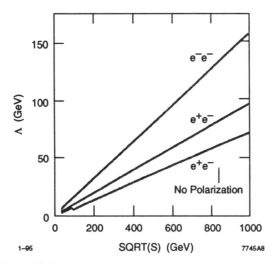

FIGURE 16. The 95% CL limits that can be obtained for the compositeness scale Λ_{LL} as a function of the e^-e^- or e^+e^- center-of-mass energy. The luminosity is given by $\mathcal{L} = 680\ pb^{-1} \times s/M_Z^2$. The polarization of the electron beam(s) is indicated in the figure.

G New Gauge Bosons: Z', Dileptons

Another standard quest, as we widen our kinematic parameter space, is the search for new gauge bosons—heavier Z' with either standard or exotic couplings, but potentially also more exotic bosons that might show up as s-channel structure in e^-e^- collisions.

The exchange of heavier Z' bosons in e^+e^- and e^-e^- scattering has been investigated in a sequential Standard Model, in LR-symmetric models, and in broader classes of models [14] – [16]. We limit ourselves to show, in Fig. 17, how sensitive Møller scattering is in comparison to Bhabha scattering when it comes to resolving the coupling parameters $v_{Z'}$, $a_{Z'}$ which enter into the cross-sections, at 95% CL, for a Z' mass of 2 TeV, and an Electron Collider energy of 0.5 TeV. For $M_{Z'} \gg \sqrt{s}$, we reach a better sensitivity in Møller scattering, where, again, an assumed 90% polarization makes a decisive difference.

More dramatic effects can be expected from dilepton (more recently renamed BIlepton) gauge bosons (with two units of charge and lepton number), such as P. Frampton has been proposing in his 331 model—which has the elegant implication of motivating the three-generation structure of the basic fermions in terms of a "natural" triangle anomaly cancellation [17]. It turns out that such states might come naturally in SU(15) grand unifying groups; with narrow widths, they would certainly lead to spectacular signals as s-channel peaks decaying into back-to-back high-transverse-momentum like-sign lepton pairs (Fig. 18).

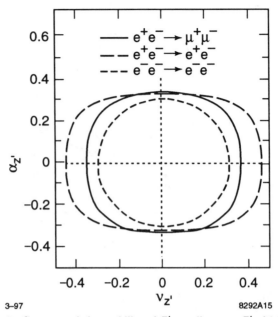

FIGURE 17. Contours of observability of Z' coupling to a Z' of 2 TeV mass.

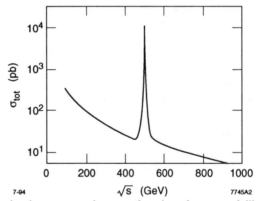

FIGURE 18. Production cross-section, as a function of energy, of dilepton gauge bosons in e^-e^- collisions. The 500 GeV mass and narrow width were assumed for illustration purposes.

H Heavy Majorana Neutrinos

One of the most tantalizing capabilities of our next Electron-Electron Collider's is quite unique, and will not easily be intruded upon by other machines: should neutrinos have Majorana mass terms, and should there be heavy (TeV-level) neutrino isosinglet states—as comes naturally in $E(6) \to SO(10) \to \ldots$ decompositions [18], a TeV e^-e^- collider might well provide spectacular signals for the process

$$e_L^- e_L^- \to W_{\text{long}}^-, W_{\text{long}}^-. \tag{9}$$

Unmistakable final-state signatures such as two unaccompanied high-p_T, like-sign leptons or two back-to-back jets that reconstruct to W masses, will be easily separated from backgrounds—and may therefore contribute to the resolution of two of the Standard Model's completely unexplained conundrums: the (SM-sanctioned) masslessness of neutrinos—through appropriate mass mixings—and the (presently absent) definition of the proper field operators for our neutral lepton sector.

The relevant scenario has been elaborated by Heusch and Minkowski [19], including the compatibility of a possible detection with present limits on the observation of neutrinoless nuclear double beta decay [20]. Again, cross-sections

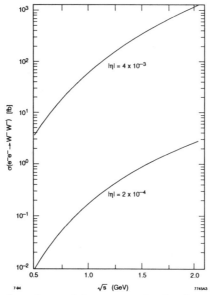

FIGURE 19. Energy dependence of the cross-section for the process $e^-e^- \to W_L^- W_L^-$ in the energy range of an NLC-type machine of a typical luminosity 10–100 fb^{-1}, for the range of neutrino mass matrix choices discussed in Ref. [18].

are detectably large only in one helicity combination; this means that even a few characteristic events will be able to establish an effect: a helicity flip in the incoming e^-e^- channel will have to eliminate the entire signal.

Obviously, the possible existence—well above any mass level that would have lent itself to direct observation—of Majorana neutrinos, will have to be examined in the framework of rare decays they might help mediate. The ensuing range of possible cross-sections for our process is shown in Fig. 19, as a function of collider energy [21]. A thorough examination of competing backgrounds [22] including only the hadronic decays of final-state W's is shown in Fig. 20. With anticipated collider luminosities, this energy range will fairly ensure detection within the first year of e^-e^- operation.

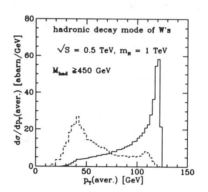

FIGURE 20. Signal (solid line) and background (dashed line) as a function of the average transverse momentum $p_\perp(aver.)$ of the jets. The hadronic invariant mass is required to be $m_{had} \geq 450$ GeV and the other parameters are chosen to be $\sqrt{s} = 500$ GeV, $m_N = 1$ TeV, $U_{eN}^2 = 4 \times 10^{-3}$ (Ref. [22]).

IV THE ELECTRON-ELECTRON COLLIDER AS A PARENT CONFIGURATION FOR ELECTRON-PHOTON AND PHOTON-PHOTON COLLISIONS

We mentioned in the introductory remarks that the Next Electron Collider will provide not only ee initial states at full energy and luminosity, but also $e\gamma$ and $\gamma\gamma$ collisions with somewhat commensurate parameters. To make this promise come true by means of the backscatter of TeraWatt-powered laser photons off the incident electrons, a number of difficult experimental and technical challenges will have to be mastered [23]. What concerns us in the present context is the need to have those laser photons scatter off highly polarized, fully accelerated electrons (or positrons).

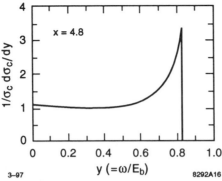

FIGURE 21. Spectrum of backscattered photons in the absence of electron polarization: an almost flat distribution for $y < 0.7$ builds up to a peak at the kinematic limit.

The motivation for this demand can easily be read off Figs. 21 and 22: Fig. 21 shows that the spectrum of the backscattered photons in the absence of electron polarization is essentially flat, with some moderate peaking at the upper end. This is clearly unacceptable as a kinematical definition of an "incoming beam." In Fig. 22(a), a fully polarized electron beam backscatters a like-helicity laser photon beam. The resulting backscattered like-helicity photon spectrum is very broad, and useless. Figure 22(b), on the other hand, has opposite-helicity laser photons backscatter off a fully polarized electron beam: a sharply peaked, highly polarized backscattered photon beam ensues, and only the small low-energy tail has the opposite polarization.

It is quite evident that there is no promise of useful photon-photon or electron-photon experimentation at the Electron Collider unless a high-quality e^-e^- version is implemented early on, and fully operational.

V CONCLUSION

As we are headed for a more closely defined proposal stage for the Next Electron Collider, it is incumbent on us to evaluate with the greatest care the physics potential of the collider configuration we believe to maximize its usefulness at unravelling the questions we will be facing some ten years from now. Given that the accelerator physics community will have no trouble whatever coming up with a design that permits us the choice of electron-electron and electron-positron initial states with comparable luminosities, identical energies, but differing helicity definitions, it is now time to scuttle the historical preoccupation with the annihilation diagram that has given us much of what the past decades taught us about the Standard Model and its signal successes. It is imperative that, in a fresh start, we take the trouble to consider what behooves us most on the exploratory trail *BEYOND* the Standard Model.

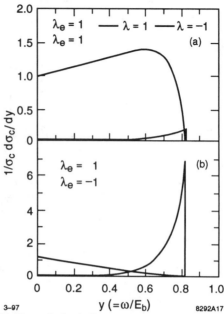

FIGURE 22. For highly polarized electrons, the spectrum and polarization of the backscattered photons is distinct. (a) For equal incident photon and electron helicities, the scattered photons have a broad distribution, and are highly polarized, retaining the incoming electron helicity. (b) For opposite incoming helicities, the spectrum is sharply peaked at the upper end, again with the incoming electron helicity. Only the broad low-energy tail has the opposite helicity.

TABLE 3. Additive quantum numbers of e^-e^- initial states.

	Q_{el}	s_z	L	L_e	I_3^W	Y^W
$e_L^- e_L^-$	-2	$1, 0, -1$	2	2	-1	-2
$e_L^- e_R^-$	-2	$1, 0, -1$	2	2	$-1/2$	-3
$e_R^- e_R^-$	-2	$1, 0, -1$	2	2	0	-4

As we do so, it is worth our while to look at the available quantum numbers of the initial state in the electron-electron configuration: Table 3 gives that information. The examples we chose above for an illustration of an e^-e^- physics program argue powerfully for a clean slate from which to choose among available incoming states when the time for decisions comes. Table 4 gives a possibly helpful overview of the principal topics we will want to attack once the new device is available some ten years hence: let us make sure our hardware—accelerator, interaction region(s), and detector(s) will permit us to make the most promising and comprehensive choice freely at that time.

TABLE 4. A listing of topics that will be investigated at a TeV-level linear collider. Check marks show the complementarity with which different initial states contribute prominently to their study. They also stake out unique contributions that the e^-e^- option will be able to make.

	QCD	$t\bar{t}$	H	SUSY	Strong WW	Anomalous Couplings	e^*	ν_M	Z'	X^{--}	Compositeness
e^-e^+		✓	✓	✓		✓		✓	✓	✓	✓
e^-e^-	✓	✓	✓	✓	✓	✓			✓		
$e^-\gamma$	✓		✓	✓		✓	✓				
$\gamma\gamma$	✓	✓	✓	✓	✓	✓					

VI ACKNOWLEDGMENTS

I wish to thank the members of our international e^-e^- Working Group for their enthusiasm and steady encouragement. Tim Barklow and Michael Peskin helped me by discussing individual items. Nora Rogers and Lilian DePorcel deserve all credit for the preparation of this manuscript.

REFERENCES

1. For a historical account, see: B. Richter, in *Proceedings of the Third International Symposium on the History of Particle Physics* (Stanford University, Stanford, CA, 1992); SLAC-PUB-6023 (1992).
2. J. E. Spencer in e^-e^- '95, Int. J. Mod. Phys. **11**, 1675 (1996).
3. P. Chen et al., in e^-e^- '95, Int. J. Mod. Phys. **11**, 1687 (1996); see also Ref. 2.
4. P. Minkowski, 1996 Snowmass, study to appear in *Proceedings of the 1996 DPF/DPB Summer Study on New Directions for High-Energy Physics* (Snowmass 96), (Snowmass, Colorado) 1997.
5. T. Rizzo, e^-e^- '95, Int. J. Mod. Phys. **11**, 1563 (1996).
6. H. Georgi, M. Machacek, Nucl. Phys. B **262**, 463 (1985).
7. F. Cuypers, G. J. van Oldenborgh, R. Rückl, Nucl. Phys. B **409**, 123 (1993)
8. M. Peskin, Lectures on Supersymmetry, Stanford University, CA (1997).
9. M. Chanowitz, LBL-PUB-32846 (1992).
10. V. Barger et al., Phys. Rev. D **50**, 6704 (1994).
11. T. Han in e^-e^- '95, Int. J. Mod. Phys. **11**, 1541 (1996).
12. D. Choudhury, F. Cuypers, Nucl. Phys. B **429**, 33 (1994); Phys. Lett. B **325**, 500 (1994).
13. T. Barklow in e^-e^- '95, Int. J. Mod. Phys. **11**, 1579 (1996).
14. F. Cuypers, PSI, Villigen Report No. PSI-PR-96-09, 1996; Int. J. Mod. Phys. A **11**, 1571 (1996).
15. A. Leike, Z. Phys. C **62**, 265 (1994).
16. D. Choudhury, F. Cuypers, A. Leike, Phys. Lett. B **333**, 531 (1994).
17. P. H. Frampton and B. H. Lee, Phys. Rev. Lett. **64**, 619 (1990); P. H. Frampton, Phys. Rev. Lett. **69**, 2889 (1992).
18. P. Minkowski in *Proceedings of the Second Int. Workshop on Physics and Experiments with Linear e^+e^- Colliders*, (Waikoloa, HI, 1993), pp. 524-537.
19. C. A. Heusch and P. Minkowski, Nucl. Phys. B **416**, 3 (1994).
20. C. A. Heusch and P. Minkowski, Phys. Lett. B **374** 116, (1996).
21. C. A. Heusch, Nucl. Phys. B (Proc. Suppl.) **38**, 313 (1995).
22. C. Greub and P. Minkowski, DESY 96-253, BUTP 96128 (1996) to be published in *Proceedings of the 1996 DPF/DPB Summer Study on New Directions for High-Energy Physics* (Snowmass 96), (Snowmass, Colorado) 1997.
23. V. Telnov, Nucl. Instrum. and Methods A **355**, 3 (1995); and elsewhere in *Proceedings of the Workshop on Gamma-Gamma Colliders*; ibid. pp. 1-184.

Ultimate luminosities and energies of photon colliders

Valery Telnov

Institute of Nuclear Physics, 630090, Novosibirsk, Russia

Abstract. A photon collider lumonosity and its energy are determined by the parameters of an electron-electron linear collider (energy, power, beam emittances) and collision effects. The main collision effect is the coherent e^+e^- pair creation. At low energies (2E< 0.5–1 TeV) this process is suppressed due to repulsion of electron beams. In this region $L_{\gamma\gamma}(z > 0.65) \geq 10^{35} \text{cm}^{-2}\text{s}^{-1}$ is possible ($10^{33} - 10^{34}$ is sufficient). At higher energies the limited average beam power and coherent pair creation restrict the maximum energy of photon colliders (with sufficient luminosity) at $E_{cm} \sim 5$ TeV. Obtaining high luminosities requires the development of new methods of production beams with low emittances such as a laser cooling.

I INTRODUCTION

Linear colliders offer the unique opportunities to study $\gamma\gamma$, γe interactions. Using the laser backscattering method one can obtain $\gamma\gamma$ and γe colliding beams with an energy and luminosity comparable to that in e^+e^- collisions or even higher (due to the absence of some beam collision effects). This can be done with a relatively small incremental cost. The expected physics in these collisions is very rich and complementary to that in e^+e^- collisions. Some characteristic examples are:

- a $\gamma\gamma$ collider provides the unique opportunities to measure the two-photon decay width of the Higgs boson, and to search for relatively heavy Higgs states in the extended Higgs models such as MSSM;

- a $\gamma\gamma$ collider is an outstanding W factory, with a WW pair production cross section by a factor of 10–20 larger than that in e^+e^- and with a potential of producing $10^6 - 10^7$ W's per year, allowing a precision study of the anomalous gauge boson interactions;

- a $\gamma\gamma$, γe collider is a remarkable tool for searching for new charged particles, such as supersymmetric particles, leptoquarks, excited states of electrons, etc., as in $\gamma\gamma$, γe collisions they are produced with cross sections larger than those in e^+e^- collisions;

- Charged sypersymmetric particles with massws higher than the beam energy could be produced with a γe collider.

The general scheme of a photon collider is shown in Fig. 1.

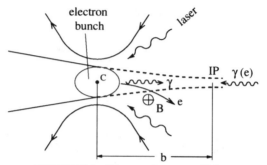

FIGURE 1. Scheme of $\gamma\gamma$; γe collider.

Two electron beams after the final focus system are traveling toward the interaction point (IP). At a distance of about 1 cm upstream from the IP, at a conversion point (CP), the laser beam is focused and Compton backscattered by the electrons, resulting in the high energy beam of photons. With reasonable laser parameters one can "convert" most of electrons to high energy photons. The photon beam follows the original electron direction of motion with a small angular spread of order $1/\gamma$, arriving at the IP in a tight focus, where it collides with a similar opposing high energy photon beam or with an electron beam. The photon spot size at the IP may be almost equal to that of electrons at IP and therefore the luminosity of $\gamma\gamma$, γe collisions will be of the same order as the "geometric" luminosity of basic ee beams.

The energy spectrum of photons after the Compton scattering for various polarization of electrons and laser photons is shown in fig. 2. At the optimum laser wave length (below the threshold of e^+e^- pair creation), the maximum energy of scattered photons is about 82% of the initial electron energy. Photons in the high energy part of spectrum can have high degree of polarization. This part of spectrum is most valuable for experiment. Below we will deal mainly with $\gamma\gamma$ luminosity produced by high energy photons.

Here we will not consider the general features of photon colliders, they are discribed in papers [2]–[5] and Proceedings of the Berkeley Workshop [6].

Preliminary studies show that the photon colliders with an energy $2E{\sim}500$ GeV and acceptable luminosity $L_{\gamma\gamma} \sim 10^{33} \mathrm{cm}^{-2}\mathrm{s}^{-1}$ (at $z = W_{\gamma\gamma}/2E > 0.7$) can be built [6,1,9]. However, physicists will be happy to have larger luminosity to study details of some processes, for example $\gamma\gamma \to Higgs$. It is also of interest to investigate main properties of the photon colliders at higher energy.

We know that e⁺e⁻ linear colliders at an energy above 1–2 TeV have serious problems. Is it easier to explore this region with a photon collider?

The present paper is focused mainly on the study of limitations on the energy and luminosity of photon colliders. We understand that linear colliders will be built not sooner than in one decade and will work another one–two decades. Therefore in discussions we will not confine ourselves by the present achievements and technologies. This concerns laser parameters (for high energy photon collider a free electron laser is needed) and emittances of electron beams. Although FELs with close parameters do not exist, but there are projects of such lasers. The way to very low emittances (laser cooling) was discussed at the first ITP symposium [10].

In the later paper it was also found how to overcome the problem of nonlinear effects in the conversion region which leads to the linear growth of the laser flash energy with the increase in the electron beam energy. In the proposed method of 'stretching' the conversion region the required laser flash energy

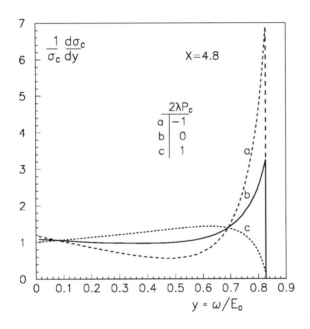

FIGURE 2. Spectrum of the Compton scattered photons for different polarizations of laser and electron beams

does not grow at all with energy.

II REQUIREMENTS TO $\gamma\gamma$ LUMINOSITY

Cross sections of the charged particle production in $\gamma\gamma$ collisions are somewhat higher than those in e^+e^- collisions. At $E > Mc^2$ the ratio of cross sections are the following:

$$\sigma_{\gamma\gamma \to H^+H^-}/\sigma_{e^+e^- \to H^+H^-} \sim 4.5;$$
$$\sigma_{\gamma\gamma \to t\bar{t}}/\sigma_{e^+e^- \to t\bar{t}} \sim 4;$$
$$\sigma_{\gamma\gamma \to W^+W^-}/\sigma_{e^+e^- \to W^+W^-}(|\cos\vartheta| < 0.8) \sim 15;$$
$$\sigma_{\gamma\gamma \to \mu^+\mu^-}/\sigma_{e^+e^- \to \mu^+\mu^-}(|\cos\vartheta| < 0.8) \sim 8.5.$$

To have the same statistics (but complementary physics) in $\gamma\gamma$ collisions the luminosity can be smaller than that in e^+e^- collisions by a factor of 5.

Cross sections decrease usually as $1/S$ ($S = E_{cm}^2$), therefore the luminosity should grow proportionally to S. A reasonable scaling for the required $\gamma\gamma$ luminosity (in the high energy peak of the luminosity distribution) at $\gamma\gamma$ collider is

$$L_{\gamma\gamma} \sim 3 \cdot 10^{33} S, \text{ cm}^{-2}\text{s}^{-1}. \tag{1}$$

With such a luminosity for the time $t = 10^7$ c one can detect
  $3.5 \cdot 10^3\ H^+H^-$,
  $2 \cdot 10^4\ \mu^+\mu^- (|\cos\vartheta| < 0.8)$;
  $2 \cdot 10^4\ t\bar{t}$;
  $2 \cdot 10^5\ W^+W^- (|\cos\vartheta| < 0.8)$;
  $2 \cdot 10^6 \cdot S(TeV^2)\ W^+W^-$.

Somewhat larger luminosity ($\sim 10^{33}$) is required for the search and study of the 'intermediate' ($M_H \sim 100 - 200$ GeV) Higgs boson which is produced as a single resonance in $\gamma\gamma$ collisions. We will see that such a level of luminosities at low energies is not a problem. The scaling (1) will be used for estimation of the maximum energy of photon colliders.

Other important problem at high luminosities is a background due to large total cross section $\sigma_{\gamma\gamma \to hadrons} \sim 5 \cdot 10^{-31}$ cm^{-2}s^{-1}. It consists of particles with $P_t \sim 0.5$ GeV uniformly (at large angles) distributed over the pseudorapidity $\eta = -\ln \tan(\vartheta/2)$ with $dN/d\eta \sim 7$ at $2E = 500$ GeV. Particle density grows only logarithmically with energy.

The average number of hadron events/per bunch crossing is about one at $L_{\gamma\gamma}(z > 0.65) = 10^{34}$ cm^{-2}s^{-1} and at the typical collision rate 5 kHz. In this paper, we are interested mainly by the luminosity in the high energy part of luminosity spectrum. However, in the scheme without deflection of used electron beams the total $\gamma\gamma$ luminosity is larger than the 'useful' $L_{\gamma\gamma}(z > 0.65)$ by a factor 5-10 due to collisions of low energy Compton photons and beamstrahlung photons. This low energy collisions increase background by a factor 2-3. At $E_{cm} = 5$ TeV with required $L_{\gamma\gamma}(z > 0.65) \sim 10^{35}$, this leads to about 30 (effectively) high energy $\gamma\gamma \to hadron$ events per bunch crossing. Similar number of hadronic events/collision is expected at LHC. However, there is important an difference between pp and $\gamma\gamma$ colliders: in the case of an interesting event (high P_t jets and leptons) the total energy of final products at photon colliders is equal to E_{cm}, while at proton colliders it is only about $(1/6)E_{cm}$. The ratio of the signal to background at photon colliders is better by a factor 6 at the same number of hadronic events per crossing. Moreover, during the reconstruction of an interesting event one can subtract smooth hadronic background and only its fluctuations are important which are proportional to $\sqrt{L}$. It means that for $L \propto E^2$ (and fixed collision rate) the ratio of signal to background is almost constant (decreases only logarithmically).

The above arguments shows that the problem of hadronic background is not dramatic for photon colliders. Of course, some increase in the collision rate with an increase in the luminosity will be useful. We will see that due to collision effects the optimum number of particles in one bunch should decrease with energy, that naturally leads to an increase in the collision rate.

III COLLISION EFFECTS. COHERENT PAIR CREATION.

There are two basic collision schemes [5]:

Scheme A ("without deflection"). There is no magnetic deflection of the spent electrons and all particles after the conversion region travel to the IP. The conversion point may be situated very close to the IP.

Scheme B ("with deflection"). After the conversion region particles pass through a region with a transverse magnetic field where electrons are swept aside. Thereby, one can achieve a more or less pure $\gamma\gamma$ or γe collisions.

During beam collision, photons are influenced by the field of the opposing electron beam. One of the important processes in this field is a conversion of photons into e^+e^- pairs (coherent pair creation) [7]. Under a certain conditions, the conversion length is shorter than the length of interaction region ($\sim \sigma_z$) and $\gamma\gamma$ luminosity is suppressed.

The probability of pair creation per unit length by a photon with the energy ω in the magnetic field $B(|B|+|E|)$ for our case) is [7], [4],.

$$\mu(\kappa) = \frac{\alpha^2}{r_e}\frac{B}{B_0}T(\kappa), \qquad (2)$$

where $\kappa = \frac{\omega}{mc^2}\frac{B}{B_0}$, $B_0 = \frac{m^2c^3}{e\hbar} = \frac{\alpha e}{r_e^2} = 4.4 \cdot 10^{13}$ G is the the critical field, $r_e = e^2/mc^2$ is the classical radius of electron.

$$T(\kappa) \approx 0.16\kappa^{-1}K_{1/3}^2(4/3\kappa) \tag{3}$$

$$\approx 0.23\exp(-8/3\kappa) \qquad \kappa < 1$$
$$\approx 0.1 \qquad \kappa = 3 - 100$$
$$\approx 0.38\kappa^{-1/3} \qquad \kappa > 100$$

In our case, $\omega \sim E_0$, therefore one can put

$$\kappa \sim \Upsilon \equiv \gamma B/B_0 \ . \tag{4}$$

The probability to create e^+e^- pair during the collision time is

$$p \approx \mu\sigma_z = \frac{\alpha^2\sigma_z}{r_e\gamma}\Upsilon T(\Upsilon) \ . \tag{5}$$

From these equations we can find Υ for a certain conversion probability p (with an accuracy higher than 25%) [4]

$$\Upsilon_m = 2.7/ln(0.1/p_1) \qquad p_1 < 0.01, \tag{6}$$
$$1.2 + 9p_1 \qquad 0.01 < p_1 < 4,$$
$$4.5\, p_1^{3/2} \qquad p_1 > 4,$$

where

$$p_1 = p\frac{r_e\gamma}{\alpha^2\sigma_z} \sim p \cdot 0.1 \frac{E[TeV\,]}{\sigma_z[mm]}.$$

For the conversion probability p the 'geometrical' $\gamma\gamma$ luminosity is suppressed approximately by a factor e^{-p}.

We will see that maximum $\gamma\gamma$ luminosity is achieved at $p > 1$. For $E = 0.5 - 2.5$ TeV and $\sigma_z = 0.1 - 0.5$ mm the parameter p_1 belongs to the second range ('transition' regime) where $\Upsilon \sim 1.2 + 9p_1$.

IV ESTIMATION OF ULTIMATE $\gamma\gamma$ LUMINOSITY

Let us find now limits posed on the luminosity due to coherent pair creation for different collision schemes.

There are three ways to avoid this effect (i.e. to keep $\Upsilon \leq \Upsilon_m$):
1) to use flat beams;
2) to deflect the electron beam after conversion at a sufficiently large distance (x_0 for $E = E_0$) from the interaction point(IP);
3) under certain conditions (low beam energy, long bunches) $\Upsilon < \Upsilon_m$ at the IP due to the repulsion of electron beams [8].

Let us consider at first requirements to the beam sizes in the case 1.

A Flat beams

The field of the beam with the r.m.s horizontal size σ_x and the length σ_z is $B \equiv |B| + |E| \sim 2eN/\sigma_x\sigma_z$. From the condition $\kappa \sim 0.8\gamma B/B_0 < \Upsilon_m$ we get

$$\sigma_x > \frac{1.6N\gamma r_e^2}{\alpha\sigma_z\Upsilon_m} = \frac{1.6N\gamma r_e^2}{\alpha\sigma_z(1.2 + 9pr_e\gamma/\alpha^2\sigma_z)} \sim \frac{40 \cdot \left[\frac{N}{10^{10}}\right]}{p + 1.3\frac{\sigma_z[\text{mm}]}{E[\text{TeV}]}} \text{ nm} \qquad (7)$$

The $\gamma\gamma$ luminosity at $z > 0.65$

$$L_{\gamma\gamma} \sim \frac{0.5k^2N^2f}{4\pi(b/\gamma)\sigma_x} \sim \frac{0.025\alpha N\sigma_z fk^2}{br_e^2}\left[1.2 + 9p\frac{r_e\gamma}{\alpha^2\sigma_z}\right]e^{-p}, \qquad (8)$$

where the coefficient 0.5 follows from the simulation for $\sigma_y = b\gamma$. It has its maximum at

$$I: \quad \tilde{p} = 0 \quad \text{at} \quad a = \frac{7.5r_e\gamma}{\alpha^2\sigma_z} = \frac{0.75E[TeV]}{\sigma_z[\text{mm}]} < 1 \ ;$$

$$II: \quad \tilde{p} = 1 - 1/a \quad \text{at} \quad a > 1 \ .$$

The corresponding luminosities for these two cases are the following

$$L_{\gamma\gamma} \sim 0.03\frac{\alpha k^2 Nf\sigma_z}{br_e^2} = 2.8 \cdot 10^{33}\left(\frac{N}{10^{10}}\right)\frac{f[kHz]}{b[\text{cm}]}k^2\sigma_z[\text{mm}], \text{ cm}^{-2}\text{s}^{-1}; \qquad (9)$$

$$L_{\gamma\gamma} \sim 0.23\frac{Nf\gamma k^2}{\alpha br_e}e^{-\tilde{p}} = 2.2 \cdot 10^{33}\left(\frac{N}{10^{10}}\right)\frac{f[kHz]}{b[\text{cm}]}k^2 E[\text{TeV}]e^{-\tilde{p}}, \text{ cm}^{-2}\text{s}^{-1}. \qquad (10)$$

Optimum horizontal beam sizes in these two cases are

$$I: \quad \sigma_x \sim \frac{1.3Nr_e^2\gamma}{\alpha\sigma_z} = 28\frac{E[\text{TeV}]\left(\frac{N}{10^{10}}\right)}{\sigma_z[\text{mm}]}, \text{ nm } ; \text{ at } a < 1 \ ; \qquad (11)$$

$$II: \quad \sigma_x \sim 0.18\alpha Nr_e = 37\left(\frac{N}{10^{10}}\right), \text{ nm at } a > 1 \ . \qquad (12)$$

The minimum value of the distance between the conversion (CP)) and the interaction regions b is determined by the length of the conversion region which is equal approximately to $b = 0.08E[\text{TeV}]$, cm (see section 6.1). For further estimation we assume that

$$b = 3\sigma_z + 0.04E[\text{TeV}], \text{ cm}. \qquad (13)$$

Let us take $N = 10^{10}$, $\sigma_z = 0.2$ mm, $f = 10$ kHz, $k^2 = 0.4$ (one conversion length) that corresponds at $E > 0.25$ TeV to the case II. For $2E = 5$ TeV we get

$$L_{\gamma\gamma} \sim 6 \cdot 10^{34} \text{ cm}^{-2}\text{s}^{-1} \text{ at } \sigma_x \sim 40 \text{ nm and } \sigma_y \sim b/\gamma = 0.3 \text{ nm}. \quad (14)$$

For a very high energy $L_{max} \sim 8 \cdot 10^{34}$ cm^{-2}s^{-1} for a chosen parameters corresponding to the beam power $P = 15 \cdot E[\text{TeV}]$ MW per beam. In the next section we will compare these approximate results with the results of simulation.

B Influence of the beam-beam repulsion on the coherent pair creation

During the beam collision electrons get displacement in the field of the opposing beam

$$r \sim \sqrt{\frac{\sigma_z r_e N}{8\gamma}}. \quad (15)$$

This estimate is obtained from the condition that at the impact parameter equal to the characteristic displacement the additional displacement is equal to the initial impact parameter.

The field at the axis (which influences on the high energy photons) $B \sim 2eN/r\sigma_z$. Then the corresponding field parameter

$$\Upsilon \sim \gamma \frac{B}{B_0} = \frac{\gamma B r_e^2}{\alpha e} \sim 5 \frac{r_e \gamma}{\alpha \sigma_z} \sqrt{\frac{\gamma r_e N}{\sigma_z}} \quad (16)$$

According to eq.(6), in the transition regime $\Upsilon_m = 1.2 + 9pr_e\gamma/\alpha^2\sigma_z$. From $\Upsilon = \Upsilon_m$ we can find the maximum beam energy when the coherent pair creation is suppressed due to the beam repulsion.

At the energy $E > 1$ TeV and bunches short enough, one can neglect the first term and get

$$\gamma_{max} \sim 3\frac{p^2\sigma_z}{\alpha^2 r_e N} \text{ or } E_{max} \sim p^2 \frac{\sigma_z[\text{mm}]}{N/10^{10}}, \text{ TeV}. \quad (17)$$

The $\gamma\gamma$ luminosity is equal

$$L_{\gamma\gamma}(z > 0.65) \sim 0.35 \frac{N^2 f k^2}{4\pi(b/\gamma)^2} e^{-p} \sim 0.1(Nf)\frac{\sigma_z \gamma p^2 k^2}{\alpha^2 r_e b^2} e^{-p}, \quad (18)$$

where the numerical factor 0.35 follows from the simulation. It has its maximum at p=2 when

$$L_{\gamma\gamma} \sim 0.05(Nf)\frac{\sigma_z\gamma k^2}{\alpha^2 r_e b^2} \sim 7\cdot 10^{33}\left(\frac{N}{10^{10}}\right)\frac{\sigma_z[\text{mm}]}{b^2[\text{cm}]}E[\text{TeV}]f[\text{kHz}]k^2\sigma_z[\text{mm}].$$
(19)

We have separated the factor (Nf) because it is a beam power. Taking in the previous example $Nf = 10^{14}$ Hz, $\sigma_z = 0.2$ mm, $k^2 = 0.4$, $b = 3\sigma_z + 0.04E[\text{TeV}]$, cm, $E = 2.5$ TeV we obtain

$$L_{\gamma\gamma}(z > 0.65) \sim 6\cdot 10^{35} \text{ cm}^{-2}\text{s}^{-1}.$$
(20)

The optimum number of particles in the beam for an energy considered (eq.(17)) is $N \sim 0.8 \cdot 10^{10}$.

These estimates show that the beam repulsion substantially influences (increases) the attainable $\gamma\gamma$ luminosity. This prediction will be checked by the simulation.

In fact, at low enough energy this effect allows to use even infinitely narrow electron beams with any reasonable number of particles in the bunch and the minimum photon spot size is b/γ, where b should be taken as small as possible. At high energies this effect also works but the number of particles in the bunch should be below some number dependent of the energy (eq.17).

V SCHEME WITH MAGNETIC DEFLECTION

Using the magnetic field B_e between the conversion and interaction regions one can sweep out electrons from the interaction point at some distance x_0 sufficient for the suppression of coherent pair creation, i.e. to satisfy condition $\Upsilon < \Upsilon_m$ given by eq.6. This distance x_0 is approximately equalto σ_x given by eq.7. The required distance b is found from relation $x_0 \sim b^2/2R = b^2 eB_e/2E$. The photon spot size at IP is b/γ and the luminosity of $\gamma\gamma$ collisions

$$L_{\gamma\gamma}(z>0.65) \sim 0.35\frac{k^2 N^2 f}{4\pi(b/\gamma)^2} \sim \frac{0.03\alpha k^2 Nf\sigma_z B_e}{er_e}(1+7.5p\frac{r_e\gamma}{\alpha^2\sigma_z})e^{-p} \quad (21)$$

The optimization over p gives

$$L_{\gamma\gamma} \sim \frac{0.03\alpha k^2 Nf\sigma_z B_e}{er_e} =$$

$$= 1.6\cdot 10^{34}\left(\frac{N}{10^{10}}\right)\sigma_z[\text{mm}]f[kHz]B_e[T]k^2 \text{ at } a = \frac{0.75E[TeV]}{\sigma_z[\text{mm}]} < 1 ; \quad (22)$$

$$L_{\gamma\gamma} \sim \frac{0.22k^2 Nf\gamma B_e}{\alpha e}e^{-(1-1/a)} =$$

$$= 1.25 \cdot 10^{34} \left(\frac{N}{10^{10}}\right) f[kHz]B[T]E[\text{TeV}]k^2 e^{-(1-1/a)}, \text{ cm}^{-2}\text{s}^{-1}; \text{ at } a > 1 . \quad (23)$$

As before, taking $N = 10^{10}$, $f = 10$ kHz, $\sigma_z = 0.2$ mm, $k^2 = 0.4$ $E = 2.5$ TeV and $B_e = 0.5$ T we get

$$L_{\gamma\gamma}(z > 0.65) \sim 2.5 \cdot 10^{34} \text{ cm}^{-2}\text{s}^{-1}. \quad (24)$$

This number is notably smaller than that in the scheme without deflection eq(20). The luminosity is proportional to B_e and one can take larger field values but it poses some technical problems. One should also remember that in the transverse magnetic field, the soft background particles produced in the forward direction (mainly e$^+$e$^-$ pairs) get a kick, begin to spiral in the detector field and to avoid backgrounds the radius of the vacuum pipe should exceed $r = 2bB_\perp/B_\parallel$.

Considering the scheme with deflection we did not consider the field created by the produced e$^+$e$^-$ pairs. These particles are closer to the beam axis than the deflected beam and their field can even exceed the field of the opposing electron beam. Therefore, our optimization of pair creation overestimates the luminosity. We will see results of simulation in next sections.

VI SIMULATION

A Assumptions

We have seen that the picture of collisions is very complicated. It is easier to get the result by simulation. The simulation code used in this work [5] includes all important processes. In the present study, the beams are collided in very ultimate conditions: very small beam sizes, high energies, too many beamstrahlung photons. In order to avoid the time consumption problem only one simplification was done: charged particles emitted the beamstrahlung photons during the beam collision but these photons were excluded from further consideration. It was assumed that:

a) the thickness of the laser target is equal to one collision length ($k = 1 - e^{-1} \sim 0.6$);

b) electrons and laser photons are polarized and $2P_c\lambda_e = -1$;

c) varying the number of particles in the beam we kept constant beam power $NfE = 15 \cdot E[\text{TeV}]$, MW;

d) the minimum distance between the CP and IP region was taken to be $b = 0.2 + 0.1E[\text{TeV}]$, cm for fig. 3 and $b = 3\sigma_z + 0.04E[\text{TeV}]$ cm for the rest figures. In the later case, the parameter ξ^2 characterizing the nonlinear effect in the Compton scattering [5] is equal to 0.6;

e) the vertical beam size is equal to $0.5b/\gamma$;

B Simulation results

Fig. 3 shows $\gamma\gamma$ luminosity as a function of σ_x at TESLA and NLC for the beam energy range 0.25–4 TeV. On the righthand graphs, the number of particles was decreased by a factor 10, while the collision rate was increased by the same factor. The distance $b = 0.2 + 0.1E[\text{TeV}]$ cm . We see on the left-hand side graphs that for 'nominal' numbers of particles in the beams the luminosity does not follow the dependence $L \propto 1/\sigma_x$ due to the conversion of photons to e^+e^- pairs. It happens at σ_x very close to our prediction, eq(12).

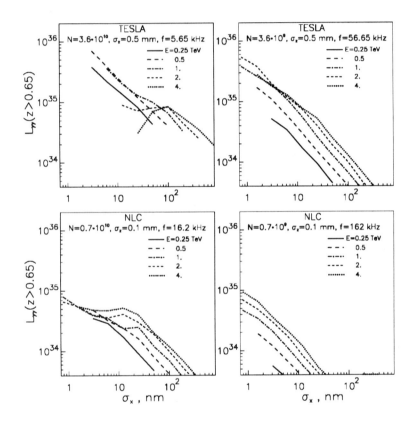

FIGURE 3. Dependence of the $\gamma\gamma$ luminosity on the horizontal beam size for TESLA and NLC beam parameters, see comments in the text.

In fig. 4 we can see the dependence of the luminosity both on N and σ_z in the case where beams are round and the conversion region is situated as close as possible: $b = 3\sigma_z + 0.04E[\text{TeV}]$ cm . The total beam power is 15

MW · E[GeV]. Looking to this pictures one can make many own observations which are clear after our theoretical consideration. Note only that the longer bunch requires the larger laser flash energy for conversion ($A \propto \sigma_z$) and not every linac (among the current projects) can accelerate a 0.5 mm long bunch due to wake fields. Let us take for further study $\sigma_z = 0.2$ mm.

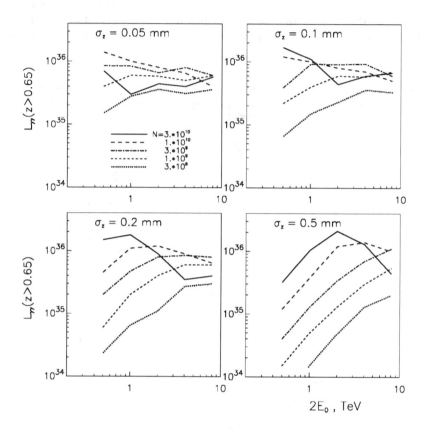

FIGURE 4. $\gamma\gamma$ luminosity for round beams at the minimum distance between interaction region and collision points, see comments in the text.

The dependence of the luminosity on σ_x (other conditions are the same as in the previous figure) is depicted in fig. 5. It is desirable to choose the working point (σ_x, N) so that not only the luminosity is large but also the corresponding curve still follows their natural behavior $L \propto 1/\sigma_x$. This means that the $\gamma \to e^+e^-$ conversion probability is still not too high. Otherwise it may happen that the low energy $\gamma\gamma$ luminosity will be much larger than that in the high energy part (low energy photons have smaller probability of conversion).

For example: at E = 2.5 TeV one can reach $L_{\gamma\gamma} \sim 7 \cdot 10^{34}$ cm^{-2}s^{-1} with all considered number of particles in the bunch.

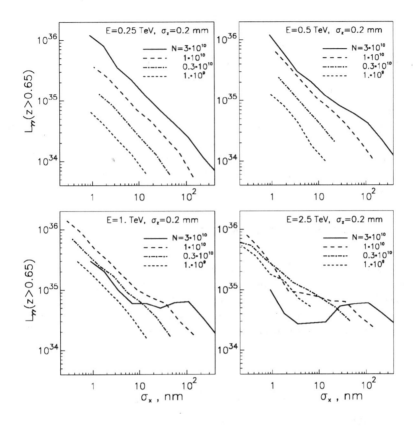

FIGURE 5. Dependence of the $\gamma\gamma$ luminosity on the horizontal beam size for $\sigma_z = 0.2$ mm, see comments in the text.

The values of σ_x and $L_{\gamma\gamma}$ where curves make zigzag are in good agreement with eqs. (10) and (12). Even higher luminosities at E = 2.5 TeV can be reached with $N = 10^9$–10^{10}, but in the case $N = 10^{10}$ it happens after zigzag on the curve that manifests that many photons have converted to e$^+$e$^-$ pairs. The best choice here is $N = 0.3 \cdot 10^{10}$, where one can reach $L = 4 \cdot 10^{35}$ without problems. The only problem here is a too small transverse beam size: $\sigma_x, \sigma_y < 1$ nm. At low energies the situation is perfect and one can dream about WW factory at the collider with an energy 2E = 500–1000 GeV.

Fig. 6 presents the result for the scheme with magnetic deflection. The strength of magnetic field is equal 0.5 T and the distance b is varied. For the comparison the case $B_e = 0$ is also shown. We see that the magnetic deflection helps at large b but the luminosity value in this region is much lower than that in the case of flat beams where conversion point is situated very close to the IP. Nevertheless, although magnetic deflection does not help to reach ultimate luminosities, in 'practical' cases (where the luminosity is far from the limit) the magnetic deflection helps to decrease the low energy $\gamma\gamma$ luminosity without degradation of the high energy part.

Comparing our luminosity scaling law (eq.1) with the results of the simulation we see that the required luminosity can be reached at least up to $2E = 4$ TeV with any (up to $3 \cdot 10^{10}$) number of particles in the bunch. Note that in our examples the collision rate is $f=10$ kHz$(10^{10}/N)$. The higher f is good for the experiment but makes some problems for the required average laser power. It seems that f=10–30 kHz is still acceptable. With N=0.3·10^{10} one can reach at 2E=5 TeV even a few times higher luminosity than that it is 'required'. The main problem here is connected with attainable beam emittances.

VII BEAM EMITTANCES

Only short remark. The normalized emittances which are written in the current projects [1] do not allow to follow the scaling low above $2E = 1$ TeV. An increase in luminosity at low energies or motion towards high energies requires serious R&D work on low emittance electron beams. One new suggestion was reported at the ITP Workshops [10]. This is laser cooling which allows to

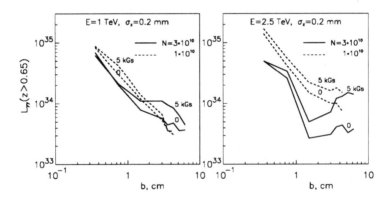

FIGURE 6. Dependence of the $\gamma\gamma$ luminosity on the distance between interaction point and conversion region in the scheme with magnetic deflection, see comments in the text.

cool beams down to values necessary for all parameters of photon colliders considered here.

VIII CONCLUSION

At $2E \sim 5$ TeV the conventional linear colliders reach their limit both in e^+e^- and in $\gamma\gamma$ mode. The number of problems grows exponentially: acceleration gradient, very small beam sizes, radiation, pair creation. The reason is common for all mode of operation:
1) the required luminosity is proportional to E^2;
2) energetic problem, because beams are used only once (but namely this feature makes possible to consider photon colliders).

Linear colliders are perfect for $2E = 0.1 - 2$ TeV and we have to use their potential with highest efficiency.

ACKNOWLEDGMENTS

I would like to thank Z.Parza, the organizer of the Program "New Ideas for Particle Accelerator" at ITP, UCSB, Santa Barbara, supported with National Science Foundation Grant NO PHY94-07194.

REFERENCES

1. Low et al., International Linear Collider Technical Review Committee Report, SLAC-Rep-471(1996)
2. I.Ginzburg, G.Kotkin, V.Serbo, V.Telnov,*Pizma ZhETF*, **34** (1981)514; *JETP Lett.* **34** (1982)491.
3. I.Ginzburg, G.Kotkin, V.Serbo, V.Telnov,*Nucl.Instr. & Meth.* **205** (1983) 47.
4. V.Telnov,*Nucl.Instr.&Meth.A* **294** (1990)72.
5. V.Telnov, *Nucl.Instr.&Meth.A* **355**(1995)3.
6. *Proc.of Workshop on $\gamma\gamma$ Colliders*, Berkeley CA, USA, 1994, *Nucl. Instr. &Meth. A* **355**(1995)1-194.
7. P.Chen,V.Telnov,*Phys.Rev.Letters*, **63** (1989)1796.
8. V.Telnov, *Proc.of Workshop 'Photon 95'*, Sheffield, UK, April 1995, p.369.
9. *Zeroth-Order Design Report for the Next Linear Collider* LBNL-PUB-5424, SLAC Report 474, May 1996.
10. V.Telnov, *Proc.of ITP Workshop 'New modes of particle acceleratios techniques and sources* Santa Barbara, USA, August 1996, NSF-ITP-96-142, SLAC-PUB 7337, e-print hep-ex/9610008.

Lepton-Hadron Collider Physics
(should there be another lepton-hadron machine in our future?)

S. Ritz[1]

Columbia University, Nevis Labs
PO Box 137, Irvington, NY 10533

Abstract. Using lessons from HERA, the physics possiblities at a new lepton-hadron collider facility are explored. The facilities considered are mainly enhancements to other proposed pp, e^+e^-, or $\mu^+\mu^-$ colliders.

INTRODUCTION AND MOTIVATION

The initial motivation for this work came from the choice of future facilities considered at Snowmass in 1996, displayed in table 1. As you see, the only ep machine on the list is not very ambitious in comparison with the other machines contemplated. HERA already has a 300 GeV center-of-mass energy, and its final luminosity will approach a significant fraction of what is shown in the table. Still, given the success of past ep experiments and the variety of important results that have already come from HERA, it is worthwhile to review what physics could be explored with a new lp collider. Realistically, if a new lp capability will materialize it will most likely come in the form of an add-on to one of the lepton-lepton or pp colliders on the list rather than as a new standalone facility. The approach, therefore, is to pair a conventional beam with one of the high energy e, μ or p beams in the list.

Why lepton-nucleon experiments? History alone is compelling: many of the results that shaped the Standard Model came from lepton-nucleon experiments, including the discoveries of partons and nucleon substructure, neutrino flavor, and neutral currents, all at $Q^2 < 300$ GeV^2. Why another lepton-proton *collider*? Recent history is compelling: the HERA program is already a great success, with important results in photoproduction (at a major leap up in photon-proton center-of-mass energy), DIS (at a major leap down in x and up in Q^2 by two orders of magnitude), and searches for new physics that are complementary to those at e^+e^- and $p\bar{p}$ colliders; and there is much more on the

---

[1] Supported by the National Science Foundation and Sloan Foundation Fellow

Machine type	E_{cm}(TeV)	L (x10^{33}cm^{-2}s^{-1})
Tevatron	2	1
LHC	14	10
NLC	0.5	5
	1	20
	1.5	20
$\mu^+\mu^-$	0.5	0.7
	0.5	5
	4	100
pp	60	10
e^+e^-	5	100
ep	1	0.1

TABLE 1. Table of Snowmass96 machine choices to study.

way [1]. HERA results have also sparked a resurgence of theoretical interest in QCD, with investigations into the interplay between hard and soft processes, BFKL tests, new classes of pQCD-calculable processes, and attempts to explain a wide variety of rapidity gap phenomena. As HERA opened large new kinematic regimes for exploration there were surprises, and there is no reason not to expect this to happen again at a new facility if it has enough of a reach into unexplored territory. Significant questions will remain after HERA that may be addressed by such a new facility.

KINEMATICS

One of the great advantages of lepton-proton collider experiments is that, for the neutral current case at least, the final state is over-constrained. The measurement of the basic kinematic variables, x and Q^2, can be made with just the scattered lepton, or with just the hadrons, or with various combinations of the two. For simplicity, the study here only considers the measurement using the scattered lepton alone. The situation is shown schematically in figure 1 with minimal assumptions about the underlying physical processes. The positive Z direction is taken as the proton beam direction, θ_e is the azimuthal lepton scattering angle, and E_e and E'_e are the incoming and outgoing lepton energies, respectively. The usual DIS kinematic variables are then given by

$$y = 1 - \frac{E'_e}{2E_e}(1 - \cos\theta_e) \qquad (1)$$

$$Q^2 = 2E_e E'_e(1 + \cos\theta_e) \qquad (2)$$

with

$$x = Q^2/sy \qquad (3)$$

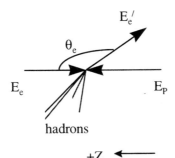

FIGURE 1. Gross features of a NC scatter. The positive, or forward, direction is defined as the direction of the proton beam.

where $s = 4E_e E_P$. (At HERA, $E_e = 27.5$ GeV and $E_P = 820$ GeV, so $s \sim 10^5$ GeV^2: a 50 TeV lepton beam would be required to do the equivalent experiment in fixed-target mode.) Combining these, one obtains

$$Q^2 = 4E_e^2 x \frac{1 + \cos\theta_e}{\frac{E_e}{E_P}(1 - \cos\theta_e) + x(1 - \cos\theta_e)} \quad (4)$$

which will be used to study the $\{x, Q^2\}$ range accessible at future machines. Another useful quantity is the virtual photon-proton center-of-mass energy, W, which is given by

$$W^2 \approx sy(1-x) \quad (5)$$

which, at low x, is simply $W \sim sy$.

F_2 STRUCTURE FUNCTION AND THE RAPID RISE OF THE $\gamma^* P$ CROSS-SECTION

That a lepton-hadron collider can open up large and important new kinematic regimes is illustrated nicely by the F_2 result shown in figure 2. The squares show the measurements from fixed target experiments and the circles show the results from ZEUS and H1. Our picture of the behavior of $F2$ is certainly very different now.

The striking rise with decreasing x at low x has important implications. $F_2(x)$ can be related to the total virtual photon-proton cross-section, $\sigma_{tot}^{\gamma^* p}$,

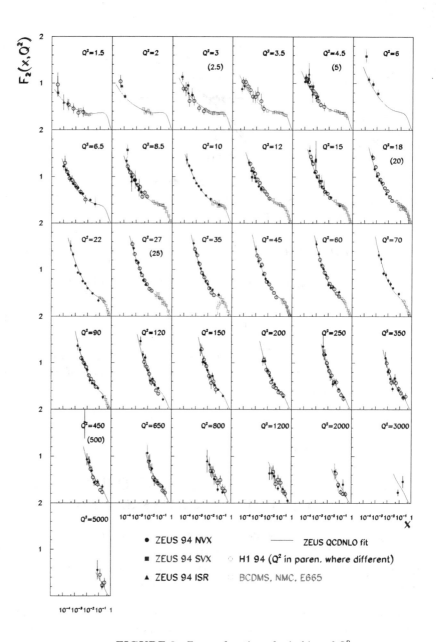

FIGURE 2. F_2 as a function of x in bins of Q^2.

as a function of W at fixed Q^2: decreasing x corresponds to *increasing* W. The results are shown for different Q^2 in figure 3. HERA data populate the large W^2 range, while fixed target experiments are at lower W^2. For photoproduction ($Q^2 \approx 0$), the rise of the cross-section with W is modest; however, for $Q^2 > 0$ the cross-section is growing rapidly with energy! At what large W will the rise slow down? At what low Q^2 will the rise decrease to match photoproduction (how, exactly, does photoproduction become DIS)? The second question has already been answered at HERA [2], but the first will likely remain even after the present HERA program is completed in 2005.

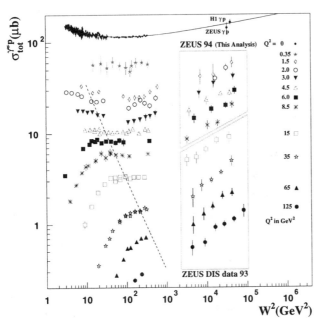

FIGURE 3. The W dependence of $\sigma_{tot}(\gamma^* p)$ for different values of Q^2.

A QCD analysis based on DGLAP evolution indicates the rise is driven by a remarkably large gluon density at low x [3]. This evolution must fail as the parton densities saturate, and the familiar QCD evolution will no longer be valid. We can use what we now know about the gluon density at low x to investigate the range of $\{x, Q^2\}$ where this will happen. An estimate of where the gluon density should saturate is given by [4]

$$xg(x,Q^2) \geq \frac{1 fm^2}{1/Q^2} \approx 25 Q^2 (GeV^2). \qquad (6)$$

Using a parametrization of the gluon density extracted from present-day data, the region in which this condition is satisfied is shown in figure 4. The experimental signature is a softening in the rise of F_2 with decreasing x in this region. Since F_2 must be rising quickly to begin with (otherwise there would be nothing to soften), the effect will likely be visible only for $Q^2 > 2$ GeV^2, indicated by the solid line.

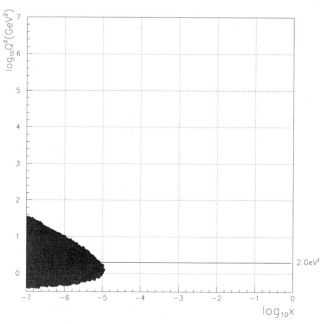

FIGURE 4. The region in the $\{x, Q^2\}$ plane in which the gluon density is estimated to show signs of saturation. The effects may only be perceptible for $Q^2 > 2$ GeV^2.

KINEMATIC COVERAGE OF VARIOUS MACHINES

We use equation 4 to explore the sensitivities of various machines to new physics. The practical range of measurements is limited. We require

1. $y > 0.01$ since the resolution on x degrades unacceptably at low y,

2. $\theta_e > 10°$ since it will be difficult to identify the scattered lepton in the hadronic splash from the proton remnant,

3. $\theta_e < 179°$ since the lepton has to leave the beamline to be measured.

Note that these limitations could be relaxed by an aggessive detector design, but the assumptions become speculative. There is already good experience at HERA doing meaurements with cuts like these, so we adopt them for the present study. It is important to state again that we are ignoring the information from the hadrons, which in some cases might expand the kinematic coverage.

The results are shown in figure 5, with the rough estimate of the gluon density saturation region overlayed as a reference. The region that satisfies the above requirements, along with the kinematic limit $y < 1$, is contained within the solid lines. Some conclusions are immediately clear:

1. To cover the low x region, the optimal configuration is obtained by pairing a very high energy proton beam with a relatively low energy electron beam. The $\theta_e < 179°$ requirement implies a lower limit on Q^2: for high energy lepton beams, even a 1 degree scatter from the incident direction implies a large Q^2, essentially independent of proton beam energy. The large relative boost of the proton system brings very low x scatters more into the central region of the detector.

2. Conversely, for such an asymmetric configuration, the high Q^2 reach is limited by the $\theta_e > 10°$ requirement. To explore the highest Q^2 physics, a more symmetric pairing of beam energies is desirable.

Thus, a very modest conventional lepton beam paired with an LHC or very high energy proton beam is useful for low x physics, while an NLC or muon collider beam would best be paired with a more conventional proton beam to explore high Q^2 physics.

LOW X, DIFFRACTION, AND ALL THAT

Why is low x interesting? There are many reasons. One perspective is obtained by considering the lifetime of the virtual photon state in the rest frame of the proton [5]

$$\tau \approx \frac{q_o}{Q^2} \approx \frac{1}{2 m_N x} \qquad (7)$$

where m_N is the nucleon mass and q_0 is the photon energy. At low x the longitudinal distances that contribute to high energy processes are HUGE (at $x \sim 10^{-4}$ this is about 10^3 fm), leading to coherent phenomena [6].

One of the early surprises at HERA was the discovery of a distinct class of events, called large rapidity gap (LRG) events. In the standard picture of DIS, color is transferred between the scattered quark and the proton remnant, as depicted schematically in figure 6. In the detector we usually see the scattered lepton, the current jet formed from the struck quark, a large hadronic energy flow in the forward region from the proton remnant, and a population of hadrons between the current jet and the remnant. However, we also found that

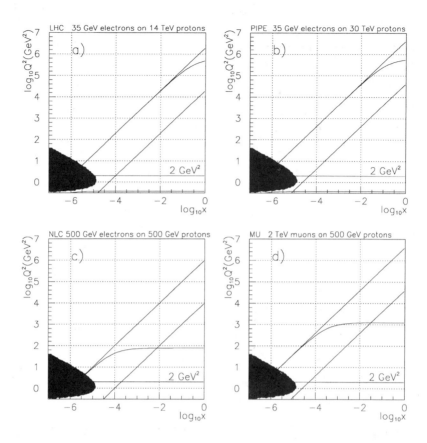

FIGURE 5. The region of the kinematic plane accessible at various lp collider configurations. The diagnal lines form a band corresponding to the requirement $0.01 < y < 1.0$.
(a) Using the LHC proton beam with a conventional 35 GeV electron beam. The limitation at very high Q^2 is due to the $\theta_e > 10°$ requirement. (b) Using an ultra-high energy proton beam (30 TeV) with a conventional 35 GeV electron beam, again limited at very high Q^2 due to the asymmetric beams. (c) Using a 500 GeV NLC electron beam with a conventional 500 GeV proton beam. The *lower* limit on Q^2 is due to the $\theta_e < 179°$ requirement. Such a configuration is limited to $Q^2 > 100$ GeV^2 and $x > 10^{-4}$, but provides access to very high Q^2. (d) Using a 2 TeV muon collider beam with a conventional 500 GeV proton beam. This configuration accesses the highest Q^2 values, but is limited at the low end to $Q^2 > 1000$ GeV^2 and $x > 10^{-3}$. The estimated zone of gluon density saturation is overlayed on each as a reference, along with a line corresponding to $Q^2 = 2$ GeV^2.

a large fraction of the time there is no remnant and no particles at all between what looks like the current jet and the forward edge of the main detector. The rate of this second type of event is large, of order 10% of the total DIS rate. It is interesting to note that a large fraction of photoproduction events also have a LRG, and there is even a charged current LRG event candidate.

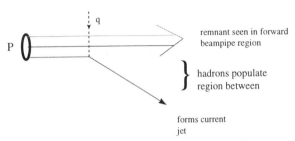

FIGURE 6. The naive expectation in DIS.

What is going on? We interpret this phenomenon as the result of a scatter involving the photon (or a Z or W) and another object with no net color, as depicted schematically in figure 7. In this figure, q is the four-vector of the photon and t is the four-vector of the other object. This is a remarkable process: the electron energy in the proton rest frame is 50 TeV, the Q^2 of the process can be large, and the energy transfer is much larger than the mass scale of the light hadrons so the overall process is quite inelastic – yet the proton emerges completely intact! The exchanged object is most likely a Pomeron, the agent required to describe pp diffractive scattering, so the process may be considered as diffractive dissociation of the virtual photon. The presence of these events leads to many important questions. What is the Pomeron? Does it "contain" partons? In what regimes can this interaction be described by QCD as two-gluon exchange? This is an area of intense investigation at HERA, and is a good example of the remarkable phenomena lurking at low x.

THE HIGH Q^2 DOMAIN

Unlike the case at low x, it is useful here to have a very high energy lepton beam. Since $s = 4E_e E_P$, the limit on Q^2 is set primarily by the beam energies. In addition, an asymmetric beam energy configuration means that high mass states have a large boost in the lab system, complicating the measurement: the highest Q^2 scatters are inaccessible at the asymmetric machines in figure 5 due to the $\theta_e > 10°$ requirement.

We can again look to the experience already obtained at HERA as a guide to the new physics potential of an lp collider at high Q^2. A recent summary

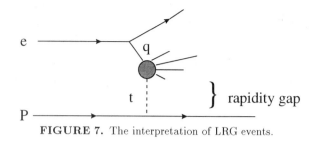

FIGURE 7. The interpretation of LRG events.

is given reference [7]. For contact interactions, leptoquarks, and R-parity violating SUSY, HERA searches are complementary to those at e^+e^- and pp machines, covering comparable ranges with different assumptions. In some cases, *e.g.*, many lepton flavor violation and excited electron searches, HERA is the best choice. For many of the new particle searches it is relatively easy to explore up to the kinematic limit (usually $\sqrt{s}$).

An exciting illustration is provided by the recent announcement [8] by both ZEUS and H1 of a statistical excess of events at high x and Q^2 over Standard Model expectation. If it persists in the new data, interest in a new lp collider will very likely continue to grow.

SUMMARY

The ep collider approach continues to be a great success. However, significant questions will remain after HERA. The F_2 structure function rises steeply with decreasing x at low x, corresponding to a fast rise of the virtual photon-proton cross-section with center-of-mass energy. This can not continue. Where does the rise slow down, and how? New QCD dynamics beyond the familiar DGLAP evolution will be required. We can not claim to understand the proton, or QCD, until this problem is understood. It is difficult to see how to explore this region without another machine[2].

For low x physics, a high energy proton beam paired with a very modest lepton beam is necessary. For high Q^2 physics, more balanced beams are desirable: a lepton beam from an electron linear collider or a muon collider, paired with a conventional proton beam, could open up more than a decade in Q^2.

As a practical matter, a new lp capability will likely appear only as an add-on to another facility. An lp program should certainly be seriously considered in any new pp machine proposal. In the nearterm, the next great proton beam

---

[2] One possiblity might be eA collisions at HERA, since the effective gluon density at low x is amplified in heavier nuclei.

will be at LHC. It is not clear that LEP will ever go back in the tunnel, but it may be worth considering creative alternatives such as colliding SpS electrons with LHC protons.

ACKNOWLEDGEMENTS

I thank Zohreh Parsa for inviting me to share some ideas on lp physics at this very enjoyable and informative meeting.

REFERENCES

1. The ZEUS and H1 homepages are http://www-zeus.desy.de/ and www.dice2.desy.de, respectively.
2. Proceedings, ICHEP96 Warsaw, 1996. Publication in preparation.
3. ZEUS Collaboration, M. Derrick et al., Phys. Lett. B345 (1995)576.,
   H1 Collaboration, S. Aid et al., Phys. Lett. B354(1995)494 .
4. A.H. Mueller, J. Phys G19(1993)1463.
5. B.L. Ioffe, Phys. Lett. 30 (1969) 123;
   B.L. Ioffe, V.A. Khoze and L.N. Lipatov, "Hard Processes", North Holland (1984) p185.
6. H. Abramowicz, L. Frankfurt, and M. Strikman, DESY-95-047, and published in SLAC Summer Inst.(1994)539.
7. Proc. 1995-96 HERA Workshop; Beyond the Standard Model Group Summary, H. Dreiner, H.-U. Martyn, S. Ritz, and D. Wyler; also HEP-PH/9610232.
8. ZEUS Collaboration, M. Derrick et al., DESY 97-025;
   H1 Collaboration, C. Adloff et al., DESY 97-024; both to appear in Z.Phys.C.

PERSPECTIVES ON FUTURE HIGH ENERGY PHYSICS

Presentation to the

Future High Energy Colliders Symposium

at the

Institute for Theoretical Physics
University of California, Santa Barbara

October 21, 1996

Burton Richter
Director
Stanford Linear Accelerator Center

A Strategic View of the High Energy Physics Program

Keep an appropriate balance among:

- present-and near-term physics programs,

- medium-term facilities additions to allow frontier physics 5–10 years downstream,

- R&D aimed at the long-term future,

- preserving strength in underlying core technologies (computing, rf power sources, electronics, etc.).

Innovate in accelerators, detectors and supporting technology.

Maintain a strong theory program closely coupled to experiments.

Outline

Present Program
    SLD
    Fixed Target
    Future of the Linac-based Program

B-Factory
    Machine
    Detector

NLC
    Technology
    ZDR
    Next Steps

GLAST — Satellite Gamma-Ray Detector

Advanced Accelerator R&D

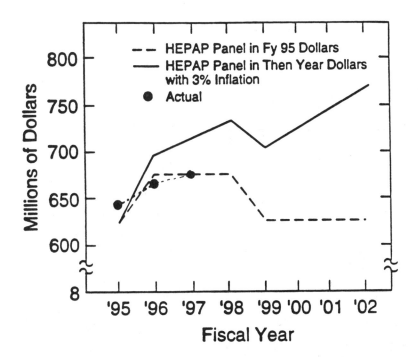

HEPAP Funding Recommendation's
(Drell Panel) May '94

The machine operates at 4×10^{10} e^+e^- per pulse.

Beam sizes are <1 micron (v) x 4 micron (h).

It is an important testbed for future colliders.

Polarization of e^- remains at about 78%.

SLD experiment continues data taking with new vertex detector.

New collision point diagnostics are now operational.

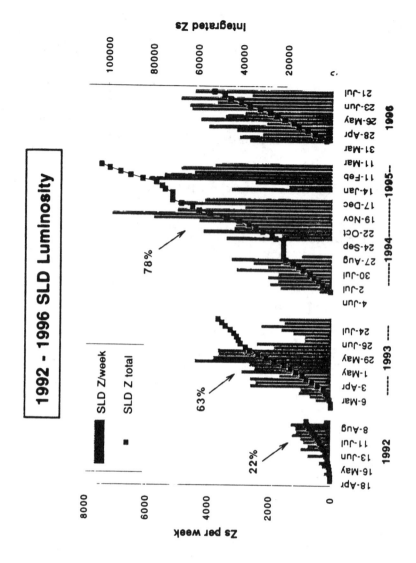

Cross Section

IP Optics

'Bouwers - Maksutov'

Catadioptric (50 years)

Meniscus cancels spherical aberration

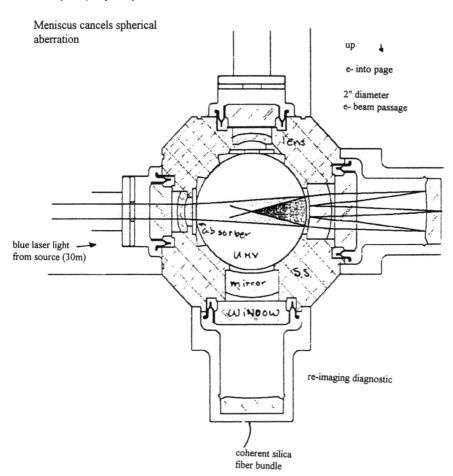

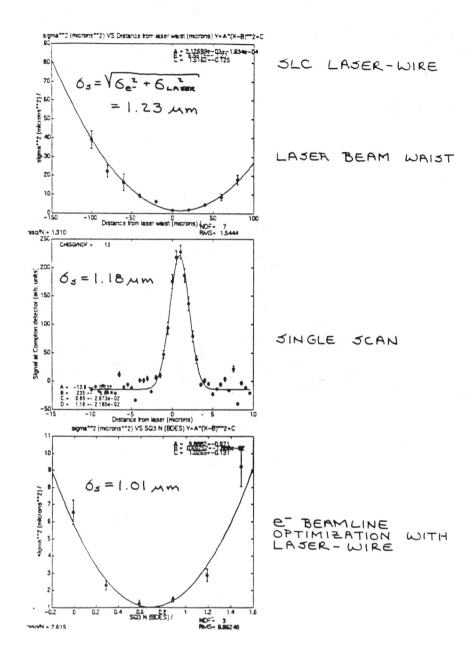

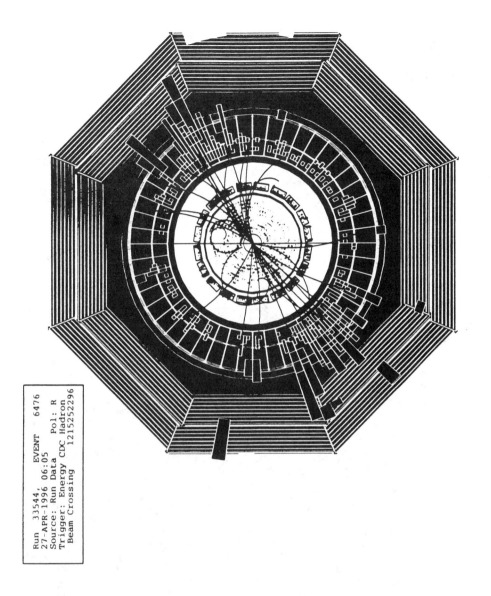

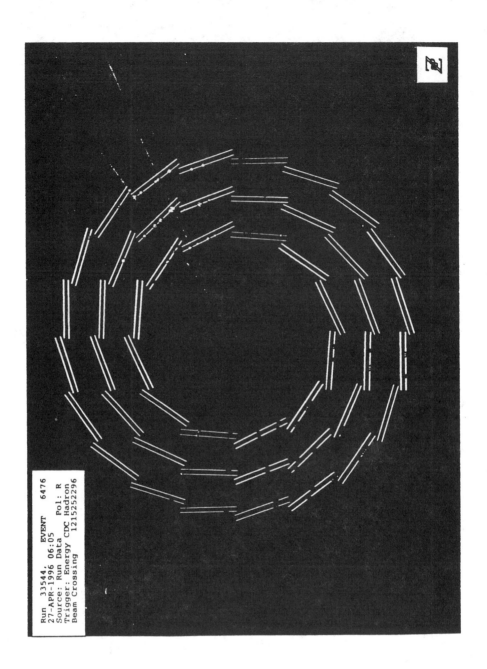

SLD Preliminary 1993-95

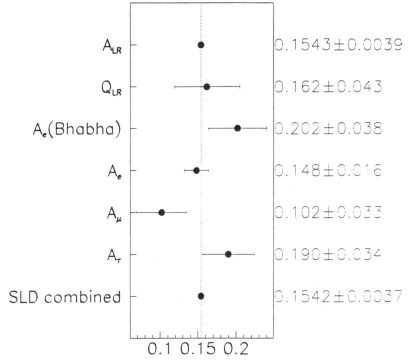

Combined Result:

$$\sin^2\theta_W^{eff} = 0.23061 \pm 0.00047$$

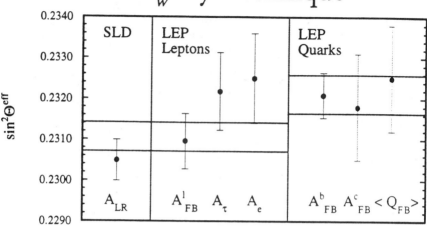

Combining our 1992 – 95 Value:

$$\sin^2\theta_W^{\mathit{eff}} = 0.23049 \pm 0.00050$$

With the LEP Lepton Average:

$$= 0.23160 \pm 0.00049$$

Results in a Lepton only Average:

$$\sin^2\theta_W^{\mathit{eff}} = 0.23106 \pm 0.00035$$

Compared to the LEP Quark Average:

$$\sin^2\theta_W^{\mathit{eff}} = 0.23211 \pm 0.00047$$
A difference of 1.79 sigma

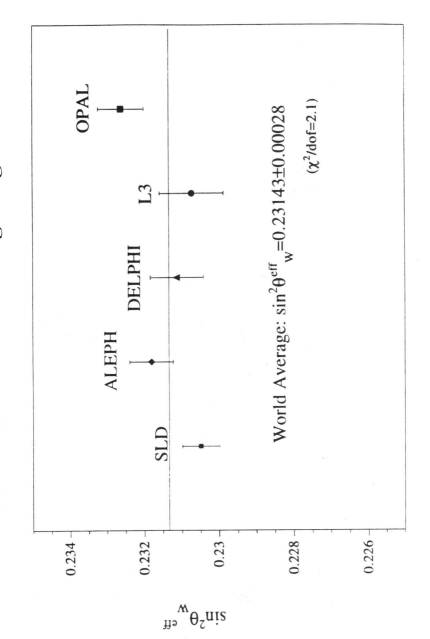

Future of the SLD R_b Measurements

More data will allow use of harder cuts to eliminate charm.

higher b-tagging efficiency from using VXD3 will assist in reducing systematics

$$dR_b \cong \left(\frac{-2R_c}{\varepsilon_b}\right)d\varepsilon_c \qquad dR_b \cong \left(\frac{-2\varepsilon_c}{\varepsilon_b}\right)dR_c \qquad dR_b \cong \left(\frac{R_b}{\varepsilon_b}\right)d\lambda_b$$

Hemisphere b-tag Performance

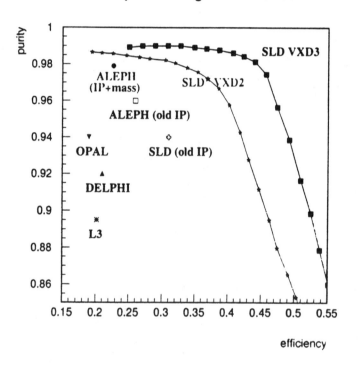

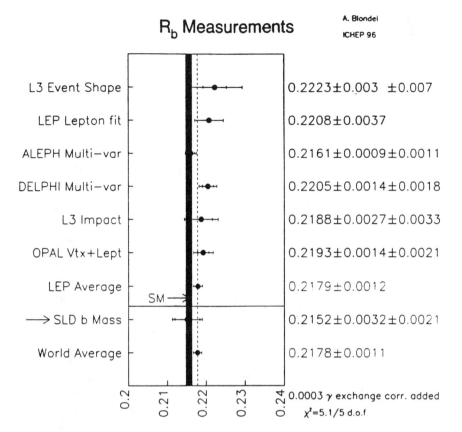

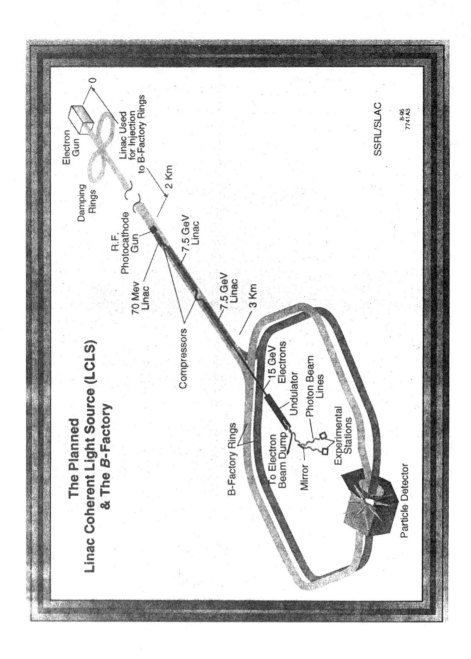

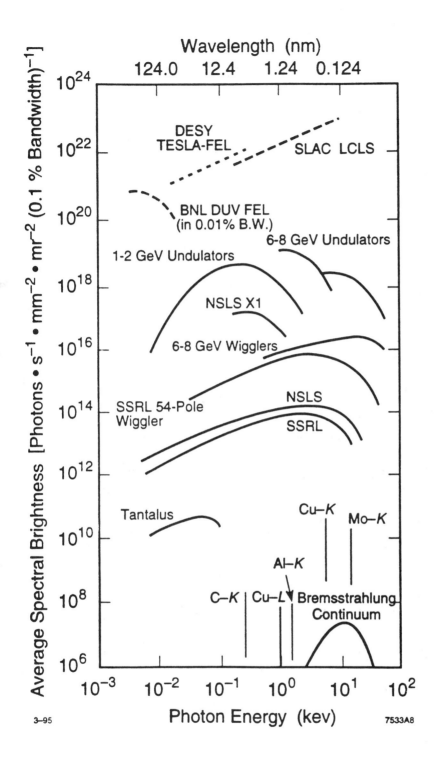

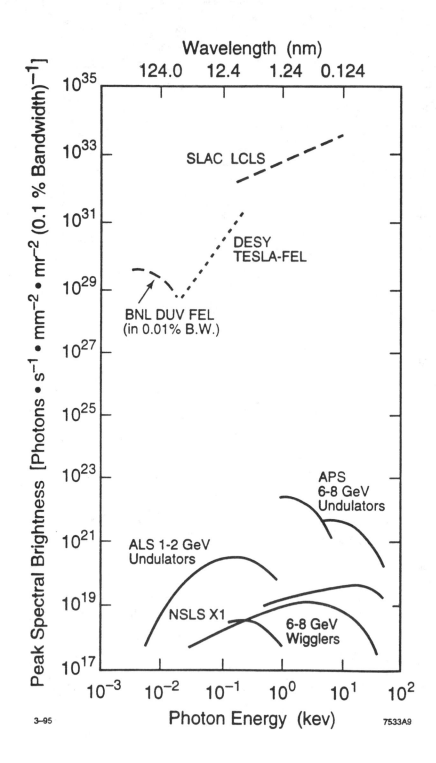

POTENTIAL LCLS SCIENCE

<u>200 femtosecond pulses</u>
- Pump-probe experiments-general
- Kinetics of cyclic solid-state reactions

<u>High brightness</u>
- Microdiffraction
- Microanalysis
- Microvolume X-ray spectroscopy

<u>200 femtosecond pulses & high brightness</u>
- Microvolume kinetics
    - e.g., phase-change memories

<u>Coherence</u>
- X-ray holography
    - e.g., percolation phenomena

<u>High peak brilliance</u>
- Non-linear X-ray phenomena

Nucleon Spin Structure Functions:

- Very high precision.

- E_e up to 50 GeV.

- Proton, neutron and deuteron targets.

- Higher beam and target polarizations.

- Less target "dilution."

- He^3 (neutron) done.

- Proton and deuteron run Spring 1997.

What carries the spin of the nucleon?

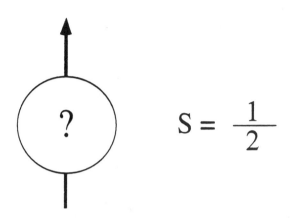

$$S = \frac{1}{2} = \frac{1}{2}\Delta q + \Delta G + \Delta L$$

quarks   gluons   orbital angular momentum

Δq is small !

Polarized Lepton-Nucleon Scattering

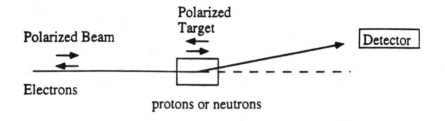

Measure Asymmetry:

$$A_1 = \frac{d\sigma^{\rightleftarrows} - d\sigma^{\rightrightarrows}}{d\sigma^{\rightleftarrows} + d\sigma^{\rightrightarrows}}$$

Extract Nucleon Spin Structure Function:

$$g_1 \approx A_1 \cdot F_1$$

QCD Sum Rule

Bjorken   (1966)

Valid at $Q^2 \to \infty$

$$\int_0^1 g_1^p(x)\,dx - \int_0^1 g_1^n(x)\,dx = \frac{1}{6}\frac{g_a}{g_v}$$

　　Proton　　　　Neutron　　　　　Beta decay

At finite Q^2,   QCD corrections:

$$\int_0^1 g_1^p(x)\,dx - \int_0^1 g_1^n(x)\,dx = \frac{1}{6}\frac{g_a}{g_v}\overbrace{\left(1 - \frac{\alpha_s}{\pi} + \ldots\right)}^{\text{Perturbative}}$$

$$+ \underbrace{\frac{C_1}{Q^2} + \frac{C_2}{Q^4} + \ldots}_{\text{Non-perturbative}}$$

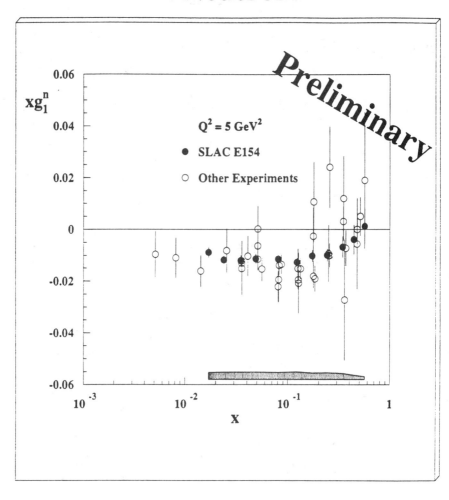

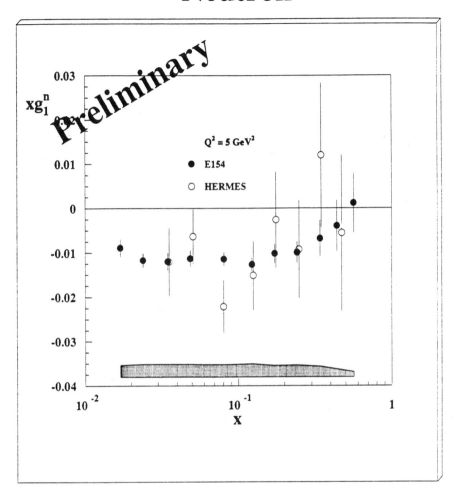

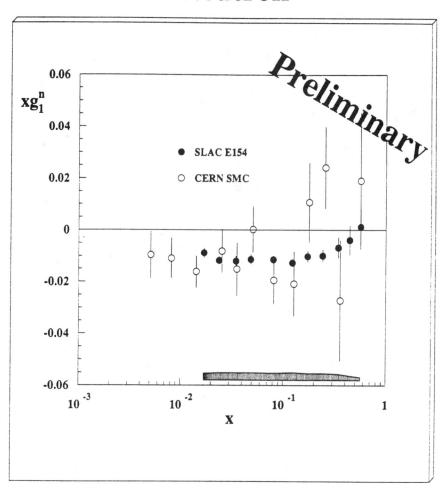

All of the present cycle of experiments will be completed in the Spring of 1998.

B-Factory commissioning begins then.

We are looking at options for other accelerator-based experiments that can run in parallel with the B-Factory in 2000 and beyond.

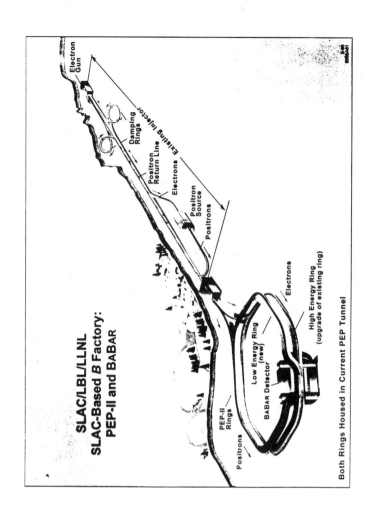

Several PEP-II Parameters

	e^+	e^-
Beam energy (GeV)	3.1	9
Beam current (A)	2.14	0.98
β_y^* (cm)	1.5	2.0
ε_x (ε_y) (nm)	64 (2.6)	48 (1.9)
σ_x (μm at IP)	155	
σ_y (μm at IP)	6.2	
σ_z (cm)	1.0	1.15
Luminosity	3×10^{33} cm^{-2}s^{-1}	
Tune shift	0.03	
Beam aspect ratio (v / h at IP)	0.04	
Number of beam bunches	1658	
Bunch spacing (m)	1.26	
Beam crossing angle	0 (head-on)	

PEP-II Commissioning Plans

Phased commissioning:

Extraction / beam transport tests: Parasitic "few" Hz extraction	Fall 1995 →
HER injection / stored beam: Work towards stored beam and high I	Spring 1997 →
LER injection / stored beam: Work towards stored beam and high I	Spring 1998 →
PEP-II collisions: Detect beam-beam collisions	Summer 1998
BaBaR collisions:	Early 1999

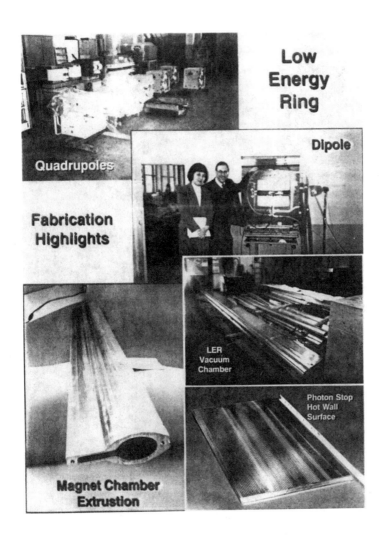

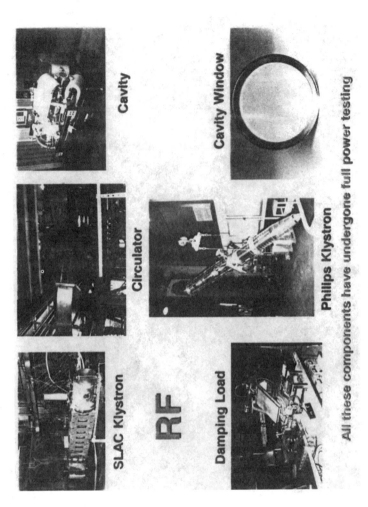

PEP-II B-Factory

HER and LER Straight Section Magnets and Vacuum Chamber on their Support Stands

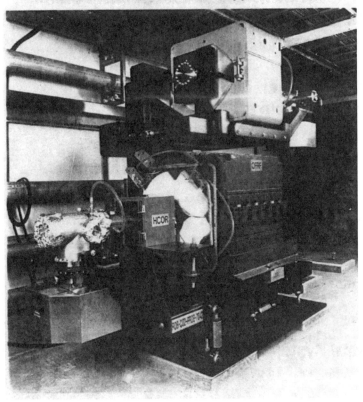

Technical Highlights

- Electron extraction and bypass beamlines installed in 1995. Checkout complete bringing beam to within 100m of ring. Positron checkout commences February '97

- Most of the High Energy Ring Magnets and Vacuum System are installed. Injection trials will commence in March '97. Stored beam commissioning will commence in April '97

- Half the Low Energy Ring Magnets are complete. The Vacuum System production is in full swing. Production rafts (holding magnets and vacuum chambers) begin installation in January '97. Ring commissioning begins April '98

- Control System - hardware and software - will be ready for both rings by April '97

- All RF components tested at or above full power. All components in production; installation ongoing

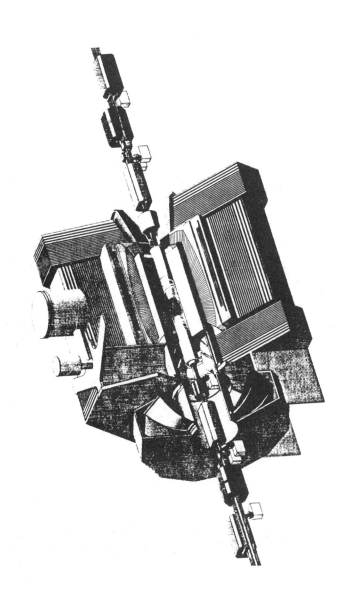

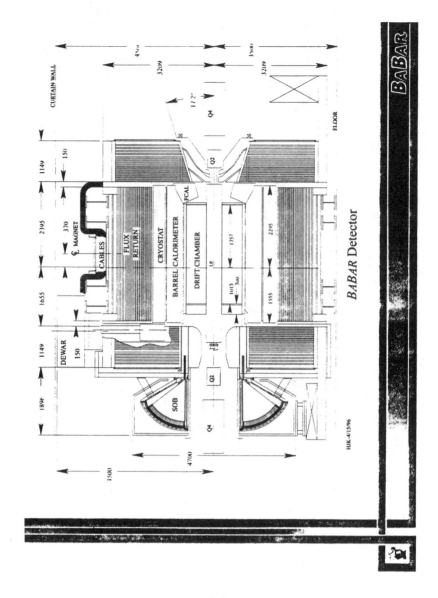

Detector System Responsibilities

System	Countries
SVT	Italy, US
Drift Chamber	Canada, US, France, Italy
DIRC PID	France, US, Russia
CsI Calorimeter	China, Germany, Norway, UK, US
Instrumented Flux Return	China, Italy, US
Magnet	Italy, US
Electronics	Canada, France, Germany, Italy, Taiwan, UK, US
Computing	Canada, France, Germany, Italy, UK, US, ...

BaBar Major acquisitions

Superconducting Solenoid	Placed with Ansaldo • Winding to begin at end of October
Magnet Flux Return	Placed with Kawasaki on May 24 • 11 ½ month delivery
Silicon strips	Placed with Micron on January 10 • Four of six detector shapes are in production
Resistive Plate Chambers	Placed with General Technica in January • First production modules in trasit to SLAC
CsI salt	Placed with VPI in October • Deliveries being made on schedule
CsI crystals	Placed with four vendors: Hilger, Kharkov Institute, SIC(Y), Crismatec • Deliveries are ramping up
Photodiodes	Placed with Hamamatsu in May • Deliveries have begun
DIRC photomultipliers	Placed with Electron Tube Ltd • Deliveries have begun
DIRC quartz	Responses to tenders received • Orders to be placed momentarily
Helium liquifier	Placed with Linde • Delivery February, 1997
Drift chamber DAQ	• Scheduled for FY97

The BABAR schedule is driven by the PEP-II schedule

- PEP-II is on schedule, from a technical point of view and a funding point of view, to begin commissioning in the Summer of 1998
- BABAR should not be on the beamline during PEP-II commissioning, but should be moved in as soon as the accelerator has reached a reasonable fraction of its design luminosity
- The BABAR schedule thus has the following goals:
  - Detector complete by the end of 1998
  - Three months of cosmic ray checkout commencing January 1, 1999
  - Move onto the beamline beginning April 1, 1999
  - Detector is ready for beam on or before July 1, 1999
- This is an aggressive schedule, necessitating a conservative design philosophy and careful management of budget and schedule
  - The funding profile from DOE, NSERC, PPARC, IN2P3, CEA, BMBF and INFN is a match to the schedule
  - We believe we can, that from both a technical and funding point of view, hold this schedule

Conclusions

- *BABAR* is concluding its final design phase and moving into the production phase
- Technical progress has been good
- Major procurements have all been placed on or close to schedule
  - Solenoid
  - RPC's
  - Quartz bars (end October)
  - Flux return
  - CsI salt
  - Silicon strip detectors
  - CsI crystals
  - DIRC photomultipliers
- Changes in the detector configuration have been made to reduce cost as well as technical and schedule risk
- Management has been able to react as required to deal with schedule problems
- The project is on schedule for completion of construction at the end of 1998, with first beam in the Spring of 1999, and the first physics run immediately thereafter

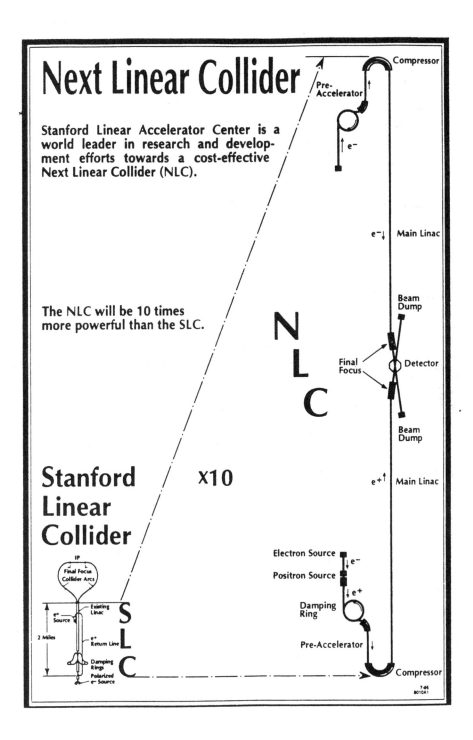

NLC Design Criteria

Collider optimized for 0.5 to 1.0 TeV
(compatible with few hundred GeV to 1.5 TeV)

First Stage 500 GeV $5 \times 10^{33} \text{cm}^{-2}\text{s}^{-1}$

Proven rf technology that exists at the outset.
All infrastructure and beamlines to support 1 TeV.

Second Stage 1 TeV $\geq 10^{34} \text{cm}^{-2}\text{s}^{-1}$

Expected improvement in rf technology of first stage.

$\Rightarrow$ Adiabatic upgrade

Expansion 1.5 TeV $\geq 10^{34} \text{cm}^{-2}\text{s}^{-1}$

First stage must be compatible with highest energy.
May require longer rf development.

e.g. gridded klystrons
cluster klystrons
TB relativistic klystrons

Polarization (e-) $\geq 80\%$

SUSY Discovery Reach at the LHC or an NLC

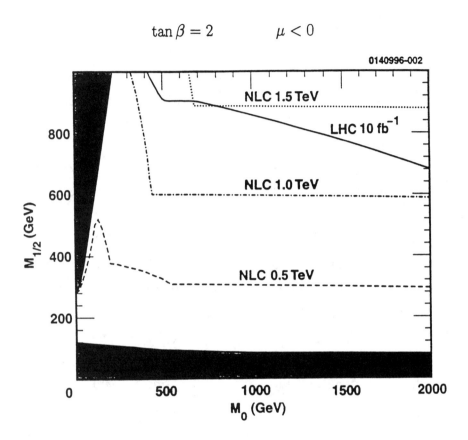

NLC ZDR Director's Review

March 18-21, 1996

Committee chaired by John Rees (SLAC) ...

G. Dugan	Cornell	H. Edwards	Fermilab
H. Frischholz	CERN	D. Gurd	LANL
T. Himel	SLAC	S. Holmes	Fermilab
N. Holtkamp	DESY	J. Ives	PBQ&D
R. Jameson	LANL	K. Oide	KEK
S. Ozaki	BNL	N. Toge	KEK

- Does the linear collider described in the ZDR include all components and subsystems needed to be comprehensive and functional?

- The cost analysis is at an early stage of development, but does the committee believe that it is "not unreasonable"?

- What areas are most in need of further R&D?

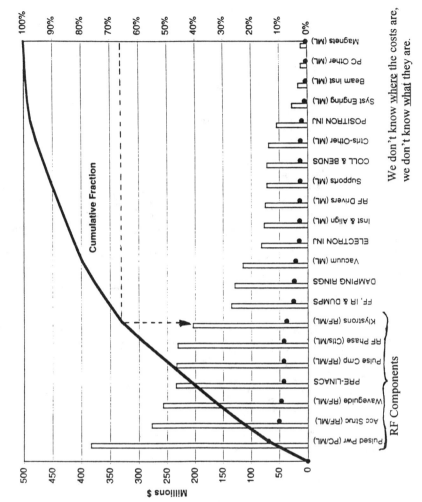

PPM Focused Klystron

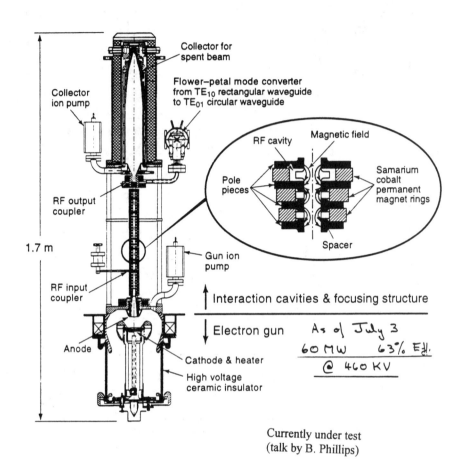

Currently under test
(talk by B. Phillips)

X-Band (11.424 GHz)

NLC Parameters

	500 GeV	1 TeV
RF Parameters		
Klystron Power (MW)	50	72
Pulse Length (µs)	1.2	1.2
Compression	3.6	3.6
Gradient (mV/m)	50	85
Linac Length (km)	17.6	19.1
Total # Klystrons	4528	9816
Wall Power (MW)	116	181
Luminosity Parameters		
Repetition Rate (Hz)	180	120
Bunches per Pulse	90	90
Particles per Bunch(10^{10})	0.75	1.1
IP σ_x (nm)	320	250
IP σ_y (nm)	5.5	4.3
Demagnification	200	200
n_γ (per e^-)	0.9	1.5
Luminosity (10^{34}cm^{-2}s^{-1})	0.6	1.4

√	65	√	75
√	1.5	√	1.1
√	4.0		
√	70		

4
150

√
3
√

70
√ 320
*

*SLC and E-144 experience

Linear Collider Test Facilities

Facility	Location	Goal	Operations
SLC	SLAC	Prototype Collider	1988-1998
ATF	KEK	Injector Damping Ring	1995 1996
SBTF	DESY	S-Band Linac	1996
NLCTA	SLAC	X-Band Linac	1996
TTF	DESY	Super C Linac	1995 (Injector) 1997 (Full Operations)
CTF	CERN	2-Beam Linac	1996 (Phase I) 1998 (Phase II)
FFTB	SLAC	Final Focus IP	1994

Linear Collider work has been international.

- Born at ICFA Workshop (1978).

- Coordinated through international workshops beginning in 1988.

- More formally through inter-laboratory M.O.U.'s since 1992.

- The first technological evaluation was done in 1995.

- The second technological evaluation is scheduled for late 1997.

I believe a large linear collider can only be realized as an international project.

To Build the International Linear Collider

- Get LHC going.

- Get a consensus of ILC Stage 1 parameters.

- Choose a technology (one, if possible).

- Hold the community together in the face of the centrifugal forces of site selection.

- Begin an international design group.

- See political agreement on construction while in the design phase.

GLAST: *Exploring the Physics and Astrophysics of Extremes using Nature's own Highest Energy Accelerators*

Stanford's interest in γ-ray detectors and sources: Hofstadter-- successful EGRET started with help from HEP.

Observational Objectives: The study of compact and extended astronomical sources of high-energy particles and radiation by measuring the source position and energy of emitted 10 MeV - 300 GeV γ-rays.

Approach: Use a large area, wide field-of-view, imaging telescope employing a pair-conversion telescope made with silicon strip detector based particle tracking; the tracker is backed by em calorimetry.

GAMMA-RAY LARGE AREA SPACE TELESCOPE

Sources: massive black holes in AGNs, galactic black holes, neutron stars, supernovae, the diffuse γ-ray background, γ-ray bursts, unidentified EGRET sources.

Mission Concept Study Team

Ames Research Center, Boston University, Columbia University (Ritz, Sciulli), Ecole Polytechnique (Fleury), France, Goddard Space Flight Center, Instituto Nazionale di Fisica Nucleare, Trieste (Barbiellini), Italy, International Center for Theoretical Physics, Trieste, Italy, Kanagawa University, Japan, Lockheed-Martin Research Laboratory, Max Planck Institut für Extraterrestrische Physik, Germany, Naval Research Laboratory, SACLAY, France, Sonoma State University, Stanford University: HEPL, Physics Dept., & SLAC (Atwood, Bloom, Godfrey), University of California, Santa Cruz (Johnson, Seiden), University of Chicago (Ong, Oreglia), University of Rome (Morselli), Italy, University of Tokyo (Kamae), Japan, University of Washington (Burnett)

10 Astronomy/Astrophysics + 10 HEP Institutions (Currently ~ 65 Collaborators)

Particle physics institutions in red

Exploring the Physics and Astrophysics of Extremes
GLAST Instrument Description

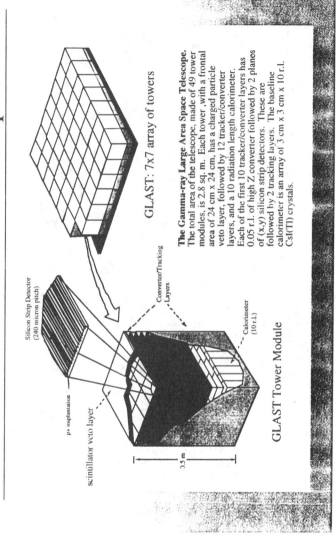

GLAST: 7x7 array of towers

GLAST Tower Module

The Gamma-ray Large Area Space Telescope. The total area of the telescope, made of 49 tower modules, is 2.8 sq. m. Each tower, with a frontal area of 24 cm x 24 cm, has a charged particle veto layer, followed by 12 tracker/converter layers, and a 10 radiation length calorimeter. Each of the first 10 tracker/converter layers has 0.05 r.l. of high Z converter followed by 2 planes of (x,y) silicon strip detectors. These are followed by 2 tracking layers. The baseline calorimeter is an array of 3 cm x 3 cm x 10 r.l. CsI(Tl) crystals.

Exploring the Physics and Astrophysics of Extremes
Scientific Objectives

Observe (and understand!) nature's highest-energy particle accelerators: active galactic nuclei (AGN), black holes, pulsars, supernova remnants, γ-ray bursts

Use these sources to probe particle physics parameters of the Universe not readily measured with other observations.

Exploring the Physics and Astrophysics of Extremes
Observing the Extragalactic γ-Ray Background

The isotropic component of the high latitude "diffuse" radiation could be either truly diffuse or the integrated flux from a large number of unresolved discrete sources, or both.

- GLAST will map the all-sky γ-ray background on finer angular scales than EGRET.

- Determine the photon energy spectrum over a broad range (10 MeV-300 GeV) with excellent source point spread capability.

- By a deep survey of extra galactic fields, resolve the background into point sources or determine if there is a truly diffuse component.

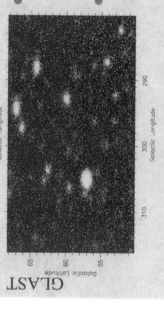

Exploring the Physics and Astrophysics of Extremes
Probing the Era of Galaxy Formation

- *GLAST will probe the extragalactic background light (EBL) through measurements of spectral cutoffs of high redshift AGNs ($Z \rightarrow 4$) in the 10 GeV - 300 GeV range.*

- *Absorption of high energy γ rays over cosmological distances via $\gamma\gamma \rightarrow e^+ e^-$ is dependent on the density of near-UV, optical, and near-IR background photons which make up the EBL.*

- *Determination of the EBL, can provide unique information on the formation of galaxies at early epochs, thereby constraining the ratio of hot dark matter to cold dark matter in the universe.*

Absorption due to EBL

unabsorbed spectra
model (2), Z=0.5
model (1), Z=0.5
model (2), Z=2
model (1), Z=2

optical depth: model (1), Z=2
optical depth: model (2), Z=2

Optical Depth

Integral Flux (counts)

E_γ (GeV)

High Energy Physics with GLAST

- The nature of the extra galactic diffuse gamma ray background at high energy. Ref.: Dar, A. and N. Shaviv, Phys. Rev. Lett. 75, 3052 (1995); Stecker, F. W. and M. H. Salamon, Phys. Rev. Lett. 76, 3878 (1996).

    --GLAST can resolve individual AGN in > 100 MeV photons with two orders of magnitude better sensitivity than current and past experiments. This will allow a significantly more sensitive measurement of the "true" diffuse extragalactic gamma ray background.

- Evidence for the origin of AGN jets by detailed observations of gamma ray spectra to cutoff energies: Ref.: Dermer, C. D., and R. Schlickeiser, ApJ. 416, 458(1993).

    --GLAST will measure the gamma ray spectrum from AGN from 10 MeV to the energy they cutoff or to about 300 GeV, whichever is lower.

- Information on the Relative Amount of Cold Dark Matter and Hot Dark Matter in the Early Universe from high energy gamma ray spectral cutoffs from AGN vs. AGN redshift. Ref.: Macminn, D. and J.R. Primack, Space and Science Reviews, 75, 413 (1996).

    --GLAST will measure the cutoff in many AGN photon spectra that have a cutoff below 300 GeV, and to a red shift of the AGN as large as about 4. Correlating the cutoff energy with redshift gives information about the ratio of HDM to CDM in the early universe.

- Extended gamma ray emission from supernova remnants evidencing the accelerators of cosmic Rays. Ref.: Drury, L. O'C., F.A Aharonian, and H.J. Volk, A&A 287, 957 (1994).

    --A typical candidate shell supernovae remnant used for detecting the signature of accelerated cosmic rays may have its flux spread out over $0.5°$. Observation of high energy photons over this diffuse source is evidence for shock acceleration by the expanding SN shock front. The collection of about 1000 photons from a SN remnant should allow GLAST to determine that the gammas are originating from an extended (rather than point) source given the excellent point spread function of the detector.

- Antimatter in the Universe: High energy photons originating in the interface between large regions of matter and antimatter over the universe. Ref. : De Rujula, A., Un Altro Modo Di Guardare L'anticielo, CERN preprint, (Sept., 1996).

    --GLAST will set limits on structure in the diffuse extra galactic gamma ray spectrum that could be explained by a variant of this model.

- Super Symmetry: The search for Neutralino (χ) annihilation signatures to gamma rays from space. Ref.: Jungman, G., M. Kamionskowski, K. Griest, to appear in Physics Reports. Can be downloaded from http://www.sns.ias.edu/~jungman.

    --$\chi\chi \Rightarrow \gamma\gamma$, GLAST will set limits for a "line" in the gamma ray spectrum at the mass of the neutralino. GLAST would be sensitive to the neutralino mass range up to about 300 GeV.

    --$\chi\chi \Rightarrow \gamma V$, where V is $c\bar{c}$, or $b\bar{b}$, GLAST will set limits on an enhancement in the gamma ray spectrum in the region of $m_\chi < 300$ GeV (with relatively low V mass).

- The search for Gamma rays from supernova due to pseudoscalar conversion. Ref.: Grifols, J. A., E. Masso, and R. Toldra, Phys. Rev. Lett. 77, 2372 (1996), and references therein.

    --If a supernova is in the GLAST acceptance when it goes off, GLAST can set limits on the coupling of a currently not observed very light pseudoscalar particle to two photons. The mean energy value of the photons to be observed is about 160 MeV. Part of the Pseudoscalar flux originating in the SN collapse would be converted to photons by the galactic magnetic field. There is a few percent chance of seeing a SN in the GLAST acceptance over a 10 year GLAST lifetime.

- Primordial Black Holes: The Search for High energy gamma rays from the evaporation of primordial black holes. Ref.: MacGibbon, J.H. and Carr, B.J., ApJ, 371, 447(1991).

    --GLAST will set limits on observation of a short burst of multiple high energy gamma rays with time resolution of about a microsecond, and essentially no deadtime.

GLAST Funding Decisions and Peer Reviews

- Spring 1992, work begins on GLAST at SLAC and Stanford Physics on initial design brought forward by W. B. Atwood.

- Winter 1992--GLAST technology study funded by NASA Scientific Research and Technology (SR&T) Program.

- March 1995 -- Selection by NASA as new mission concept in Astrophysics for possible flight in the next decade.

- February 1996 -- DOE SAGENAP committee: "GLAST project has high scientific merit and should be supported." Encouraged DOE funding of GLAST R&D to bring the project to the proposal stage using existing SLAC funds. Need for future review of the GLAST proposal by the DOE was emphasized.

- December 1995 -- GLAST beam test for summer 1996 approved.

GLAST DOE & NASA Reviews (continued)

- October 1996 -- SR&T funding renewed and augmented to support a three year hardware R&D program.
- October 1996 -- NASA institutes the "Gamma-ray Large Area Space Telescope Science Working Group (GSWG)". This committee includes Astronomers, Astrophysicists and Particle Physicists.
- November 1996 -- SLAC SPC encourages SLAC participation in GLAST.
- April 1997--GLAST chosen as new start for FY 2000-4 by SEUS (astrophysics) advisory committee to NASA.
- May 1997 -- Will GLAST make it into NASA's strategic plan??

Advanced Accelerator R&D

Electron-positron LC's can be pushed to 5 TeV (Siemann) to 8 TeV (Richter) without "tricks."

Four-beam systems can perhaps go further, though charge and current neutralization cannot be complete (non-constant logitudinal distribution).

Muon colliders are in an interesting early state with problems to be solved.

SLAC is beginning to look at very high-gradient acceleration for the long-term future.

What Limits Gradient?

Wakefield Instrumentation

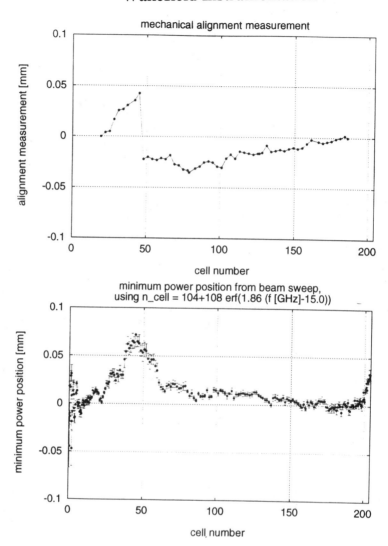

MM-Wave Structure Fabrication

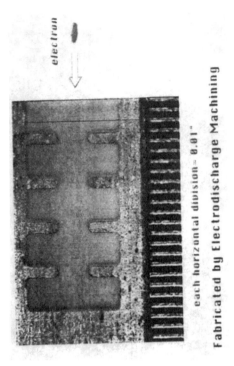

Summary

- SLD and spin structure experiments stop in 1998.

- The B-Fctory will be completed in 1998 and colliding beam trials begin mid-1998 with the detector off line. The physics program will begin in 1999.

- Use of the linac for other physics in the B-Factory era is under review.

- Critical decisions on the GLAST program will come in 1999.

- NLC R&D is moving ahead rapidly at SLAC and other labs.

- The ILC design team should begin work in 1998.

Future High Energy Colliders

October 21–25, 1996

Coordinator: Z. Parsa

Advisory Committee: D. Cline, G. Jackson, W. Marciano and P. Wilson

Schedule

Monday, October 21, 1996

Time:	Speaker:	Title:
8:00 am	Registration	ITP Lobby

Convener: Z. Parsa, BNL

8:40	J. Hartle, ITP Director	Welcome
	Z. Parsa, ITP/BNL	Welcome & Introduction

Defining Perspective Presentations:

9:00	D. Gross, Princeton Univ.	Perspectives on Future High Energy Physics
9:55	Refreshment Break	ITP Front Patio

Convener: G. Jackson, FNAL

10:25	J. Peoples, FNAL Director	Perspectives on Future High Energy Physics
11:20	B. Richter, SLAC Director	Perspectives on Future High Energy Physics
12:15 pm	Lunch	ITP Front Patio

Convener: R. Pecci, UCLA

1:45	N. Samios, BNL Director	Perspectives on Future High Energy Physics

View From Washington:

2:40	M. Krebs, DOE	Perspectives on Future High Energy Physics
3:35	Refreshment Break	ITP Front Patio

4:05	B. Kayser, NSF	Perspectives on Future High Energy Physics
5:00	Session ends	
5:15	Wine & Cheese	ITP Front Patio
5:45	Buffet Dinner	ITP Front Patio

Tuesday, October 22, 1996

Convener: V. Barger, U. Wisconsin

9:00 am	W. Marciano, BNL	Physics of the Standard Model and Beyond
9:45	S. Willenbrock, UIL	Higgs Physics
10:30	Refreshment Break	ITP Front Patio
11:00	J. Gunion, UCD	Supersymmetry
11:45	T. Appelquist, Yale	Strongly Interacting New Physics
12:30 pm	Lunch	ITP Front Patio

Convener: M. Harrison, BNL

2:00	L. Rolandi, CERN	LEP Status Report
2:55	D. Amidei, UMI	Collider Physics at FNAL & TeV2000, Physics Issues
3:45	Refreshment Break	ITP Front Patio
4:15	G. Jackson, FNAL	TeV2000, Accelerator Issues
5:30	Session ends	

Wednesday, October 23, 1996:

Convener: N. Lockyer, U. Pennsylvania

9:00 am	E. Keil, CERN	Status of LEPII and LHC
9:45	I. Hinchliffe, LBL	Large Hadron Collider (LHC) Physics
10:30	Refreshment Break	ITP Front Patio
11:00	M. Harrison, BNL	Big Hadron Collider
11:45	H. Murayama, LBL	e+ e- Physics

12:30	Lunch	ITP Front Patio

Convener: D. Silverman, UC Irvine

1:45	F. Paige, BNL	Complementarity of Lepton and Hadron Colliders
2:25	D. Burke, SLAC	Overview and outlook for Linear Colliders and Next Linear Collider SLAC ZDR Design
3:30	Refreshment Break	ITP Front Patio

Convener: L. Rolandi, CERN

4:05	N. Toge, KEK	Japan's Status on Next Linear Collider
4:45	R. Brinkman, DESY	Superconducting Collider (TESLA)
5:35	Session ends	

Thursday, October 24, 1996

Convener: A. Tollestrap, FNAL

8:30 am	P. Wilson, SLAC	Scaling Linear Collider to 5 TeV & above
9:10	Refreshment Break	ITP Front Patio
9:45	R. Palmer, BNL	Overview of Muon Collider Design Simulation & Detector
10:40	D. Cline, UCLA	Concepts for $\mu^+\mu^-$ Cooling & polarized Sources
11:20	V. Barger, UWI	$\mu^+\mu^-$ Collider Physics capabilities
12:05	Lunch	ITP Front Patio

Convener: R. Sawyer, UCSB

1:30	C. Heusch, UCSC	e- e- Collider
1:50	V. Telnov, INP	Photon-Photon Collider
2:30	S. Ritz, Columbia U.	Lepton-Hadron Collider

3:05	Refreshment Break	ITP Front Patio

Convener: C.. Pellegrini, UCLA

3:35	R. Siemann, SLAC	Snowmass 96 - summary
4:15	Z. Parsa, R. Palmer, M. Harrison (BNL), G. Jackson, A. Tollestrup, (FNAL), J. Irwin, P. Wilson (SLAC) C. Pellegrini, (UCLA) and other speakers	New Reports* & Round Table Discussion on Basic Issues in Accelerator Physics & Technology of: Linear Accelerators, Muon, Big Hadron Collider, etc.
5:30	Light Buffet Dinner	ITP Front Patio

Friday, October 25, 1996

Convener: M. Muggee, LBL

9:00 am	J. Irwin, SLAC	Fundamental Limitations of Particle Accelerators
9:40	Refreshment Break	ITP Front Patio

Futuristic Technology:

10:15	T. Katsouleas, USC	Highlights of "New Modes of Particle Acceleration Techniques and Sources" Symposium, ITP (96)
10:55	S. Chattopadhyay, LBL	A Report on the Advanced Accelerator Concepts Workshop (Oct 96)
11:35	Z. Parsa, BNL	Summary and Closing Talk
12:20 pm	Lunch	ITP Front Patio

LIST OF PARTICIPANTS*

Dante Amidei	University of Michigan
Thomas Appelquist	Yale University
Katsushi Arisaka	University of California, Los Angeles
Tofigh Azemoon	CERN, Switzerland
Mark Bachman	University of California, Irvine
Vernon Barger	University of Wisconsin, Madison
Frederick Bernthal	University Research Association, Inc.
Reinhard Brinkmann	DESY, Germany
Carl Bromberg	Michigan State University
David Burke	Stanford Linear Accelerator Center
Swapan Chattopadhyay	Lawrence Berkeley Laboratory
David Cline	University of California, Los Angeles
Andrej Czarnecki	Universität Karlsruhe
Robert Diebold	U.S. Department of Energy
Alex Dragt	University of Maryland, College Park
Anton Eppich	University of California, Santa Barbara
Sergio Ferrara	CERN, Switzerland
Jorge Fontana	Santa Barbara
Steve Giddings	University of California, Santa Barbara
Henry Glass	Fermi National Accelerator Laboratory
Vladimir Gorev	Kurchatov Institute
Aaron Grant	University of California, Los Angeles
David Gross	Princeton University
John Gunion	University of California, Davis
Michael Harrison	Brookhaven National Laboratory
James Hartle	University of California, Santa Barbara
Clem Heusch	University of California, Santa Cruz
Ian Hinchliffe	Lawrence Berkeley Laboratory
John Irwin	Stanford Linear Accelerator Center
Gerald Jackson	Fermi National Accelerator Laboratory
Basim Kamal	Brookhaven National Laboratory
JooSang Kang	Korea University
Thomas Katsouleas	University Southern California, Los Angeles
Boris Kayser	National Science Foundation
Eberhard Keil	CERN, Switzerland
Robert Koppensteiner	Johannes Kepler University
Martha Krebs	U.S. Department of Energy
Yoshitaka Kuno	KEK, Japan
James Langer	University of California, Santa Barbara
David Lange	University of California, Santa Barbara
Roy Lee	University of California, Irvine

*This may not include the names of the late resistants.

Nigel Lockyer	University of Pennsylvania
William Marciano	Brookhaven National Laboratory
Scott Metzler	Fermi National Accelerator Laboratory
William Molzon	University of California, Irvine
Andrew Morgan	University of California, Los Angeles
Rollin Morrison	University of California, Santa Barbara
Marshall Mugge	Lawrence Livermore National Laboratory
Hitoshi Murayama	Lawrence Berkeley Laboratory
Tim Nelson	University of California, Santa Barbara
Frank Paige	Brookhaven National Laboratory
Robert Palmer	Brookhaven National Laboratory
Zohreh Parsa	Brookhaven National Laboratory
Roberto Peccei	University of California, Los Angeles
Claudio Pellegrini	University of California, Los Angeles
John Peoples	Fermi National Accelerator Laboratory
Changqing Ziao	University of California, Santa Barbara
Burton Richter	Stanford Linear Accelerator Center
Steven Ritz	Nevis Laboratory, Columbia University
Luigi Rolandi	CERN, Switzerland
Joel Rozowsky	University of California, Los Angeles
Francesco Ruggiero	CERN, Switzerland
Nicholas Samios	Brookhaven National Laboratory
Raymond Sawyer	University of California, Santa Barbara
Steven Schnetzer	Rutgers University
Bertram Schwarzschild	Physics Today
Robert Siemann	Stanford Linear Accelerator Center
Dennis Silverman	University of California, Irvine
Sunil Somalwar	Rutgers University
Ray Stefanski	Fermi National Accelerator Laboratory
Andrew Strominger	University of California, Santa Barbara
Robert Sugar	University of California, Santa Barbara
Valery Telnov	Institute of Nuclear Physics
Nobukazu Toge	KEK, Japan
Alvin Tollestrup	Fermi National Accelerator Laboratory
Anthony Veletto	University of California, Los Angeles
Scott Willenbrock	University of Illinois, Urbana
Perry Wilson	Stanford Linear Accelerator Center
Michael Witherall	University of California, Santa Barbara
José Wudka	University of California, Riverside
Herng Yao	University of California, Irvine
Pavel Zenkevitch	Institute for Theoretical and Experimental Physics, Moscow

Author Index

A

Amidei, D., 95
Appelquist, T., 65

B

Barger, V., 219
Berger, M. S., 219
Brinkmann, R., 173

C

Cline, D. B., 203

G

Gunion, J. F., 41, 219

H

Han, T., 219
Harrison, M., 139
Heusch, C. A., 235
Hinchliffe, I., 129

K

Keil, E., 113
Krebs, M., 9

M

Marciano, W. J., 11
Murayama, H., 143

P

Paige, F. E., 157

R

Richter, B., 287
Ritz, S., 275
Rolandi, L., 81

S

Samios, N. P., 1

T

Telnov, V., 259

W

Willenbrock, S., 27
Wilson, P. B., 191

AIP Conference Proceedings

Title	L.C. Number	ISBN
No. 334 Few-Body Problems in Physics (Williamsburg, VA 1994)	95-76481	1-56396-325-6
No. 335 Advanced Accelerator Concepts (Fontana, WI 1994)	95-78225	1-56396-476-7 (Set) 1-56396-474-0 (Book) 1-56396-475-9 (CD-Rom)
No. 336 Dark Matter (College Park, MD 1994)	95-76538	1-56396-438-4
No. 337 Pulsed RF Sources for Linear Colliders (Montauk, NY 1994)	95-76814	1-56396-408-2
No. 338 Intersections Between Particle and Nuclear Physics 5th Conference (St. Petersburg, FL 1994)	95-77076	1-56396-335-3
No. 339 Polarization Phenomena in Nuclear Physics Eighth International Symposium (Bloomington, IN 1994)	95-77216	1-56396-482-1
No. 340 Strangeness in Hadronic Matter (Tucson, AZ 1995)	95-77477	1-56396-489-9
No. 341 Volatiles in the Earth and Solar System (Pasadena, CA 1994)	95-77911	1-56396-409-0
No. 342 CAM -94 Physics Meeting (Cacun, Mexico 1994)	95-77851	1-56396-491-0
No. 343 High Energy Spin Physics Eleventh International Symposium (Bloomington, IN 1994)	95-78431	1-56396-374-4
No. 344 Nonlinear Dynamics in Particle Accelerators: Theory and Experiments (Arcidosso, Italy 1994)	95-78135	1-56396-446-5
No. 345 International Conference on Plasma Physics ICPP 1994 (Foz do Iguaçu, Brazil 1994)	95-78438	1-56396-496-1
No. 346 International Conference on Accelerator-Driven Transmutation Technologies and Applications (Las Vegas, NV 1994)	95-78691	1-56396-505-4
No. 347 Atomic Collisions: A Symposium in Honor of Christopher Bottcher (1945-1993) (Oak Ridge, TN 1994)	95-78689	1-56396-322-1
No. 348 Unveiling the Cosmic Infrared Background (College Park, MD, 1995)	95-83477	1-56396-508-9

Title	L.C. Number	ISBN
No. 349 Workshop on the Tau/Charm Factory (Argonne, IL, 1995)	95-81467	1-56396-523-2
No. 350 International Symposium on Vector Boson Self-Interactions (Los Angeles, CA 1995)	95-79865	1-56396-520-8
No. 351 The Physics of Beams Andrew Sessler Symposium (Los Angeles, CA 1993)	95-80479	1-56396-376-0
No. 352 Physics Potential and Development of $\mu^+\mu^-$ Colliders: Second Workshop (Sausalito, CA 1994)	95-81413	1-56396-506-2
No. 353 13th NREL Photovoltaic Program Review (Lakewood, CO 1995)	95-80662	1-56396-510-0
No. 354 Organic Coatings (Paris, France, 1995)	96-83019	1-56396-535-6
No. 355 Eleventh Topical Conference on Radio Frequency Power in Plasmas (Palm Springs, CA 1995)	95-80867	1-56396-536-4
No. 356 The Future of Accelerator Physics (Austin, TX 1994)	96-83292	1-56396-541-0
No. 357 10th Topical Workshop on Proton-Antiproton Collider Physics (Batavia, IL 1995)	95-83078	1-56396-543-7
No. 358 The Second NREL Conference on Thermophotovoltaic Generation of Electricity	95-83335	1-56396-509-7
No. 359 Workshops and Particles and Fields and Phenomenology of Fundamental Interactions (Puebla, Mexico 1995)	96-85996	1-56396-548-8
No. 360 The Physics of Electronic and Atomic Collisions XIX International Conference (Whistler, Canada, 1995)	95-83671	1-56396-440-6
No. 361 Space Technology and Applications International Forum (Albuquerque, NM 1996)	95-83440	1-56396-568-2
No. 362 Two-Center Effects in Ion-Atom Collisions (Lincoln, NE 1994)	96-83379	1-56396-342-6
No. 363 Phenomena in Ionized Gases XXII ICPIG (Hoboken, NJ, 1995)	96-83294	1-56396-550-X
No. 364 Fast Elementary Processes in Chemical and Biological Systems (Villeneuve d'Ascq, France, 1995)	96-83624	1-56396-564-X

	Title	L.C. Number	ISBN
No. 365	Latin-American School of Physics XXX ELAF Group Theory and Its Applications (México City, México, 1995)	96-83489	1-56396-567-4
No. 366	High Velocity Neutron Stars and Gamma-Ray Bursts (La Jolla, CA 1995)	96-84067	1-56396-593-3
No. 367	Micro Bunches Workshop (Upton, NY, 1995)	96-83482	1-56396-555-0
No. 368	Acoustic Particle Velocity Sensors: Design, Performance and Applications (Mystic, CT, 1995)	96-83548	1-56396-549-6
No. 369	Laser Interaction and Related Plasma Phenomena (Osaka, Japan 1995)	96-85009	1-56396-445-7
No. 370	Shock Compression of Condensed Matter-1995 (Seattle, WA 1995)	96-84595	1-56396-566-6
No. 371	Sixth Quantum 1/f Noise and Other Low Frequency Fluctuations in Electronic Devices Symposium (St. Louis, MO, 1994)	96-84200	1-56396-410-4
No. 372	Beam Dynamics and Technology Issues for + - Colliders 9th Advanced ICFA Beam Dynamics Workshop (Montauk, NY, 1995)	96-84189	1-56396-554-2
No. 373	Stress-Induced Phenomena in Metallization (Palo Alto, CA 1995)	96-84949	1-56396-439-2
No. 374	High Energy Solar Physics (Greenbelt, MD 1995)	96-84513	1-56396-542-9
No. 375	Chaotic, Fractal, and Nonlinear Signal Processing (Mystic, CT 1995)	96-85356	1-56396-443-0
No. 376	Chaos and the Changing Nature of Science and Medicine: An Introduction (Mobile, AL 1995)	96-85220	1-56396-442-2
No. 377	Space Charge Dominated Beams and Applications of High Brightness Beams (Bloomington, IN 1995)	96-85165	1-56396-625-7
No. 378	Surfaces, Vacuum, and Their Applications (Cancun, Mexico 1994)	96-85594	1-56396-418-X
No. 379	Physical Origin of Homochirality in Life (Santa Monica, CA 1995)	96-86631	1-56396-507-0
No. 380	Production and Neutralization of Negative Ions and Beams / Production and Application of Light Negative Ions (Upton, NY 1995)	96-86435	1-56396-565-8
No. 381	Atomic Processes in Plasmas (San Francisco, CA 1996)	96-86304	1-56396-552-6

Title	L.C. Number	ISBN
No. 382 Solar Wind Eight (Dana Point, CA 1995)	96-86447	1-56396-551-8
No. 383 Workshop on the Earth's Trapped Particle Environment (Taos, NM 1994)	96-86619	1-56396-540-2
No. 384 Gamma-Ray Bursts (Huntsville, AL 1995)	96-79458	1-56396-685-9
No. 385 Robotic Exploration Close to the Sun: Scientific Basis (Marlboro, MA 1996)	96-79560	1-56396-618-2
No. 386 Spectral Line Shapes, Volume 9 13th ICSLS (Firenze, Italy 1996)		1-56396-656-5
No. 387 Space Technology and Applications International Forum (Albuquerque, NM 1997)	96-80254	1-56396-679-4 (Case set) 1-56396-691-3 (Paper set)
No. 388 Resonance Ionization Spectroscopy 1996 Eighth International Symposium (State College, PA 1996)	96-80324	1-56396-611-5
No. 389 X-Ray and Inner-Shell Processes 17th International Conference (Hamburg, Germany 1996)	96-80388	1-56396-563-1
No. 390 Beam Instrumentation Proceedings of the Seventh Workshop (Argonne, IL 1996)	97-70568	1-56396-612-3
No. 391 Computational Accelerator Physics (Williamsburg, VA 1996)	97-70181	1-56396-671-9
No. 392 Applications of Accelerators in Research and Industry: Proceedings of the Fourteenth International Conference (Denton, TX 1996)	97-71846	1-56396-652-2
No. 393 Star Formation Near and Far Seventh Astrophysics Conference (College Park, MD 1996)	97-71978	1-56396-678-6
No. 394 NREL/SNL Photovoltaics Program Review Proceedings of the 14th Conference— A Joint Meeting (Lakewood, CO 1996)	97-72645	1-56396-687-5
No. 395 Nonlinear and Collective Phenomena in Beam Physics (Arcidosso, Italy 1996)	97-72970	1-56396-668-9
No. 396 New Modes of Particle Acceleration— Techniques and Sources (Santa Barbara, CA 1996)	97-72977	1-56396-728-6
No. 397 Future High Energy Colliders (Santa Barbara, CA 1997)	97-73333	1-56396-729-4

Future high energy colliders : Santa Barbara, California, October 1996 / editor, Zohreh Parsa.
POA Circulating Books
QC787.C59 F87 1997

PHYSICS

Non-Circulating Until
JUL 1 4 1998